GEOMORPHOLOG TS

General Editor: K. M. CLAYTON, U. of East Anglia

5

GLACIAL AND FLUVIOGLACIAL LANDFORMS

Many areas of the earth's surface have been covered by glaciers during the last three million years; glacial geomorphology is concerned with the landforms created over such areas after the invasion and subsequent wastage of glaciers and ice sheets. Dr Price concentrates on trying to reach an understanding of past processes through the study of the margins of glaciers existing today. Such an approach, which assumes that processes of landform development observable in association with existing glaciers are similar to those which accompanied glaciation during the Pleistocene, is necessarily descriptive rather than quantitative.

There are two introductory chapters, the first indicating the historical background of glacial geomorphology, and the second giving a general description of ice and meltwater as agents of erosion, transportation and deposition. The following four chapters discuss the morphology, internal characteristics, classification and mechanics of formation of landforms produced by glacial erosion and deposition, and fluvioglacial erosion and deposition. A major theme in these chapters is the explanation of specific landforms in terms of the processes responsible for their development. In Chapters VII and VIII models of glaciation and deglaciation are constructed.

As with all the books in this series, there are excellent illustrations, both line and photographic, including many important aerial photographs.

R. J. Price is Lecturer in the Department of Geography at the University of Glasgow.

OTHER TITLES IN THIS SERIES

1	*Geomorphology*	F. MACHATSCHEK	1969
2	*Weathering*	CLIFF OLLIER	1969
3	*Slopes*	A. YOUNG	1972
4	*Geographical variation in coastal development*	J. L. DAVIES	1972

R. J. PRICE
Geography Department
University of Glasgow

GLACIAL AND FLUVIOGLACIAL LANDFORMS

Edited by K. M. Clayton

OLIVER & BOYD · EDINBURGH

OLIVER AND BOYD
Tweeddale Court
14 High Street
Edinburgh EH1 1YL
A Division of Longman Group Limited

ISBN 0 05 002646 1

First published 1973

Printed in Great Britain by Cox & Wyman Ltd
London, Fakenham and Reading

CONTENTS

PREFACE AND ACKNOWLEDGEMENTS

The importance of glacier ice and meltwater as agents responsible for the form of the earth's surface over extensive areas in middle and high latitudes has been widely recognized for over one hundred years. This book stems from my desire to understand the processes responsible for the landforms and deposits that occur in areas that have experienced glaciation during the Pleistocene. In order to understand the processes associated with past glaciations I have examined the margins of existing glaciers in Alaska, Iceland, Sweden and Switzerland. This book, therefore, is strongly influenced by my own field experience, and because that experience has been limited to a small sample of the many complex glacial and fluvioglacial environments that exist, the contents of this book represent a biased approach to the study of glacial and fluvioglacial processes. The rigour and quantification that have characterized the developments in fluvial geomorphology over the past twenty years have not been paralleled by similar developments in glacial geomorphology. The observation and measurement of glacial and fluvioglacial processes is often extremely difficult and even accurate descriptive data are either sparse or nonexistent. I have attempted to take a process-oriented approach, but in many instances observations and measurements of some processes are completely lacking and the mechanisms involved in the development of some landforms can only be broadly interpreted.

I am deeply indebted to the many people who have guided my research work and assisted me in the field. Dr. E. Watson of the University College of Wales, Prof. J. Ross MacKay of the University of British Columbia and Dr. J. B. Sissons of Edinburgh University provided stimulating teaching and research supervision. Numerous people greatly assisted me in obtaining experience in areas of existing glaciers and I am particularly indebted to Prof. R. P. Goldthwait and the staff of the Institute of Polar Studies, Ohio State University, to Prof. S. Thorarinsson of the University of Iceland, and to Prof. R. Miller and the staff of the Department of Geography, University of Glasgow. To administrators and granting bodies who have ensured financial support for fieldwork, to fellow academics who have given advice or help in the field or laboratory, to the undergraduate and postgraduate students who have been participants in field projects and to the local inhabitants of the areas in which I have carried out fieldwork, I am most grateful.

Thanks are due to the following who read parts of the manuscript and provided helpful comments: Dr. J. T. Andrews, Dr. B. Bluck, Dr. C. Clapperton, Dr. A. Gemmell, Dr. W. S. B. Paterson, Dr. J. B. Sissons, Dr. D. Sugden, Dr. C. Swithinbank. The series editor, Prof. K. Clayton, was a constant source of guidance throughout the preparation of this book.

My wife has been a great deal of help as typist, research assistant and constant source of encouragement. Not only has she assisted with the preparation of this book but also with the preparation entailed in my fieldwork in Alaska and Iceland.

Mr. A. Kelly drafted most of the line drawings and the photographic work was done by Mr. I. Gerrard. Miss S. Luther-Davis designed Figs. 93, 94 and 95.

I wish to thank the following for permission to reproduce illustrations: Pergamon Press and W. S. B. Paterson for Figs 3, 7, 8, 85; The Geographical Review and R. P. Sharp for Figs 6 and 32; Journal of Glaciology and J. Nye for Figs 8 and 9; The Royal Geographical Society for Fig. 12; U.S. Geological Survey for Fig. 13; Journal of Glaciology and G. Boulton for Fig. 15; Scottish Journal of Geology and R. Kirby for Fig. 17; Journal of Geology and R. P. Goldthwait for Fig. 18; Journal of Arctic and Alpine Research for Figs 20-28 and 76-78; Glasgow University for Figs 29 and 59; Geological Society of London for Figs 30 and 31; National Air Photo Library, Canadian Department of Energy Mines and Resources for Fig. 33; Geological Society of America and J. Ives and R. Kirby for Fig. 34; Geografiska Annaler and V. Schytt for Fig. 36; Scottish Geographical Magazine and J. B. Sissons for Fig. 38; Geografiska Annaler and C. Mannerfelt for Fig. 40; Institute of British Geographers for Figs 45, 60-66, 72-76, 80 and 81; Jökull and P. J. Howarth for Figs 46, 47, and 48; R. J. Rice for Fig. 51; Geografiska Annaler for Figs 52-58; U.S. Department of Agriculture Forest Service for Fig. 69; Landmaelingar Islands for Figs 75 and 76; Endeavour and G. de Q. Robin for Fig. 86.

R. J. PRICE

I INTRODUCTION

Glacial geomorphology is concerned with the landforms that result from the invasion by, and the subsequent wastage of, glaciers and ice sheets on an area of the earth's surface. The fact that large parts of the earth's surface have been covered by glaciers during the last three million years has only been widely accepted since the 1860's, and therefore the development of glacial geomorphology, glacial geology and Pleistocene studies in general is only a little over a century old. This chapter is concerned with the history of the development of the 'Glacial Theory' and of glacial geomorphology, the type of data used by glacial geomorphologists, the methods of analysis they employ, the terminology they use and the relationships that exist between the subdisciplines of glacial geomorphology, glacial geology and glaciology.

HISTORY OF THE GLACIAL THEORY AND GLACIAL GEOMORPHOLOGY

Up to 1862. To trace the original source of a new idea can be both time consuming and exasperating. Claiming that one man was responsible for a major breakthrough in scientific progress is usually over generous to the man concerned and often indicates a lack of understanding of the circumstances in which the so-called breakthrough was achieved. Jean Louis Rodolphe Agassiz (1807–1873) has usually been credited with the introduction of the Glacial Theory. Agassiz was Professor of Natural History at Neuchâtel and a distinguished specialist in fossil fishes. As Chorley, Dunn and Beckinsale point out (1964, p. 195), '. . . glaciology benefited quite as much from the use of Agassiz's name as from the application of his ideas.' This distinguished palaeontologist relied heavily on the contributions of little-known men.

In 1723 J. J. Scheuchzer, a Swiss naturalist, presented a theory of ice movement and also suggested that glacier ice was capable of transporting sand and large stones. He was followed in 1744 by P. Martel, a Swiss engineer, who suggested that glacier ice had been the transporting agent for the stones seen lying beyond the present limits of glaciers thus inferring the former greater extent of the glaciers. It should be noted that both these authors were Swiss and therefore very well acquainted with existing glaciers. Although Hutton was unable to visit the Alps he specifically accredited glaciers with the power to move large boulders: 'For the moving of large masses of rock, the most powerful engines without doubt which nature employs are the glaciers. . . . These great masses are in perpetual motion, . . . impelled down the declivities on which they rest by their own enormous weight together with that of innumerable fragments of rock with which they are loaded.' (Playfair, 1802, p. 388).

In 1815 J. Perraudin, a Swiss guide, suggested the former extension and increase in volume of glaciers, to account for boulders occurring high up on valley sides above existing glaciers. In 1821 I. Venetz, in a paper read to the Société Helvétique des Sciences Naturelles (published 1833), drew attention to Perraudin's ideas and pointed out that moraines similar to those seen near the snouts of existing glaciers could be seen far away from these glaciers and, therefore, the glacier ice must have been much more extensive in the past. By 1829 Venetz was explaining all erratic blocks in Switzerland and northern Europe as due to transport by ice.

By the 1820's, therefore, the concepts of glacier movement, the ability of glacier ice to transport debris and the former greater extent of the Alpine glaciers had been clearly stated. J. Esmark, working in Norway, had reached similar conclusions by 1827: '. . . we find the large stones lying separated from one another, surrounded by sand and gravel, a circumstance which cannot be explained in another way but by supposing that the whole has formerly been filled up with ice, which has pushed the whole mixed mass up the slope of the mountain. The water of the ice, afterwards thawing, carried off by its rapid streams a part of the stones and gravel which were then heaped together, deeper down in the valley: these heaps resemble entirely those which glaciers carry before them.' (Esmark 1827, p. 116). This quotation clearly represents an understanding of the importance of both glacial and fluvioglacial erosion, transportation and deposition. Esmark was also convinced that immense masses of snow and ice formerly existed in Norway, probably down to sea-level and he attributed the deep inland troughs and coastal fiords to glacial erosion.

Two further developments in the 1830's in Germany are worthy of note before looking at the part played by Agassiz in the development of the Glacial Theory. In 1832, Professor A. Bernhardi published the following statement (p. 257-9): 'The polar ice once reached as far as the southern limit of the district (in Germany) which is still marked by the erratics. This ice, in the course of thousands of years, shrank to its present proportions, and the deposits of erratics must be identified with the walls or mounds of rock fragments which are deposited by glaciers large and small, or in other words, are nothing less than the moraines which the vast sea of ice deposited in its shrinkage and retreat.' Similarly, in 1837, Karl Schimper, after studying erratics in Bavaria concluded that Europe had been covered by "*ein grosses Eisfeld*." It can be seen from the above comments that the concepts of a former ice cover over Switzerland, Germany, Norway and Northern Europe in general had been propounded in several papers by the 1830's. However, the proponents of these ideas were not leading figures in the scientific world of the time, and the more widely held view of the significance of the Great Deluge as a means of explaining erratics and spreads of sand and gravel was little affected by the clearly stated deductions which demanded that glacier ice be recognized as a significant agent in moulding landforms and depositing erratics.

In 1835 J. de Charpentier read a paper in Lucerne in which he gave credit to Venetz's pioneer work (1833) and went on to ascribe striations to glacial erosion. It was this paper that inspired a friend of de Charpentier to take an interest in the Glacial

Theory. That friend was Agassiz, and in his Presidential Address to the Swiss Society of Natural Sciences in 1837 he said, 'I wish to talk about glaciers, moraines and boulders'; he went on to postulate that the extensions of the glaciers were not just local phenomena but part of a climatic change which affected the whole of Europe. For the first time a distinguished scientist had put forward a claim that glacier ice had been much more extensive in the past and that the former ice cover was responsible for striations and the transport and subsequent deposition of erratics. These conclusions were published by Agassiz in 1840 in a book entitled, *Études sur les Glaciers*. The relationship between Agassiz and Venetz, de Charpentier and Schimper is well documented by North (1943, p. 18-19). The contributions made by the various scientists other than Agassiz have been overshadowed by the fact that Agassiz was the man who publicized the Glacial Theory and realized its possible application in areas such as Britain where, although no glaciers existed in the nineteenth century, considerable attention was being paid to the erratics and other superficial deposits that covered large areas. Until Agassiz came to Britain in 1840 these erratics and superficial deposits were generally attributed to the Deluge (Biblical Flood) or floating icebergs. North (1943, p. 20), states that, 'There can be no doubt, however, that it was Agassiz's championship of the new theory that brought matters to a head, and that finally gave the death stroke to the theories associating the phenomena in question with the Deluge and (except in certain instances) with floating icebergs, for as Sir H. H. Howarth somewhat cynically remarked in this very connection, scientific idolatry is determined chiefly by the size of the idol, and as a naturalist Agassiz was undoubtedly famous.'

The first public presentation in Britain of the Glacial Theory took place at the September meeting of the British Association in Glasgow in 1840. The presentation of Agassiz's paper was probably made possible and even encouraged by his friend, Buckland. Buckland had been in the field in Switzerland with Agassiz in 1838 and at that time was not convinced by Agassiz's arguments. However, after the Glasgow meeting, Agassiz travelled in Scotland with Buckland and Murchison and it must have been the result of that excursion which convinced Buckland that glacier ice had once existed in Britain. On November 18, 1840, at a meeting of the Geological Society in London, Agassiz, Buckland and Lyell all read papers demonstrating the former existence of glacier ice in northern Britain. These papers were strongly criticized by Murchison, Greenough and Whewell, and such criticism of the Glacial Theory continued in Britain for many years to come.

One quotation from Agassiz's paper to the Geological Society in 1840 clearly illustrates the refinements that had already taken place in the understanding of the effects of a former ice cover: 'The common origin of moraines, and of accumulations of rounded pebbles and of blocks, cannot be doubted. The former are simple ridges formed on glaciers; the latter materials rounded and polished under glaciers or great masses of ice, and are disposed by the water thus produced.' (Agassiz, 1840 c, p. 330). It can be seen that a distinction was already being made between glacial and fluvioglacial processes, and great use was also being made of observations obtained in areas of existing glaciers in solving the problems encountered in areas with a former ice cover.

After the introduction of the Glacial Theory into Britain in 1840 many British geologists remained opposed to it, and there was no great application of it until Ramsay used it in his work in Wales in 1862 and Jamieson in Scotland in the same year. Buckland was coming to the end of his active career as a geologist and became Dean of Westminster in 1845. Agassiz went to Harvard in 1847. However, the work of a new generation of geologists made great use of the Glacial Theory, and the period 1862-1900 could well be called the most productive with regard to the development of British glacial geomorphology. It is not surprising that after Agassiz's paper at Glasgow and the three papers by Agassiz, Buckland and Lyell in London, all of which referred to examples of the effect of glacier ice in Scotland, that the most significant developments in glacial geomorphology in Britain, and maybe even in the world, took place in Scotland during the period 1860-1890. During the same period Agassiz's presence in the United States also led to significant developments in that country.

1860-1890. Between 1860 and 1865 the concept of a great ice sheet covering northern Britain in the past and the ability of that ice to erode, transport and deposit was clearly stated in the British literature. In 1860, Jamieson, writing about the drifts of north-east Scotland stated, (Jamieson, 1860, p. 370), '. . . an extensive development of glaciers and land ice, which polished and striated the subjacent rocks, transported many of the erratic blocks, destroyed the pre-existing alluvium and left much boulder earth in various places.' In his famous paper on 'Ice Worn Rocks of Scotland' in 1862, Jamieson elaborated these ideas, and after further fieldwork in the west of Scotland was able to suggest that there was evidence of ice erosion from the tops of some of the highest mountains down to sea level. He even produced a map showing the probable direction of ice movement. This paper contains some excellent evidence and reasoning and the major hypothesis is supported by reference to the descriptions of the great ice sheets of Greenland and Antarctica which Jamieson had read in the reports of the early explorers of those regions (e.g., Ruik, P. C. Sutherland, Sir James Ross). However, even though Jamieson was convinced of a former ice sheet over Scotland his awareness of the raised beaches along the coast and the occurrence of marine shells in these deposits and in the boulder drift in some areas also convinced him that a great submergence of the country had taken place after the ice sheets had disappeared.

Also in 1862, the then President of the London Geological Society, A. C. Ramsay stated that (1862, p. 204): 'I am therefore constrained to return, at least in part, to the theory many years ago strongly advocated by Agassiz, that . . . great parts of North America, the north of the continent of Europe, a great part of Britain, Ireland and the Western Isles were covered by sheets of true glacier ice in motion, which moulded the whole surface of the country, and in favourable places scooped out depressions that subsequently became lakes.'

Apparently Ramsay's work had convinced A. Geikie of the significance of a former ice cover over Scotland, because in his monograph on the Glacial Drift in Scotland (1863), Geikie disposes of the 'Deluge' and 'Ice-berg' theories; he goes on to describe the till in great detail and concludes that it had been deposited by moving ice. He also described and accounted for striae and *roche moutonnée* and stresses that the ice cover was not simply confined to the valleys but that it was '. . . one wide sheet covering the whole or nearly the whole of the country.' (A. Geikie, 1863, p. 78). He also compared this former ice sheet with the ice sheet described by Sutherland, Rink and Kane during their visits to Greenland. Geikie believed the Scottish ice sheet to have been at least 2000 feet thick over parts of Scotland, and he described the main centres of ice dispersal and lines of ice movement.

Papers by Jamieson on 'The Parallel Roads of Glen Roy' (1863) and 'The History of the Last Geological Changes in Scotland' (1865), and by Young on 'The Former Existence of Glaciers in the High Grounds of the South of Scotland' (1864), are excellent examples of the detailed analytical fieldwork undertaken by these early workers. The quality of observation and interpretation embodied in these papers is of a high order, and the basic conclusions put forward still stand, even in the face of recent research.

Summaries of these early papers and of his own extensive fieldwork as an officer of the Geological Survey appeared in Archibald Geikie's book *The Scenery of Scotland*, published in 1865. In the preface Geikie refers to the fact that Agassiz, Ramsay, Lyell, Chambers and Jamieson had all been convinced that Scotland had been affected by land ice. He then goes on to devote three whole chapters and parts of others in demonstrating the effects of this former ice cover. It was only another eight years until James Geikie produced an outstanding volume entitled, *The Great Ice Age* (1873). That in ten years two brothers could bring glacial geomorphology and glacial geology from a situation of uncertainty to a position of wide respect and great interest and activity is quite remarkable. The table of contents of *The Great Ice Age* makes it hard to believe that ten years previously only a handful of persons would even accept the theory that large areas of the earth had once been covered by glacier ice and few workers had attempted to explain the origins of the landforms and deposits resulting from this former ice cover. However, by 1873 J. Geikie was discussing theories of ice movement, the Greenland ice sheet, the causes of the Ice Age and the origin and stratigraphy of glacial and post-glacial deposits in Scandinavia, Switzerland, Italy, Britain and North America.

No apology need be made for stressing the history of the development of the Glacial Theory in Britain, and in Scotland in particular. These developments were both dramatic and well-documented and somewhat remarkable in that Britain was so far removed from areas of existing glaciers. Similar developments were taking place in Europe and in North America. After the arrival of Agassiz in the United States in 1847 geologists in that country soon took an interest in the Glacial Theory, and during the 1860's and 1870's numerous papers relating to glacial geology and glacial geomorphology were written. The equivalent of J. Geikie's *The Great Ice Age*, was G. F. Wright's *The Ice Age in North America*, published in 1889. It was not only based on the author's own work in the eastern United States over the period 1874-1887, but included his own observations of ice conditions, landforms and deposits in Muir Inlet, Glacier Bay, Alaska in the summer of 1886. Wright's book was very comprehensive and includes a chapter on glaciology, four chapters on existing glaciers in various parts of the world, seven chapters on glacial geology and geomorphology and various chapters on the causes, extent and date of the Ice Age and its effects in both North America and Europe.

It can be stated, therefore, that the basic principles and content of glacial geomorphology had been well established by 1890. It is interesting to note two very distinct characteristics of this early period in the establishment of glacial geomorphology. First, the prime methodology adopted by the early workers was one of analogy. Glaciers were known to exist in certain areas and it was established that they had once been more extensive. Landforms and deposits similar to those observed near existing glaciers were also observed in areas without any existing glaciers, and it was argued that they were produced by similar processes and that therefore there must have been an ice cover in the past. Secondly, the major developments in the field of glacial geomorphology were associated with men who either had first-hand knowledge of glacierized areas or who were very familiar with the literature relating to glacierized areas either in Alpine or Polar regions. This second characteristic is particularly interesting because the rate of development of glacial geomorphology has probably been directly related to the geomorphologists' awareness and knowledge of glacierized environments. The statement that the most satisfactory developments in the field of geomorphology have resulted from the study of actual processes, which have then led to a fuller understanding of landforms, is hard to challenge. Whenever geomorphologists have relied solely on the landform to suggest process or environment of formation the results have often been misleading and confusing. This is not surprising in that it is now generally accepted that a particular landform may be developed by one of several different processes. The difficulty of deducing the environment and process of development of a specific landform from the end product, i.e. the landform, was clearly understood by the early glacial geomorphologists because they were very much aware of the complexity of the glacierized environment.

This awareness of the conditions to be found near existing glaciers on the part of American geomorphologists explains why, after the British had been at least at the same stage of development as the Americans in 1890, they were left behind during the early years of the twentieth century.

1890-1920. The period 1890-1920 was extremely rich from the point of view of the glacial geomorphologist seeking information about landforms and deposits associated with existing glaciers. During this period Alaska was the focus of attention for American earth scientists. Papers or books by Russell (1892, 1893), Reid (1896), Tarr and Martin (1906), Tarr (1909), Tarr and Butler (1909), von Engeln (1912), Tarr and Martin (1914) and Muir (1915), provide detailed descriptions of wide ice fronts at the terminus of long valley glaciers as well as the classic piedmont lobes to be found in Alaska. Much information was provided about glacial and fluvioglacial processes and stratigraphy in environments that were probably not very different from those which had existed in the coterminous United States or in Europe, some 15000 years earlier. The information collected from Alaska was extensively used as a basis for the interpretation of landforms and deposits observed by glacial geomorphologists working in areas of former glaciation in the United States (e.g., Stone 1899, Dryer 1901, Crosby 1902, Trowbridge 1914).

This period was also significant for British exploration of high Alpine and Polar environments. However, the expeditions sent out from Britain were not primarily concerned with geological and glaciological matters, and the relevance of the publications of these expeditions to the geomorphologist was minimal. Conway (1898) and Garwood (1899), writing about the glaciers of Spitzbergen, did provide some information of use to the geologist and geomorphologist. Although quite a large number of papers on the glacial geology and geomorphology of Britain were written between 1890 and 1920 most of them did not take into account the growing literature relating to areas of existing glaciers.

The one person to dominate British glacial geomorphology during this period was P. F. Kendall. His paper of 1902 on 'A System of Glacier-Lakes in the Cleveland Hills' had a profound (and some would argue a detrimental) effect on British geomorphology. Kendall's paper is detailed and well-argued, and it set a pattern of analysis which was to dominate the British literature for another 30 years. Kendall's thesis was that ice was watertight and could therefore act as a dam and thus permit the development of large lakes in valleys not actually occupied by ice. It is interesting to note that the literature on areas of existing glaciers which had appeared in the last decade of the nineteenth century presented evidence of meltwater drainage on, into and under glaciers. In the introduction of his work on the Cleveland Hills, Kendall mentioned the theoretical possibility of englacial and subglacial drainage. Although aware of the work on existing glaciers he did not apply it to the Cleveland Hills and nowhere does he refer to drainage into or beneath the ice. The record of the discussion that followed the reading of Kendall's (1902) paper indicates that it was very well received. Referring to Kendall's work on the Cleveland Hills, Harmer (1907, p. 470), wrote, 'It is clear moreover, that such cases (of ice-dammed lakes) must be typical and not anomalous.' The general acceptance of the development of ice-dammed lakes and the cutting of overflow channels dominated British glacial geomorphology until the 1950's. This is a particularly interesting example of the almost circular reasoning adopted by glacial geomorphologists during this period. A particular set of landforms, the meltwater channels, were explained in terms of an ice dam blocking normal drainage routes. The limits of the ice cover and the condition of the ice were determined from the existence of the channels. The development of the channels was explained in terms of the limits of the ice and the condition of that ice. Whenever the glacial geologist or glacial geomorphologist has been tempted to interpret the genesis of a particular landform, or the nature of the glaciological environment, in areas of former glaciation without any reference to known processes and environmental conditions in areas of existing glaciers he has often followed paths that have led him, and others who have followed him, into confusion.

1920-1945. Papers on regional glaciation and deglaciation and on the origins of landforms and deposits in areas formerly covered by ice, formed by far the largest part of the literature of the period 1920-1945. By 1945 the number of publications relating to glacial geomorphology was very large (see Charlesworth, 1957). Whereas the period 1890-1920 was characterized by the contributions of workers studying areas of existing glaciers the following period was characterized by the theoretical contributions of workers studying areas of former glaciation. The main results of all the detailed descriptions of the forms and deposits in areas of former glaciation were new concepts of deglaciation, and these in turn started another major controversy amongst geomorphologists. Flint (1929, p. 256) outlined two methods of ice dissipation; 'These are, 1) retreat of ice due to excess of melting over alimentation, with the preservation of a well-defined glacier front; and 2) dissipation of the ice as a "dead" or stagnant mass resulting from total loss of forward motion while at its maximum extent.' Flint maintained that these two methods of ice dissipation each produced characteristic landforms. Normal retreat resulted in recessional moraines channelled by proglacial streams, and dissected outwash terraces traceable into specific recessional moraines. The concept of downwastage implied the appearance of nunataks and the washing of debris by meltwaters on to the ice occupying the lowlands, thus protecting it from melting. The shrinkage of the ice was inward from the valley sides, resulting in the formation of marginal lakes and ice-contact kame terraces. It is possible that paired terraces would develop on either side of a valley due to hydrostatic connection through crevasses in the ice linking two marginal lakes. Kettles would occur in the almost horizontal terraces and some terraces would be related to spillways that drained the marginal lakes in which the deposits accumulated.

These detailed models of the landforms and deposits which result during deglaciation had a profound effect on workers in the field. Flint himself applied the concept, or model, of the downwastage of an ice mass with resultant development of characteristic fluvioglacial forms, sometimes related to an englacial water-table, to the deglaciation of central Ireland, (Flint, 1930). In 1931 Andersen published similar opinions based on work in Denmark and he enlarged the concept by suggesting that when the ice surface became so thin that the water-table was high enough to cause the ice to lift, there was a rushing and irregularly streaming meltwater drainage that would produce kames and pitted plains.

In a paper on the deglaciation of New England, R. J. Lougee (1940) presented strong criticism of the concepts of stagnation and downwastage put forward by Flint and Andersen. Lougee pointed out that features which, on the basis of their morphology, had been deduced as being formed in contact with stagnant ice could have been formed in contact with active ice; he cited examples of kettles, reticulated ridges and long ice-contacts formed in association with active ice in New England.

In 1942 Flint and Demorest published a more detailed account of the evidence that glacier thinning occurred during deglaciation and of the resulting forms. Evidence from existing glaciers indicated that thinning does take place but it is not necessarily accompanied by stagnation, and that the ice surface downwasted in a particular area will be indicated by the presence of a succession of lateral moraines or kame terraces although these forms are not necessarily associated with stagnant ice. The authors conclude (p. 132) '. . . the available evidence indicates that separation or stagnation of vanished ice can rarely be proved and that at best it can be inferred to have affected only small areas at any one time.' Rich (1943) also favoured local stagnation without regional stagnation. He envisaged the burial of ice, under aggrading outwash at the margin of a continental glacier, that became detached and stagnant as a result of its burial.

Active ice sheared over the top of the stagnant masses but when the stagnant ice melted, features indicating stagnation would be formed.

It is interesting that such a debate should have taken place about the manner of ice wastage when reference to observations made in areas of existing glaciers would have clearly indicated that downwastage of the ice surface and backwastage of the ice front can both take place simultaneously. The only explanation is that once again glacial geologists and geomorphologists were attempting to deduce glaciological conditions from morphology and deposits instead of turning to glaciological sources for information to explain the landforms and deposits.

Between 1920 and 1945 a large number of publications in Britain, in the form of memoirs of the Geological Survey and papers by Gregory (1926), Charlesworth (1926a, 1926b), Trotter (1928), Hollingworth (1931), Simpson (1933), and Linton (1933), all portray the wide acceptance of the concept of normal or horizontal retreat of an ice front as opposed to downwastage. The subglacial origin of eskers had long been accepted by British geomorphologists, but the other forms of fluvioglacial deposition had always been associated with the proglacial environment and frequently with ice-dammed lakes, the only evidence for which had been the existence of 'overflow channels.'

1945-1960. During 1945-1960 some major changes in attitudes took place amongst glacial geomorphologists and several new methods and techniques were introduced. The immediate post-war years saw the publication of papers which in many instances were based on projects at least initiated prior to the war, and it may well be that the delay in publication resulted in far more mature and stimulating presentation. Undoubtedly the period 1945-57 represents the most stimulating phase in the development of glacial geomorphology since the turn of the century.

Scandinavian workers made the biggest impact. Two papers by Mannerfelt (1945, 1949) were the forerunners of many others which applied the observations made in Alaska, Spitzbergen and other areas of existing glaciers, to the features of fluvioglacial erosion and deposition in the Swedish–Norwegian mountains. Mannerfelt concluded, on the basis of detailed morphological mapping, that the ice sheet that once covered the Scandinavian mountains had downwasted so that the higher peaks and ridges emerged from beneath the ice. The meltwater channels occurring in the cols between the highest peaks had developed as the ice downwasted. The channels were not generally the result of outlet drainage from ice-dammed lakes but were formed, '. . . by streams which flowed along through the mountain passes at the bipartition of the shrinking ice mass.' (Mannerfelt 1945, p. 224). Channels were also cut between the ice and rock of emerging spurs and ridges but Mannerfelt pointed out that (p. 223), 'A typical feature is that the lateral meltwater tries to take a sublateral or subglacial course under the ice margin.' He also realized that the ice remained longest in the valleys and that along the margins of downwasting ice masses, marginal channels were cut that in places were traversed by subglacial chutes. Mannerfelt suggested that the occurrence of flights of marginal channels separated by fairly constant vertical distances, indicated that the ice surface was lowered by ablation between three and five metres each year.

Since these channels were formed at the ice margin they would also indicate the slope of the ice surface. Both of these ideas put forward by Mannerfelt have been subjected to criticism. To determine the gradient of the ice surface and the rate of downwastage (i.e. annual ablation) by using data derived from meltwater channels, there must be no doubt about the true marginal origin of the channels. Mannerfelt himself stated (1949, p. 197), 'There is always a natural tendency for the lateral meltwaters to undermine the margin of the ice and find their way into the subglacial chutes. Sublateral drainage gullies are the commonest.' It has proved almost impossible to attribute with certainty a marginal position to any channel during its formation, as even channels that are parallel to the present contour pattern could be submarginal. Therefore it is usually impossible to use meltwater channels to determine the slope of the ice surface and the rate of downwastage.

Hoppe (1950) also demonstrated the progressive thinning of the ice sheet in Scandinavia and generally supported Mannerfelt's conclusions. Publications by Gillberg (1956) and Holdar (1957) continued the methodology adopted by Mannerfelt and Hoppe. This methodology was primarily one of detailed morphological and stratigraphical mapping by means of both fieldwork and photo interpretation and photogrammetric techniques. The stress laid by these Scandinavian workers on the significance of landforms produced by fluvioglacial erosion and deposition led to a resurgence of interest both in Britain and North America in those areas where moraines and till sheets were neither well developed nor well exposed. The distribution and orientation of channel systems became much more significant in studies of deglaciation as a result of the detailed work carried out in Scandinavia. The publication of a large and controversial volume by Gjessing (1960) represented the culmination of this period of emphasis on fluvioglacial features.

During this period of intense activity in Scandinavia other Scandinavians were working in areas of existing glaciers. Okko (1955) and Arnborg (1955) were working in Iceland and Schytt (1956) in Greenland and the Antarctic. The close liaison between workers studying areas of former glaciation and those working in areas of existing glaciers and in some cases of workers having had wide experience in glaciology before attempting studies in areas of former ice cover, has undoubtedly been one of the greatest strengths of Scandinavian glacial geomorphology.

Hollingworth (1952) was the first British worker to apply the conclusions of the Scandinavian workers in Britain, but it was Sissons (1958a, 1958b, 1958c, 1960a, 1960b, 1961a, 1961b, 1961c) who really followed the Scandinavians' methodology and applied their concepts of downwastage and subglacial and englacial drainage with great success. Sissons' work is characterized by very detailed morphological and stratigraphical mapping, usually on a scale of 1:10000. He used aerial photographs extensively and mapped thousands of meltwater channels in central and southern Scotland, which he used not only to prove that the occurrence of meltwater channels need not indicate the former existence of an ice-dammed lake but also to interpret the minimum height of the former ice surface, the probable slope of the surface of the ice sheet and the mode of dissipation of the ice. Until these publications by Sissons, Kendall's 1902 paper still dominated British glacial geomorphology. Sissons (1960b,

1961a) provided a critique of Kendall's work and the majority of workers accepted the fact that ice-dammed lakes were not as common as had been generally believed and that evidence other than the existence of channels had to be sought before ice-dammed lakes could be postulated.

Two other developments in Britain during the 1950's are worthy of mention. First, papers by Glen (1952, 1955, 1956) and Nye (1951, 1952, 1957, 1959a, 1959b) on ice movement led to new developments in glaciology that had an effect on most aspects of glacial geomorphology. Secondly, in 1957, Charlesworth published his two volumes entitled *The Quaternary Era*. These volumes contain a magnificent summary of the vast literature in many languages that had accumulated since the general acceptance of the Glacial Theory. As a source book it has no parallel.

Developments in the United States between 1945 and 1960 were of three main types. There was a continued interest in areas of existing glaciers. R. P. Sharp's papers (1947, 1949, 1951a, b) contain a great deal of relevant information about ablation, meltwater drainage and supraglacial debris. Goldthwait (1951) working in east-central Baffin Island described moraines in the process of formation. The strong link between geomorphology and geology in the United States and Canada is reflected by the dominance in the North American literature of this period of studies of till and till fabric. Papers by Holmes (1941), Dreimanis *et al.* (1957), Harrison (1957) and Gravenor and Kupsch (1959) laid the foundations for a field of investigation which has received much attention in recent years.

Two text books were published in the United States during this period. In 1947 Flint published *Glacial Geology and the Pleistocene Epoch* and this was followed in 1956 by Thwaites' *Outline of Glacial Geology*. Both these books reflected the strong geological interests of their authors but they both contain a great deal of information relating to glacial geomorphology. Flint's work was very comprehensive and included sections on glaciology, glacial geomorphology, proglacial drainage and aeolian phenomena, changing sea-levels and the Pleistocene stratigraphy and chronology of North America and Europe. This text was really the first modern equivalent of those written by J. Geikie (1873) and G. F. Wright (1889).

1960-1971. It is probably unwise to look for trends in the field of glacial geomorphology over the last decade. The volume of literature has continued to grow and it is only possible to keep track of it with the aid of abstracting journals and bibliographies (e.g. *Geographical Abstracts*—series A; Glaciological literature in Journal of Glaciology). Since 1960 there have probably been three major developments. First there has been increased interest and activity in areas of existing glaciers. Work in Alaska (Goldthwait 1963, Haselton 1966, Clayton 1964, Reid and Callender 1965, Price 1964, 1965, 1966, Petrie and Price 1966, Lindsay 1966), Axel Heiberg Island (Müller 1962), Baffin Island (Andrews 1965), Iceland (Howarth and Price 1969, Price 1969, 1970), Spitzbergen (Boulton 1967, 1968) and other areas has contributed to a greater awareness of the complexity of glacial and fluvioglacial processes and environments. Secondly, there has been a growing concern for quantification in glacial geomorphology. Numerous papers have been based on measurements made in the field or laboratory, of the sedimentary character of the glacial and fluvioglacial landforms or of their shape and altitudinal relationships. It is interesting that by far the greatest attention has been given to the analysis of the end-products rather than the processes themselves as observed in the field or simulated in the laboratory. It is questionable if the refined statistical and computer methods used in recent years by glacial geomorphologists have achieved more than a more precise description of phenomena. There are few instances where such methods have contributed to a greater understanding of glacial and fluvioglacial processes.

The third development in the field since 1960 has been the appearance of several systematic and regional texts devoted to various aspects of the Pleistocene and summarizing literature relating to glacial geomorphology. Embleton and King's book *Glacial and Periglacial Geomorphology* (1968), is an excellent summary of the modern literature and is remarkable in that it is the first text book to include the words 'glacial geomorphology' in the title. The 1965 Congress of the International Association for Quaternary Research, held in Boulder, Colorado, resulted in numerous publications containing results of work undertaken by glacial geomorphologists from all over the world. The general volume on *The Quaternary of the United States* edited by Wright and Frey (1965) is a most useful survey of the state of knowledge in the United States. The equivalent in the European literature is two volumes edited by Rankama entitled *The Quaternary*. Volume one deals with Denmark, Norway, Sweden and Finland (1965) and volume two deals with the British Isles, France, Germany and the Netherlands (1967). These two volumes are primarily concerned with Quaternary stratigraphy and chronology. A text by West (1968), entitled *Pleistocene Geology and Biology*, although containing chapters on glaciology and glacial geomorphology, is also strongly oriented to stratigraphy and chronology. The one regional account of Quaternary events in Europe which emphasizes glacial geomorphology is Sissons', *The Evolution of Scotland's* Scenery (1967), which deals in a systematic manner with the landforms and deposits produced by glacial and fluvioglacial processes in Scotland.

The last decade has seen great increases, both in the number of workers in the field of glacial geomorphology and in the number of publications. Several general texts summarizing the post-war literature have helped to overcome the increasing problem of coping with the ever-growing literature. However, there is as yet little sign that glacial geomorphologists are in general facing up to the problem that has dominated the field since its early days in the nineteenth century. Until a greater understanding is reached of the mechanisms responsible for glacial and fluvioglacial deposits and landforms, by studying the processes at work under, in, on, and at the margins of existing glaciers, the interpretation of past glacial environments and the development of landforms and deposits in areas of former ice cover will remain a very difficult and often unrealistic exercise.

This book uses what is known about these mechanisms in an attempt to highlight the weaknesses in the interpretations which currently dominate the literature. The fact that the available information about glacial and fluvioglacial processes is at best limited and at worst entirely absent for certain processes, means that this contribution will be rather unbalanced.

DATA, METHODOLOGY AND TERMINOLOGY

The glacial geomorphologist is primarily concerned with the origin of a particular group of landforms. These landforms result from the erosional, transportational and depositional work of glacier ice and meltwater. It is necessary to examine the nature of the data studied by glacial geomorphologists and the methodologies adopted for its analysis. The purpose of most glacial geomorphological work is to classify landforms on the basis of their origin which, in turn, facilitates meaningful description and allows the interpretation of the environmental conditions prevailing at the time of formation of the landforms. Establishing a terminology or a classification system can either be an end in itself, or a means to another end. One of the biggest problems in glacial geomorphology is the terminology which has developed in a rather haphazard manner. Unfortunately, the literature is full of examples of terms which have various uses depending upon the author concerned. This confusion largely stems from a lack of knowledge about the processes responsible for certain landforms and the adoption of a generic classification for these landforms.

The geomorphologist's data are basically of two types. First, he is concerned with 'form'. That is, he requires a description of morphology in terms of length, breadth, height, plan shape and profile shape. This type of data is best obtained when the landform is new or little modified by processes other than those directly responsible for its formation. It is therefore obvious that the glacial geomorphologist seeking data relating to glacial and fluvioglacial landforms is well advised to study those areas presently being deglacierized, or alternatively those areas only recently exposed from beneath an ice cover. The relationships between the value of morphological data and the age of the landform can probably be expressed by an exponential curve in which the value of the data declines at an ever-increasing rate with the passage of time. It can be argued that the data obtained from a specific landform exposed from beneath the ice cover some ten years ago would be five times more valuable than the data obtained from the same landform after 1000 years and 100 times as valuable as data obtained after 10000 years.

It has long been established that the same morphology can be produced by different processes. The glacial geomorphologist can not, therefore, rely solely on morphology as a basis of classification or as his only source of data for interpreting origins. The character of the materials constituting a landform must also be analysed. In glacial geomorphology there are two basic classes of landforms: those produced dominantly by depositional processes and those produced dominantly by erosional processes. The two classes are in turn subdivided on the basis of the type of material upon which these processes have been at work, that is, whether the processes have been working on solid rock or on unconsolidated ('drift') materials.

In the study of glacial and fluvioglacial landforms eroded from solid rock the geomorphologist can obtain data about the lithology and structure of the rocks concerned. He may be able to ascertain the extent to which these lithological and structural factors have influenced the development of the morphology resulting from the erosional processes. However, it may be very difficult to ascertain the morphological character of an area prior to any ice cover or prior to the last ice cover. The influence of inherited characteristics from a previous condition probably constitutes the largest unknown in any type of geomorphological analysis.

With regard to erosional landforms developed on unconsolidated materials, the same type of problem exists. Unstratified and stratified sediments represent the two end members of what is actually a continuum. It is generally accepted that unstratified sediments in a glaciated area indicate deposition by ice itself, in contrast to the stratified sediments which are the result of deposition by meltwaters. However, glacial deposition can be accompanied by the release of meltwater and it is not uncommon to find lenses and even extensive strata of stratified deposits incorporated within unstratified sediments. On the other hand, very poorly stratified or even unstratified materials are known to result from rapid fluvioglacial deposition. However, in terms of the erosional forms developed on both stratified and unstratified unconsolidated sediments, the environment of deposition of the sediments is no more significant than the sedimentological or thermal origins of solid rocks. The major problems for the glacial geomorphologist are to ascertain the effect of the internal character of sediments and also the effect of surface form prior to erosion.

Depositional forms are generally much more susceptible to analysis by the glacial geomorphologist than are erosional forms. The data available from the former are not limited to their morphology and their resistance to erosion. Although morphological data relating to depositional landforms can be of great value it is usually the character of the sediments which constitute the landform that is the most useful data source. By studying the particle size, shape, orientation and sedimentary structures it is often possible to interpret the nature of the environment of deposition and the processes involved. One of the major difficulties encountered in collecting data about the character of glacial and fluvioglacial sediments is the limitation of access to the interior of sediment accumulations. The data obtained by glacial geomorphologists in such circumstances are usually only a very small sample of the whole accumulation.

Data relating to the actual geomorphological agents involved in the production of both erosional and depositional forms are largely derived from the work of two groups of specialists, who are usually only incidentally interested in glacial and fluvioglacial landforms. The two agents involved are ice and water. The study of glacier ice has developed into a distinctive discipline, that of glaciology. The mechanics of ice accumulation, ice movement and ice wastage is a complicated field, largely dominated by physicists, meteorologists and hydrologists. The glacial geomorphologist has to turn to these specialists to derive data regarding the processes involved in glacial erosion and glacial deposition and the discharge of meltwater. In the same way the glacial geomorphologist has to turn to the hydrologist, engineer, sedimentologist and fluvial geomorphologist for data regarding the processes involved in meltwater erosion, transportation and deposition. Unfortunately, the glaciologist and hydrologist have not always been concerned with those problems of most interest to the glacial geomorphologist and frequently the data required by the latter is simply not available. Data relating to the hydrology of glacierized environments have been, until very recently, almost completely lacking. The inaccessibility of many areas of existing glaciers, the rapid fluctuations of discharge and debris load and the frequent abandonment of one stream course in favour of another has made the collection of data in such environments extremely difficult.

The collection of data relating to ice accumulation, movement and ablation has also been very difficult. The developments in glaciology have been remarkable considering the environmental difficulties and the relatively small number of workers involved. The one major weakness in the data that are available is that much of it has been derived either from small high Alpine glaciers or from selected sites on the polar ice caps. The amount of data relating to the type of ice sheets that covered so much of North America and Europe during the Pleistocene is limited.

Some data about ice and meltwater can be derived from areas of former glaciation. The former extent, the direction of movement and some information about the minimum thickness of the Pleistocene ice sheets can be determined from the location of glacial and fluvioglacial landforms and deposits. Attempts have also been made to interpret at least the mode of ice dissipation and even rates of ice wastage. However, these interpretations have been based on limited information and a considerable amount of circular argument. Assumptions are often made about the mode of origin of certain landforms and deposits on the basis of a preconceived idea of the nature of the ice, and ice wastage, and these same features are then used as a basis of the interpretation of the condition of the ice and the mode of ice dissipation.

Laboratory experiments have also been used as a source of data relating to ice and meltwater. However, they have been largely concerned with theoretical studies of ice deformation and ablation, and water movement, transportation and deposition. The problems of transferring both spatial and time scales from the glacierized environment to the laboratory are considerable.

Data relating to the actual processes of glacial and fluvioglacial erosion, transportation and deposition are largely derived from two types of sources; areas of existing glaciers and areas of former ice cover. The prime difference between these two sources of information is that in areas of existing glaciers certain processes can be actually observed at work and the resultant forms and deposits can be observed in a very fresh condition, whereas in areas of former ice cover the processes can only be deduced from their much altered end products.

There is little data available based on actual observation and measurement of glacial erosion and deposition in areas of existing glaciers. This is not so very surprising because the mechanisms of glacial erosion and deposition largely occur beneath the ice. The only way such processes can be observed is by tunnelling into the ice and, although this has been done in a few locations, there are two major problems. First, the act of cutting the tunnel inevitably changes the subglacial environment in which these processes work. Secondly, such tunnels that have been cut represent very selective and limited samples of subglacial environments.

The most successful studies in areas of existing glaciers have dealt with the transportation of material on or near the ice surface, the redistribution of this material during ice wastage and the accumulation of supraglacial and englacial debris at the ice margin. Apart from these observations of the transport and deposition of debris by glacier ice the next most important source of data regarding glacial processes is from those areas that have only been deglacierized during the last 100 years. It is in such localities that the end products of glacial processes can be most easily studied without the interference of a thick vegetation cover or the extensive modification of form and internal character which results from long periods of weathering and mass movement. Even the interpretation of processes from fresh end-products can be difficult because fluvioglacial erosion and deposition may have removed or buried part of the evidence and very similar forms could have been produced by different mechanisms. However, the freshly deglacierized area has probably been the most useful source of data relating to processes of glacial erosion and deposition.

In areas of former glaciation a time lapse of many thousands or, in a few cases, a couple of million years may have occurred since the landform or deposit was originally produced. The environmental changes which have taken place in the area of study may have been very great. The extent to which the form and internal character of the feature being studied have been modified by processes other than those directly responsible for it is a very big unknown. In these circumstances great care has to be taken when adopting refined techniques of analysis to ensure that the data being used are worthy of such sophisticated methods.

Considerably more data have been obtained about fluvioglacial erosion, transportation and deposition from areas of existing glaciers, than about glacial processes. The work of water on the surface, at the sides and in front of existing glaciers is relatively easy to observe, but once again the englacial and subglacial environments are largely inaccessible. Data are available about the amount of discharge in a drainage basin resulting from ablation, the importance of precipitation in the non ice-covered parts of the basin, the development of ice-dammed and proglacial lakes and the great fluctuations in discharge and debris load which occur in glacierized areas. Meltwater streams have been observed in the process of eroding and depositing on the ice surface, along ice margins and in the immediate proglacial area. In areas of very recent deglacierization the results of these various fluvioglacial processes can be most clearly seen. The availability of aerial photography has enabled studies to be made of the evolution of the proglacial area over a period of 20 years. Changes that take place over such a period as a result of changes in the proglacial drainage system and the melting out of buried ice from beneath fluvioglacial deposits can be quite dramatic.

In areas of former ice cover the detailed description of features produced by fluvioglacial erosion and deposition has contributed some significant information. It must be remembered however, that the hypotheses built up on this data are limited in their value by the incompleteness of the data. For example, a system of meltwater channels and a series of fluvioglacial deposits only represent a part of the drainage system that was functioning at the time of their formation. The remainder of the system, for which there may be no evidence, was contained in the ice.

Another major problem in the interpretation of fluvioglacial processes using evidence obtained from areas of former glaciation is to ascertain the relative importance of meltwater activity as opposed to normal fluvial processes. It can be argued that any stream channel or valley system which existed in an area of glaciation prior to the invasion of the area by ice and which survived the processes of glacial erosion and deposition, almost certainly carried meltwater during deglaciation. Separation of the work done by meltwater from that done by the pre-glacial and post-glacial streams involves a large amount of subjective judgement.

The form and internal character of landforms produced by fluvioglacial processes can be modified considerably by post-formational weathering and mass movement. The major source of information about landforms produced by fluvioglacial deposition is undoubtedly the sedimentary and structural character of the deposits. However, the interpretation of the environments of deposition, using evidence obtained from exposures in the deposits, is limited by the small size of the samples revealed by exposures, and by the controversial statements made by sedimentologists on the basis of the evidence available.

It can be seen from the above comments that the data used by the glacial geomorphologist vary widely in their value. The lack of observation and measurement of processes at work reflects both the difficulty of the environment in which these processes can be observed and the problems of designing suitable instrumentation. Data obtained from areas of very recent deglacierization are largely concerned with the results of the various processes rather than the processes themselves. Data obtained from areas of former glaciation are not only incomplete as a result of post-glacial destruction but are subject to widely differing interpretations, because similarities in both form and internal character can result from different sets of processes. Another major difficulty is a lack of knowledge about the pre-glacial character of the areas invaded by ice and about the prevailing environmental conditions during ice advance and ice wastage.

Glacial geomorphology in its present state of development is, therefore, a very inexact science. The landforms to be studied are the results of processes which are controlled by the laws of physics and chemistry. The emphasis during the last 100 years has been on the interpretation of processes from their end products. Major progress can only be made when observation and measurement of the processes themselves become the focus of attention of glacial geomorphologists.

It is of interest to examine the methods of analysis adopted by glacial geomorphologists when using the data that has been discussed above. In areas of existing glaciers the description and measurement of both landforms and processes have been undertaken. In areas of former ice cover the emphasis has been on description in terms of form, internal character and distribution of landforms and deposits. The origins of these landforms and deposits have usually been postulated by means of analogy with observed or postulated processes in areas of existing glaciers. These analogies have been of three types.

In the first type of analogy emphasis is laid upon detailed description of the observed phenomena followed by the construction of an hypothesis or model. Examples of this methodology are the stagnant-downwasting-ice model of Flint (1929), and Hoppe's model of the development of hummocky-moraine (1952). This type of model building is limited, first, by the incompleteness of the data available, and secondly, by the necessity of assuming that a set of conditions known to be true in one area of existing glaciers can be transposed successfully to another situation, maybe thousands of miles away, and of much greater age. The quality of the models constructed in these circumstances is a direct function of the amount of information available and the universality of the constraints built into the model.

The second type of analogy entails detailed description of the phenomena involved followed by direct comparison with similar phenomena in the process of formation. This task is particularly suitable in areas which may have been deglacierized 50 to 100 years ago but which contain forms very similar to those being produced by the glaciers at present (see, for example, Price, 1966, 1969).

The third type of analogy involves the classic scientific method of observation, description and experiment. Laboratory experiments have been used only rarely by glacial geomorphologists. There are a few examples, such as the simulation of the formation of eskers by a subglacial stream issuing into a body of water (Hanson 1943, Lougee 1954), and experiments on glacial erosion (Lister *et al.* 1967) and deposition (Harrison 1957). The main problems encountered with the experimental method are associated with scale and the paucity of data relating to measurements of the actual processes.

A major hindrance to any person examining the literature produced by glacial geomorphologists is associated with terminology. The confusion that prevails is largely the result of attempts to develop a genetic classification of glacial and fluvioglacial landforms and deposits without an adequate knowledge of their genesis. The problem is accentuated by the fact that similar landforms and deposits can probably be formed by different processes, and this has led to considerable difficulties when morphology has been used as the single most important criterion for classification. For example, if an esker is simply defined as a narrow ridge of sand and gravel without any reference to the relationship of this feature to the former position of the ice edge at the time of its formation, or to the nature of its internal stratification or structure, the feature could in fact be a crevasse filling, a moraine or a remnant of a dissected outwash plain. A genetic terminology is certainly the most satisfactory, but it is completely dependent upon an adequate understanding of the processes involved.

A solely descriptive terminology also has its problems. If such a terminology is based only on morphology confusion quickly arises, because similar forms frequently have very different internal character and relationships with other forms. It would be absurd to suggest that all linear ridges of certain external dimensions should be given the same name. If this descriptive terminology is extended to include both form and internal character, confusion once again prevails. In such a system, all ridge-like features consisting of sand and gravel would have the same name, although of differing origin, while it is again necessary to include in this classification the position of the feature in question in relationship to the ice margin at the time of its formation. Thus the classification ceases to be solely descriptive.

Attempts by glacial geomorphologists to bring together the elements of both descriptive and genetic classification have certainly led to considerable confusion. In some instances this confusion has been unavoidable, because of the lack of information about the detailed character of the form or deposit involved and its genesis. On the other hand, much confusion could have been avoided if authors had been more ready to state the basis of their classification system and had indicated the limitations of their evidence. There is, for example, no excuse for those authors who insist on calling a ridge or a series of mounds of well-sorted and stratified sand and gravel a moraine, when there is abundant evidence that meltwater rather than ice is primarily responsible for its deposition.

This chapter has dealt with the development of the field of glacial geomorphology and the nature of the data, methodology and terminology used by geomorphologists, and provides the background to what follows. Before discussing in a systematic manner the glacial and fluvioglacial processes involved in the development of landforms and deposits it is necessary to examine both ice and water, the two major agents, in terms of their own characteristics and the work they are capable of doing. These two agents are the prime concern of three separate disciplines: glaciology, hydrology and glacial geology (including sedimentology).

In the first part of this chapter it has been demonstrated that by far the most significant contributions to glacial geomorphology between 1860 and 1920 came from geologists, both in Europe and North America. During the last fifty years, at least in Europe, there has been an increase in the contribution made by physical geographers and a relative decrease in the contribution of geologists, who have largely concerned themselves with stratigraphical and chronological aspects of the Pleistocene as opposed to the geomorphological aspects. In the United States, and in Canada to a lesser extent, glacial geomorphology has largely developed at the hands of geologists. There is no major philosophical difference between the glacial geomorphology undertaken by geologists and that undertaken by physical geographers. No one would deny the necessity of examining both morphology and internal character of glacial landforms at the same time, and the training of the physical geographer and the glacial geologist must permit him to describe and analyse morphology, areal distribution and internal constituents. Both the physical geographer and the glacial geologist must have an understanding of glaciology if their interpretations are to be of value. It is not in terms of a lack of communication between glacial geologists and physical geographers that problems arise, but in a lack of communication between glaciologists and glacial geomorphologists and a lack of interest on the part of many glacial geomorphologists in becoming familiar with presently glacierized environments and developments in glaciology. Glaciology and hydrology have a great deal to contribute to the understanding of problems in glacial geomorphology. It is beyond the scope of this book to provide detailed coverage of these fields, but a great deal of relevant information can be found in the pages of the Journal of Glaciology and such texts as those by Lliboutry (1964), Shumskii (1964), Kingery (1963) and Paterson (1969). In the following chapter an examination will be made of the two agents of prime concern to the glacial geomorphologist, namely ice and meltwater.

II GLACIER ICE AND MELTWATER

THE geomorphological processes associated with glacier ice and meltwater are the major concern of this book, but before discussing those processes and the landforms and deposits they produce, it is necessary to discuss the character of glacier ice and meltwater. Such a discussion involves an examination of two disciplines that are quite separate from geomorphology but which are very relevant to the work of geomorphologists.

Ice in all its aspects is the object of study of glaciology. This discipline is not only concerned with glacier ice but with sea and lake ice and ice on rivers and in the soil. However, there is now a large body of literature concerned with glacier ice, much of it to be found in the Journal of Glaciology. In addition, several excellent summaries already exist (e.g., Sharp 1960, Lliboutry 1964, Embleton and King 1968, Paterson 1969, Flint 1971). Glaciology has developed rapidly during the last ten years and the subject now has a substantial body of data, some distinctive techniques and some well-tried theories. In the following sections those aspects of glaciology of direct relevance to the glacial geomorphologists will be discussed.

The meltwaters released when glacier ice melts are responsible for many of the landforms and deposits that occur in an area that has been glaciated, with meltwater acting as a geomorphological agent in broadly the same manner as any flowing water in stream channels. Landforms and deposits produced by fluvial processes are the separate concern of another branch of geomorphology (see Leopold, Wolman and Miller 1964, and Morisawa 1968). However, many of the processes studied by fluvial geomorphologists are relevant to the study of landforms and deposits produced by meltwaters. Bearing in mind the very distinctive characteristics of glacial environments it is necessary to discuss the ways in which meltwater streams differ in the work they perform when compared with streams in other environments. For this reason a section of this chapter will deal with meltwater streams as geomorphological agents.

THE PRESENT AND FORMER EXTENT OF GLACIERS AND ICE SHEETS

Glacier ice at present covers 15 million km^2 or about 10% of the earth's land area. Only 4% of the ice occurs outside Antarctica and Greenland. The Antarctic ice sheet is the largest mass of ice, covering an area of 12·6 million km^2; it has an average thickness of 2·2 km and a maximum thickness of 4·25 km (Donn *et al.*, 1962). The volume of ice in the Antarctic ice sheet is probably of the order of 31 million km^3 and if all that ice melted world sea-levels would rise by some 40 m (allowing for an isostatic recovery of 20 m). Net snow accumulation on the ice sheet ranges from 0 to 85 cm of water equivalent per year but a common figure is 12 cm. Most estimates indicate a positive mass balance for the ice sheet as a whole at present and it is believed that it was probably initiated in the Tertiary period and existed throughout the Pleistocene (Hollin 1962).

The Greenland ice sheet covers an area of 1·73 million km^2, some 83% of which is above an altitude of 1400 m (the firn line). It has an average thickness of 1·6 km, a maximum thickness of 3·4 km and an approximate volume of 2·8 million km^3. This ice sheet has a mean accumulation of 31 cm of water equivalent per year and a balanced budget.

The remainder of the earth's present ice cover is mainly located in areas of high altitude where a high percentage of the annual precipitation occurs as snow. Most of the earth's great mountain belts contain glaciers or ice caps. Large glaciers occur in the Western Cordillera of North America, the Andes, the European Alps, Scandinavia, the Urals, the Himalayas and New Zealand. Although the ice masses of these areas are very significant local phenomena, they only constitute 4 per cent of the earth's ice cover and have a total volume of 500 000 km^3.

The fact that a much larger area of the earth's surface had been covered by glacier ice in the past was only established in the middle of the last century. This idea was concerned with events commonly referred to as the Great Ice Age, which was known to have been a very recent event in terms of earth history. Much earlier periods of ice cover have subsequently been established. Glaciations occurred in the Precambrian and the Permo-Carboniferous (Kilburn, Pitcher and Shackleton, 1965). There is considerable information available about the extent of the Permo-Carboniferous glaciation but the period over which it lasted is not known at all accurately.

The largest amount of information about periods of glaciation during the history of the earth refers to the last 3 million years. Mean annual temperature in Central Europe during the Eocene was 21°-22°C (Woldstedt, 1954). During the rest of the Tertiary period mean temperatures dropped about 10°C and the beginning of the cold period known as the Pleistocene is usually dated as somewhere between 2 and 3 million years BP. The Pleistocene period was characterized by extensive ice sheets that waxed and waned several times. There were several separate cold phases characterized by an extension of the ice sheets separated by periods of milder climate during which the ice sheets either became much smaller or disappeared altogether.

At their maximum extent, the Pleistocene ice sheets (Figs 1 and 2) covered an area of 47 million km^2 (Donn *et al.*, 1962), which represents about 30 per cent of the earth's land area. The major ice sheets outside of the Antarctic and Greenland were the Laurentide ice sheet of North America (including the Ellesmere-Baffin ice sheet) and the Scandinavian-Great Britain ice sheet. The North American ice sheet covered an area of 18·45 million km^2 and attained a maximum thickness of 3·5 km. The Scandinavian ice sheet covered an area of 6·67 million km^2 and had a maximum thickness of 3 km. A third, non-polar ice sheet of much smaller dimensions occurred in North Central Siberia and at its maximum it covered 3·73 million km^2. These three ice sheets contained half of the world's ice by volume, at their maximum extent during the

FIG. 1. Maximum extent of Pleistocene glaciers and ice sheets in the northern hemisphere.

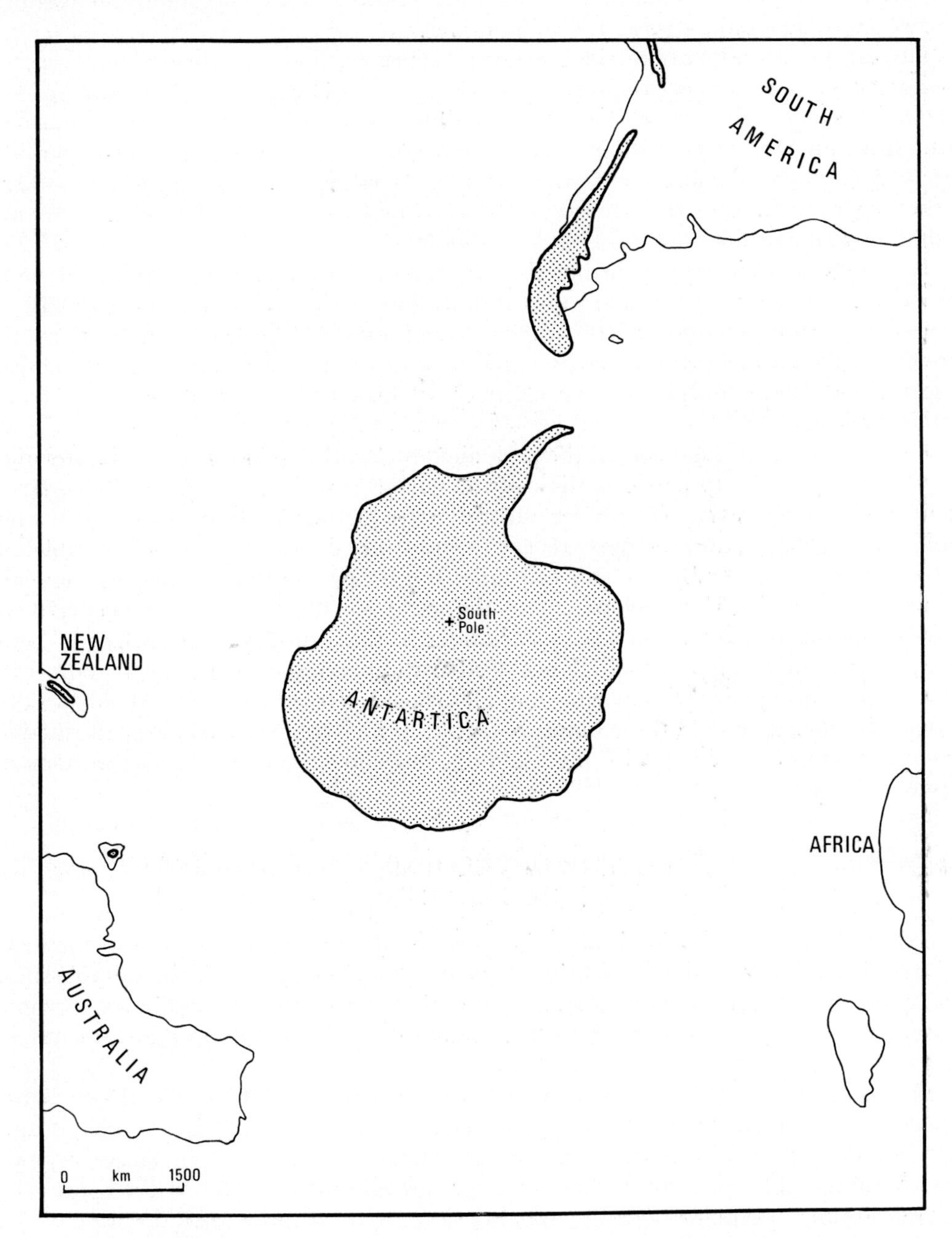

FIG. 2. Maximum extent of Pleistocene glaciers and ice sheets in the southern hemisphere.

Pleistocene. The total volume of ice on the earth's surface at the maximum extent of the ice sheets has been estimated as 100 million km^3.

The various extensions of the Pleistocene ice sheets seem to have covered more or less the same areas on each occasion and probably had similar patterns of build up, ice flow and wastage. The greatest extent of the Pleistocene ice sheets probably occurred in the Illinoian or Saale glaciations. Average world temperatures in the glacial periods were 5°-7°C cooler than at present while the interglacials may have been rather warmer. In the European Alps the snow-line was depressed by 1300 m, in Britain by 1200 m and in the Western Cordillera of the U.S.A. by 800 m.

This increase of ice cover during the Pleistocene Period as compared to the area presently covered by ice has important implications as far as the geomorphologist is concerned. An explanation of the origins of landforms and deposits that occur on 20 per cent of the earth's surface can be found in terms of the processes currently active beneath, and at the margins of, the ice masses which now cover 10 per cent of the earth's surface.

The question as to what caused the periodic deterioration in the earth's climates and the subsequent build up and extension of the ice cover has been considered by earth scientists for many years. The older theories are summarized by Brooks (1949) and Charlesworth (1957), and the more recent theories (i.e. the last 20 years) by Embleton and King (1968). The debate still continues but it appears that no single cause can satisfactorily explain the development of the great ice sheets that have occurred at various times in the earth's history. Variations in the altitude of the surface of land areas, short-term changes in the balance of solar radiation and longer-term changes in the quantity and quality of solar emission, all probably play their part. Whatever the cause of the glacial periods, the geomorphologist is primarily interested in the distinctive processes that are associated with the build up, movement, and subsequent wastage of masses of glacier ice.

THE DEVELOPMENT OF GLACIERS AND ICE SHEETS

Glaciers and ice sheets develop in situations where snow accumulation exceeds melting. This requires suitable climatic conditions as well as landforms that provide suitable sites. The combination of climatological and morphological factors conducive to snow accumulation frequently occurs above the annual snow-line either on plateau surfaces or in valley heads.

The transformation of snow to glacier ice is accomplished by changes in the constituent crystals, involving crystal displacement, changes in size and shape of crystals and their internal deformation (Paterson, 1969). These changes result in larger crystals that are more tightly packed, so that air spaces are eliminated. When the ice is near the pressure melting point these changes are accomplished more easily because of the presence of meltwater. The processes involved in the transformation of snow to glacier ice are compaction, sintering, recrystallization and movement along the internal glide planes of crystals. All these processes produce an increase in density of the material.

Fresh snow has a density of 0·1 to 0·2 and is a loose aggregate of snow flakes that can be regarded as an unconsolidated aeolian sediment accumulated in layers. It is possible for small rock fragments, volcanic dust and pollen grains to be trapped in the snow layers. Loose snow changes to granular snow, with a density of 0.3, over a period of a few days or, at most, a few months. The almost spherical granules of snow, frequently referred to as 'old snow', are a loose and highly pervious granular aggregate.

After the formation of granular snow, densification continues by modification of individual grains and by packing which reduces pore spaces and permeability and increases density and compactness. When a density of 0·5 is reached the material is known as 'firn'. When the remaining pore spaces are sealed off and the mass is impermeable and a density of 0·85 has been reached the transformation to glacier ice has taken place. Further densification can take place, so that common densities for glacier ice are 0·89 and 0·90. The density of pure ice is 0·917.

The rate of transformation of snow to ice varies from place to place because the process is sensitive to temperature and the rate of snow accumulation. The presence of meltwater speeds up the process, so that the transformation proceeds more rapidly in

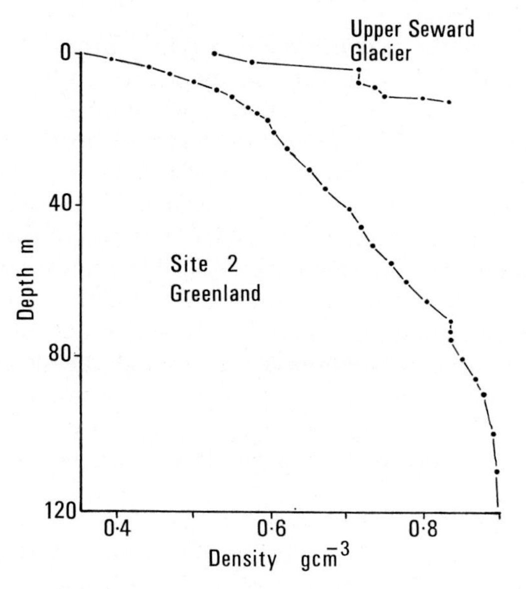

FIG. 3. Variation in firn density with depth in the upper Seward Glacier (Sharp, 1951) and in the Greenland ice sheet (Langway, 1967). (After Paterson, 1969).

areas of wet snow than areas of dry snow. Such differences in the rate of transformation can be illustrated by two examples (Paterson, 1969), one based on data from the Yukon Territory and the other from Greenland (Fig. 3). Firn becomes ice (density=0·85) at a depth of about 13 m on the Upper Seward Glacier in the Yukon, but not until a depth of 80 m at Site 2 in Greenland. In terms of the known rates of snow accumulation in each area, snow is transformed into ice in three to five years in the Upper Seward Glacier, whereas the process takes more than 100 years at Site 2 in Greenland.

The original snow accumulation can be regarded as an aeolian sediment which is converted to a sedimentary rock by diagenetic processes. When the glacier ice begins to move it is converted into a metamorphic rock with structures and fabrics. Before discussing the mechanisms of ice movement it is necessary to consider the problem of the classification of glaciers and ice sheets.

THE CLASSIFICATION OF GLACIERS AND ICE SHEETS

There are various ways to classify glaciers and ice sheets. A classification can be based on the temperature of the ice forming the glacier or ice sheet, the dynamic characteristics of the ice or the morphology of the ice.

H. W. Ahlmann's classification (1935b, 1948), based on the temperature of the ice and the amount of surface melting, still remains the most widely used. However, more recent investigations (Müller, 1962) indicate that not all glaciers and ice sheets can be fitted into any one of Ahlmann's three types and that several zones of different type can occur on one glacier or ice sheet. Ahlmann's classification allowed for three basic types: temperate, subpolar and polar (cold).

In a temperate glacier the ice is at the pressure melting point throughout, except for a surface layer about 10 to 20 m thick in which the temperature can be below 0°C for at least part of the year. This cold wave must be eliminated by the end of summer for a glacier to be classed as temperate. Meltwater can penetrate to the base of a femperate glacier. This means that the rate of movement of the ice is usually faster in a temperate glacier than in subpolar and polar or cold glaciers. The rate of transtormation of snow to firn is also greater on temperate glaciers because of the summer meltwater produced in the accumulation zone which speeds up the firnication process. In cold glaciers and ice sheets the temperature is below the pressure melting point. The main difference between subpolar and polar glaciers is that on subpolar glaciers some melting takes place in the accumulation area in summer even though, below a depth of 10 to 20 m, the ice is well below the pressure melting point. In contrast, there is no melting in the accumulation area of the high polar type, the transformation from snow to glacier ice takes much longer and most of the glacier is well below the pressure melting point.

The dynamic classification of glaciers is based on their internal characteristics and on this basis it is possible to define three types: active, passive, and dead. Active glaciers usually have a positive mass balance, but in some cases movement is maintained even though a negative mass balance exists. In general, very active glaciers tend to have large mass budgets, i.e. a large annual accumulation accompanied by large amounts of ablation. A passive glacier still receives nourishment but displays very little activity. A dead glacier no longer receives nourishment from the accumulation zone and movement is restricted to that produced by local slopes. This is not the most useful classification of glaciers and ice sheets, because it is also related to the thermal character of the ice as well as to the mass balance of the glacier or ice sheet. Cold glaciers are frozen to their bases and require a much greater shear stress to induce movement. In general, cold glaciers move much more slowly than temperate glaciers. A glacier with a large mass budget will have more rapid rates of movement than a glacier with a small mass budget. Since cold glaciers usually have much smaller budgets than temperate glaciers they can be expected to move at slower rates. Rapidly moving temperate glaciers with a high budget can have rapid rates of movement even in terminal lobes, which are experiencing large amounts of ablation, and even if the ice-margin is retreating rapidly.

Linked with the dynamic classification of glaciers are the concepts of thickening and thinning of an ice mass and the associated changes in its volume and areal extent. A positive mass balance, that is with accumulation in excess of ablation, produces a thickening of an ice mass. This in turn is usually reflected in a forward movement of the terminus and a steepening of the ice front, although there may be a considerable time lag. A negative mass balance, i.e. ablation in excess of accumulation, produces a thinning of the ice mass which is reflected in a retreat of the terminus and a flattening of the ice front. The thinning of an ice mass can also result in the cutting off of the source of supply of ice to a certain part of the glacier or ice sheet. When this occurs that part of the ice mass which has lost its ice supply becomes 'dead'. An area of dead ice may not become immobile for a considerable time after the supply is cut off, but when all accretion and movement cease such an ice mass is termed stagnant.

The morphological classification (Fig. 4) is based on the size, shape and position of ice masses as well as the relative proportions of the ice masses that occur at different altitudes. In very general terms small ice masses can indicate either very early or very late stages in development (i.e. in terms of time), whereas very large ice masses have usually been in existence for a relatively long period of time.

Cirque Glaciers occupy valley heads or depressions on mountain slopes. The mass of ice in a cirque glacier may only have a length and width measured in hundreds of metres and a thickness measured in tens of metres. Cirque glaciers may occupy rock basins with steep back walls and frequently occur at high altitudes in mountainous areas. They are usually the first to develop and the last to disappear during a period of glaciation. A fundamental characteristic of a cirque glacier is that its shape is determined by the rock walls that surround it (Fig. 4*a*).

Valley Glacier—Alpine type. During a period of deteriorating climate and therefore of falling snow-line, cirque glaciers develop a positive mass balance and begin to extend down-valley. Several cirque glaciers often combine and feed the main valley glacier, which is characteristically long and narrow and confined between the valley walls and terminates in a narrow tongue of ice (Fig. 4*b*). Valley glaciers are common in areas of high mountains, for example, the European Alps and the coastal mountains of Alaska. If a large number of cirque glaciers at high altitudes and with positive mass balances combine to form a valley glacier the latter can descend to very low altitudes. Valley glaciers can be subdivided on the basis of the percentage of their areas in different altitudinal zones.

Valley Glacier—outlet type. This type is similar in its lower reaches to the Alpine valley glacier but instead of being fed by a cirque glacier or a series of cirque glaciers, it is fed by an ice cap or ice sheet (Fig. 4*c*). Numerous valley glaciers of this type descend from the Vatnajökull ice cap in Iceland and the Greenland and Antarctic ice sheets.

Transection Glacier. It is possible for a number of glaciers to occupy a series of radiating valleys in a mountain massif in which the accumulation area is not large enough to permit the development of an ice cap. Frequently the mountain massif is too deeply dissected to allow an ice cap to form. In this situation it is possible for the

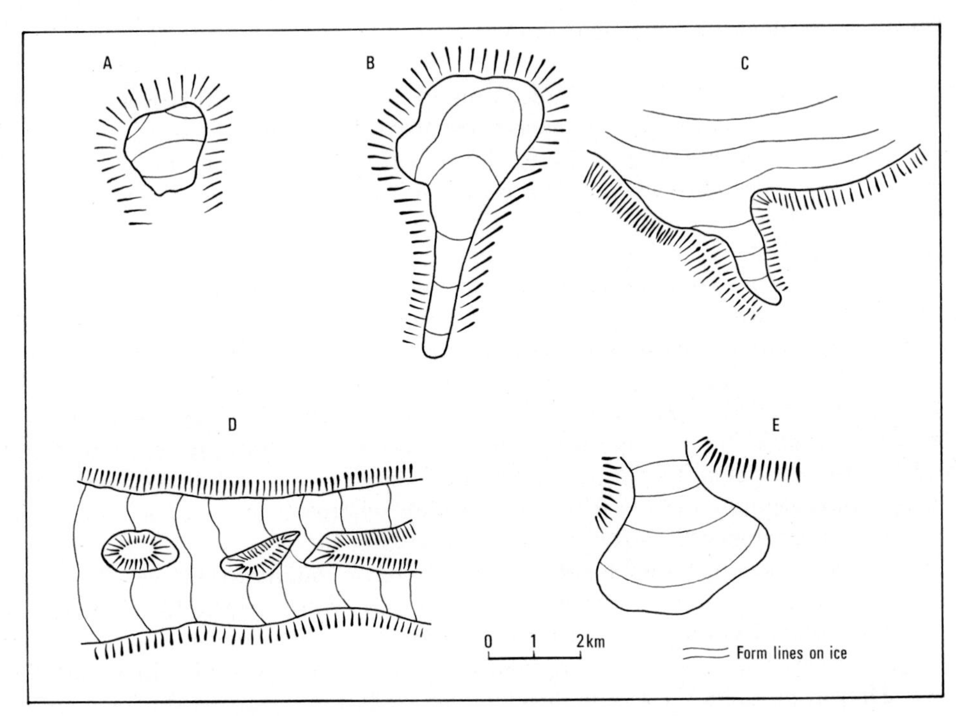

FIG. 4. Morphological classification of glaciers. *A*—Cirque glacier; *B*—Valley glacier, Alpine type; *C*—Valley glacier, outlet type; *D*—Transection glacier; *E*—Piedmont glacier.

valley glaciers to become so thick that they overspill across the divides between their confining valleys (Fig. 4*d*). It is in such circumstances that transfluent ice breaches and cols can develop.

Piedmont Glacier. When a valley glacier extends beyond its confining valley walls on to a lowland it spreads out to form a piedmont glacier (Fig. 4*e*). In such a situation a relatively large percentage of the glacier surface is at a low altitude.

Ice Cap. The development of an accumulation area at a relatively high altitude on an upland surface can produce an ice cap. Unlike cirque glaciers or valley glaciers, which are confined by valley walls, the ice in an ice cap moves in directions largely determined by the slope of the ice surface, which is in turn related to rates of accumulation. Vatnajökull in south-east Iceland is a typical ice cap that has attained a thickness of 600-700 m on a surface 1000-2001 m above sea level. The only difference between an ice cap and an ice sheet is that ice sheets are much larger.

Lowland Ice Cap. This term is usually applied to relatively small ice masses at low altitudes in the Arctic, e.g., Barnes ice cap. These ice masses are usually not very active because of their low total budget and low underlying relief.

Ice Sheet. Both during the Pleistocene and at present large areas of the earth's surface have supported large continuous areas of ice cover. These ice sheets completely submerge the underlying topography and develop lines of movement unrelated to the submerged topography. Only two ice sheets exist at present. The Greenland ice sheet contains 11 per cent of the world's ice and the Antarctic ice sheet 85 per cent. During the Pleistocene the Laurentide ice sheet in North America covered about the same area as the present Antarctic ice sheet but the Scandinavian ice sheet only covered half that area.

ICE MOVEMENT

When walking across a glacier it is often hard to believe that the mass of material beneath one's feet, which may be hundreds or even thousands of metres thick, is constantly moving. The flow of glaciers has been compared with the flow of rivers and although both glaciers and rivers consist of water, albeit in different states, the actual flow mechanisms and the factors that determine the rate of flow in each case are very different.

Altmann (1751), who recognized that gravity was the cause of glacier movement, believed that the movement was accomplished by the glacier sliding over its bed. Bordier (1773) suggested that ice flowed as a viscous fluid. However, it has only been in the last twenty years that a proper understanding of the mechanics of ice flow has been obtained with major contributions from such workers as Sharp, Nye, Glen, Lewis, Weertman and Lliboutry. Rapid progress was made after the realization that ice as a crystalline solid would deform, like other crystalline solids such as metals, at temperatures near its melting point. The application of modern ideas in solid state physics and metallurgy to problems in glaciology has produced a strong theoretical approach that has been largely substantiated by field observations (Paterson, 1969, p. 2).

A glacier moves in three ways: by internal plastic deformation, by sliding over its bed and by faulting. Paterson, (1969, p. 26) states, 'The ice in an active glacier is a mass of interlocking, roughly equigranular crystals. Relative movement between crystals is limited, but each crystal may deform internally. However, the fact that the general shape of the crystals does not change greatly with time shows that deformation is accompanied by recrystallization.'

Plastic deformation in a single crystal is achieved by the gliding, one over another, of layers parallel to the basal plane, the process being affected by the shape and orientation of the individual crystal. The plasticity of ice is demonstrated by the way in which glacier ice is often moulded to the form across which it moves. J. G. McCall (1960) observed in a tunnel cut through Vesl-Skautbreen, Norway, that grooves were impressed in the ice as it moved across its rock bed and retained their form for some 50 m beyond the obstacle. Since the glacier was moving at the rate of 3 m per year it appears that such grooves could survive for about 15 years. Such observations suggest that the ice behaves plastically under pressure but remains rigid after the pressure is released.

Rigsby (1960) studied crystal characteristics in 8000 samples from three temperate and three polar glaciers. Crystals in samples taken from regions of high stress were small but crystals from stagnant ice were up to 10 cm in diameter. Crystals from near the centre line of a valley glacier were randomly oriented whereas those from ice near valley walls, where high stresses result from the drag of the valley wall, had strong preferred orientations. Kamb and Shreve (1963) studied crystals in a core in the Blue Glacier, U.S.A. Crystal size increased slightly down to 70 m and then decreased towards the bottom at 137 m. The fabric diagrams varied with depth but below 100 m they had three or four maxima.

The actual mechanism and resultant crystal characteristics associated with plastic deformation are much better understood than the mechanism of basal sliding. On average, sliding comprises about half the total movement of glaciers at the pressure melting point. If the basal ice is below the pressure melting point sliding will not take place.

Weertman (1957, 1964a) has suggested two mechanisms of glacier sliding. The first involves pressure melting. The up-stream sides of irregularities on the glacier bed provide a resistance to ice movement and there is a resultant increase in pressure on the up-stream sides of obstacles. Melting then takes place and the meltwater flows around the obstacle and refreezes on the down-stream side in an area of lower pressure. This process is maintained by the flow of latent heat of fusion, released on refreezing on the down-stream side, through the obstacle and surrounding ice to the up-stream side. This process does not work when the ice is below the melting point, or when obstacles on the glacier bed are of the order of more than one metre in length because the amount of heat conducted through such obstacles is small.

Weertman's second process involves enhanced plastic flow. All ice is deformed plastically but near an obstacle the longitudinal stress in the ice is greater than average, so the strain rate will be above average. Velocity is proportional to strain rate times distance, and the larger the obstacle the greater the distance over which stress is increased and therefore the greater the velocity. This mechanism is thus more effective for large obstacles. It is probable that a glacier moves over its bed by a combination of these two mechanisms.

Weertman (1964a), also studied the effect on sliding velocity of a water layer on the glacier bed. If a water layer submerged small obstacles, they would not support any basal shear stress; the larger obstacles would, therefore, support larger stresses and plastic flow and sliding velocity near the larger obstacles would increase. Weertman calculated that a water depth of about one tenth of the controlling obstacle size would increase the sliding velocity by about 20 per cent. This is a possible mechanism to account for the observed increase in glacier movement during the melt season and may help to explain glacier surges.

There is little information available about the nature of water in and at the base of a glacier (Paterson 1969, p. 126). The water may occur in channels, as a sheet or in cavities. Lliboutry (1965, p. 647-53) thinks water-filled cavities are important. Such cavities will develop on the lee side of obstacles and if they become filled with water they will enlarge and tend to increase basal sliding.

Weertman and Lliboutry used different models for the glacier bed. In Weertman's model the glacier bed is rough in both directions and in Lliboutry's model the bed has a 'washboard' form. In both cases the bed is supposed to be impermeable to water.

The conditions at the glacier bed are probably very variable and only a very small sample of these conditions has been observed, in tunnels, crevasses and caves (e.g., Carol, 1947; Haefeli, 1951; McCall, 1952; Kamb and LaChapelle, 1964), but some of these observations do indicate the existence of both pressure melting and plastic deformation. The nature of the form of the bed, the presence or absence of till or gravel, the permeability of the underlying rock and the distribution and amount of any water all play their part in determining the rate of basal sliding. According to Paterson (1969), basal slip and the related topic of the causes of glacier surges are the most important unsolved problems in glacier dynamics today.

Having discussed the mechanisms of ice movement, it is now necessary to examine the factors that cause ice to move and which influence its rate of movement. Ice movement is directly related to the thickness, the temperature and the surface slope of the ice mass. These factors are all related to the mass balance of the ice mass, that is, the balance between accumulation and ablation. Equilibrium is rarely achieved for long and the relationship between mass balance and movement of a glacier is a complex one. Since changes in the mass balance affect the volume of the ice and therefore the form of the mass, they also affect the rates of flow in the ice mass which in turn affect the position of the ice margin.

In an ideal temperate valley glacier of uniform width throughout its length, some ice and all the snow accumulated in the previous winter is lost during the summer (i.e., in the ablation zone). In the upper part of the glacier (accumulation zone) the amount of snow added to the surface exceeds the amount lost by melting. If the profile of a glacier is to remain unchanged, '. . . the amount of ice flowing, in a year, through any cross section (perpendicular to the direction of flow) in the accumulation area must equal the total amount of snow which has collected during the year on the area of the glacier above the cross section. Similarly, in the ablation area, the amount of ice flowing through a cross section in a year must equal the amount of ice lost from the glacier between the cross section and the terminus. Thus the amount of ice flowing through any cross section must increase steadily from zero at the head of the glacier to a maximum at the "equilibrium line" (the boundary between the accumulation and ablation areas) and from there, decrease steadily towards the terminus.' (Paterson, 1969, p. 64). So long as ice thickness and width remain constant throughout the length of the glacier, velocity will vary in the same manner as discharge, that is, maximum velocities will occur near the equilibrium line. Velocity vectors will be inclined downwards relative to the surface (Fig. 5) in the accumulation area and upwards relative to the surface in the ablation areas.

In terms of a simple theoretical model, ice velocities are mainly determined by ice thickness and surface slope. Ice tends to flow in the direction of maximum surface slope even if that is up-hill in terms of the slope of the subglacial surface. Velocity is proportional to the fourth power of ice thickness and the third power of surface slope. The product of thickness times surface slope is fairly constant so that where the ice is thin the ice surface tends to be steep and where it is thick the ice surface tends to have a low angle of slope.

A sudden increase in the mass balance of a glacier is often propagated through the length of the glacier by an increase in thickness which moves along the length of the

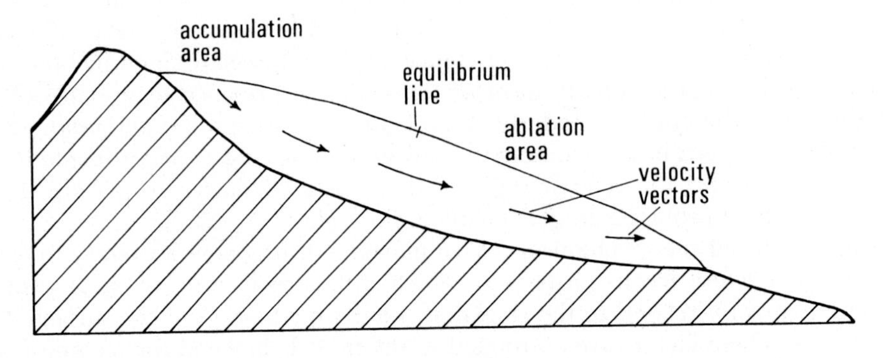

FIG. 5. An ideal valley glacier.

glacier in the form of a kinematic wave. 'In the accumulation area, the surface rises towards its new steady state value at an exponentially decreasing rate. In the ablation area, the surface initially rises and at an increasing rate.' (Paterson, 1969, p. 205). This increase in thickness can be relatively very large near the terminus and so the glacier will advance. In the reverse situation involving a decrease in mass balance, the effect at the terminus can be equally large, so that a relatively small reduction in mass balance can produce a substantial retreat.

Since many of the physical properties of ice are dependent on temperature it is necessary to consider the temperature of the ice in glaciers and ice sheets when discussing ice movement. From a geomorphological point of view it is important to know if a glacier is frozen to its bed; if so, it is hard to see how much erosion could be accomplished.

Temperatures in an ice mass are affected by two sources of heat, the geothermal heat flux at the base of the ice and the temperature at the ice surface. Heat transfer through the ice mass is achieved by conduction, advection and the movement of meltwater (if any). In the upper 15 m the temperature of the ice varies with changes in air temperature at the surface. Seasonal variations in ice temperature are not detected at depths greater than 15 m and at 10 m their amplitude is unlikely to exceed 1°C. Annual mean air temperatures and snow temperatures below the level of seasonal fluctuations in the firn in dry snow areas are very similar, e.g., Camp Century, Greenland, Lat. 77°10'N, Long. 61°08'W, firn temperature at 10 m below the surface was −24·0°C and air temperature was −23·6°C (Weertman, 1968). At the same site the negative temperature gradient commonly encountered in polar ice masses has been measured (Hansen and Langway, 1966).

Depth in metres	Temperature °C
10	−24·0
154	−24·6
1387	−13·0

From this example it can be seen that there is a temperature difference of 11°C between the surface ice and the basal ice. This difference can theoretically amount to as much as 30° in a very large thick ice sheet. It is therefore not possible to regard large ice sheets as isothermal. The highest temperatures and the greatest shear stresses occur at the base of ice sheets and the change of horizontal velocity with depth should be large near the bottom but small elsewhere.

The steady state thickness of an ice sheet is probably very insensitive to its mass balance. The colder the ice sheet, the thicker it should be. It has been estimated that a decrease in temperature of about 20°C will increase thickness by about 35 per cent. The shape of an ice sheet is a function of the plastic properties of ice. Where ice is thick the surface slope is small, but where ice is thin the surface slope will be relatively steep, therefore irregularities in the ice surface reflect irregularities of the rock floor.

Weertman (1961a) has concluded that if a small ice sheet exceeds a certain critical width, it may be unstable and become large as a result of only a moderate increase in accumulation. Similarly, a moderate decrease in accumulation, or increase in ablation, if maintained, might cause a large ice sheet to decrease in size rapidly. Such a theory might explain the repeated rapid growth and decay of the Pleistocene ice sheets (Paterson, 1969, p. 162).

In very general terms it can be stated that cold ice will move more slowly than temperate ice and maximum shear will take place near the base of cold ice masses where temperatures are likely to be highest. If a cold ice mass thickens as a result of increased accumulation, slow rates of internal movement and low rates of ablation, it is possible for the temperature of the basal layers to increase because of the blanket effect of the ice cover and the continued supply of geothermal heat. Once the basal ice reaches the pressure melting point rates of flow increase and the ice sheet will extend itself over a larger area by outward movement. After such a period of rapid expansion, shrinkage sets in until either a new equilibrium condition is reached or the ice sheet disappears. Weertman (1964b) has calculated, by assuming a net accumulation of 20-60 cm per year and assuming perfect plasticity, that it would take 15000 to 30000 years to build up a Pleistocene ice sheet. Also assuming ablation rates of one to two metres per year over an appreciable area of the ice sheet it would only take 2000 to 4000 years to destroy the ice sheet.

It can be seen therefore, that the rate of ice movement in large ice sheets is closely related to the temperature of the ice. Problems associated with the build up, movement and wastage of large ice sheets must be considered in terms of the temperature of the basal ice; many problems can only be solved if the basal ice, even in otherwise cold ice sheets, is at the pressure melting point at least during part of the time that a large ice sheet exists. Not only is it necessary to regard different zones of individual glaciers as having different internal temperature characteristics at the same time, but it is highly likely that any given ice sheet will have different temperature characteristics both at different places at the same time and at different times throughout its history. For a flow line through Byrd Station in West Antarctica, Budd, Jenssen and Radok (1970, p. 303) have calculated that 'melting starts in a deep trough where the ice thickness is over 3,500 m. The meltwater refreezes on to the base of the ice as the ice flows up over a ridge, and this is followed by further melting and finally refreezing as the ice shelf is approached.' The map accompanying this paper indicates considerable areas around the periphery of the Antarctic ice sheet that are covered by ice, the basal temperature of which is at the pressure melting point. Zotikov (1963) has estimated that the basal ice may reach the melting point over about half of the area of the Antarctic ice sheet. The evidence of landforms and deposits developed in subglacial environments associated with the wastage of the North American and European Pleistocene ice sheets is very considerable. Such evidence suggests that at least during their period of decay the marginal zones of those ice sheets were at the pressure melting point, but this does not deny a former polar (i.e. cold) character for the same ice sheets at an earlier stage.

All the ice in a temperate glacier or ice sheet is at the pressure melting point except for that portion of the ice mass near the surface, which is subject to seasonal variations. It is sometimes assumed that all ice masses in the polar regions are cold and all ice masses outside the polar regions are temperate. This is hardly justified on the basis of the limited number of measurements of ice temperature that currently exist. The best conditions for the development of temperate ice masses are heavy winter snow falls and intensive summer melting. As Paterson (1969, p. 177), points out, 'Few temperature measurements have been made at depth in ice near the melting point. To the normal problems of deep drilling is added the fact that drilling may affect the temperature by introducing meltwater and by changing the pressure in the ice.' Indirect information about ice temperature can be obtained by the behaviour of meltwater streams in the ice mass. If meltwater streams penetrate to the base of the ice mass and appear in tunnels at the base of the terminus, then the ice mass is at least temperate in its basal layers. If on the other hand meltwater streams are restricted to channels in the ice surface, or to tunnels within the ice, then the ice mass is likely to be cold.

The movement of ice when the whole mass is at the pressure melting point is enhanced by the presence of meltwater. This is illustrated by seasonal variations in velocity recorded on temperate glaciers. Elliston (1963) has shown that on the Gornergletscher, Switzerland, winter velocity is 20 to 50 per cent slower than the average annual velocity and the summer velocity exceeds the average by 20 to 80 per cent. These changes can be understood if meltwater acts as a lubricant at the base of the glacier.

Most valley glaciers move at rates of between 10 and 200 m per year although in ice falls the rate may increase to 1000 to 2000 m per year. Rates of movement on the Antarctic ice sheet are of the order of 30 m per year, but outlet valley glaciers draining from the ice sheet move at rates ranging between 300 and 1400 m per year (Paterson 1969, p. 75). Velocity rates across the surface of a valley glacier (Fig. 6) are usually fastest near the centre line and decrease towards the valley sides. Velocities in temperate valley glaciers also tend to decrease towards the base of the glacier (Fig. 7). Savage

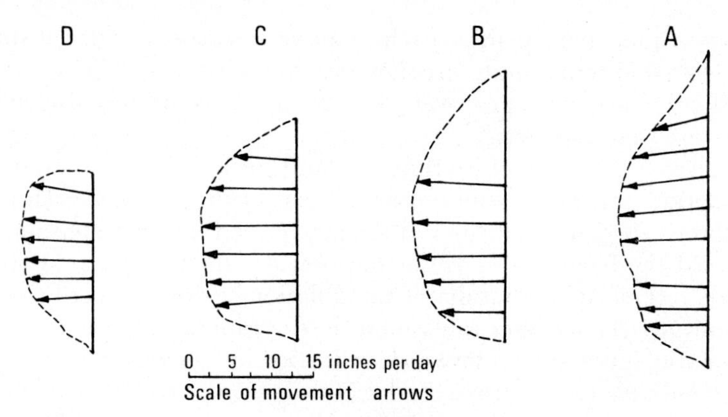

Fig. 6. Horizontal exponent of surface velocity measured on the Blue Glacier (1957-58) along transverse profiles near the firn line (*A*) and near the terminus (*D*). (After Sharp, 1960.)

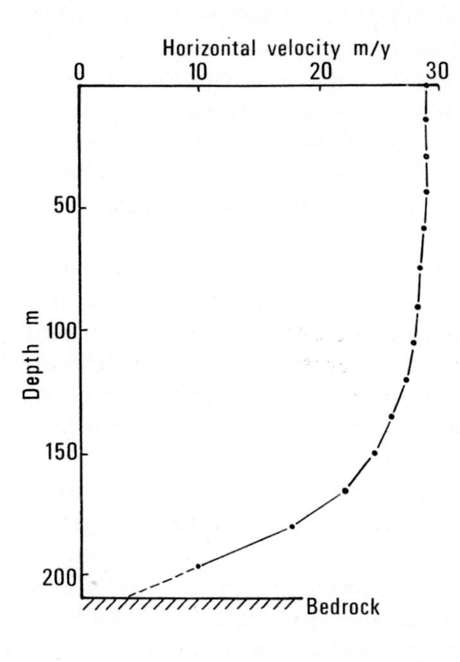

Fig. 7. Variation of horizontal velocity with depth: Athabasca Glacier. (After Paterson, 1969.)

and Paterson (1963) have published data based on borehole measurements in the Athabasca Glacier in Canada which show that velocity varies very little with depth in the upper half of the glacier but in the lower half velocity decreases at an increasing rate as the bottom is approached.

Seasonal variations in glacier velocities can be considerable. Summer and winter velocities in the same glacier can differ by as much as 10 to 20 per cent. Monthly average velocities may differ by as much as 40 per cent and hourly differences by as much as 100 per cent. Paterson (1969, p. 78) states 'Measurements over periods of a few hours suggest that movement at each point proceeds in a series of small jerks and that the jerks at different points are not synchronized. . . . The great majority of measured velocities are averages over periods of weeks or months, in which these discontinuities are smoothed out.'

Since velocity is a function of ice thickness and surface slope it should be at a maximum in late spring and at a minimum at the end of the summer. However, since it appears that lubrication by meltwater at the glacier bed is very important in ice masses in which at least the basal layers are at the pressure melting point, velocities in such glaciers will be largest at the middle of the ablation season. There can also be temporary increases in velocity after a period of heavy rain.

Very dramatic increases in forward velocities in glaciers associated with rapid advance of ice margins have attracted a lot of attention in recent years (see *Science 162*, (3852), 1968, and *Canadian Journal of Earth Sciences*, *6*, (4), 1969). These glacial surges can be regarded as rare events although there is a mounting record of their total occurrence (Tarr and Martin, 1914; Glen, 1941; Desio, 1954; Lliboutry, 1958; Hattersley-Smith, 1964; Post, 1960 and 1965). These surges are usually completed in one or two years and may involve a forward movement of the ice front of 10 to 20 km in a matter of a few years. Ice velocities during a surge can average 20 m per day, e.g., The Muldrow Glacier, (Post, 1960), but maximum ice velocities may be one hundred times greater than the velocities which occur in the glacier when it is not surging. During a glacier surge the upper part of a valley glacier may be lowered by 50 or 60 m and tributary glaciers may be left hanging, while in the lower part of the glacier the ice may thicken by 50 or 60 m.

Numerous theories have been put forward to explain glacier surges. Increased avalanche activity related to the occurrence of earthquakes has been suggested as a cause of the very rapid increase in accumulation and therefore of an unstable situation resulting in rapid increase in velocity. Temperature changes at the base of the glacier as a result of an increase in velocity in turn produced by an increase in net mass balance, have also been suggested. Other workers have related surges to local increases in geothermal heat. Whatever the cause, a glacier surge has very important geomorphological implications, not only in terms of rapid changes in the location of ice margins but in terms of the increased powers of erosion, transportation and deposition that the increases in ice velocity produce.

It can be seen from the above discussion that glacier movement is a very complex subject. It is closely related to mass balance studies and therefore to climatological factors. The mechanisms and changes in rate of ice movement are strictly the province of the glaciologist but the results of ice movement are of great concern to the geomorphologist. Not only does the balance between accumulation, movement and ablation determine the surface form and areal extent of glaciers and ice sheets but the temperature of the ice affects the mechanism and rate of ice movement and the penetration of meltwater into the ice mass.

The form and extent of an ice mass is of great concern to the geomorphologist. A great many of the landforms and deposits to be discussed in later chapters of this book are affected by the form of the ice mass with which they were associated. Variations in the thickness of the ice mass and changes in the profile, shape and position of the ice terminus are of fundamental importance in the development of glacial and fluvioglacial landforms and deposits. The relationship between rates of ice movement, rates of ablation and oscillations in the ice margin will be discussed with reference to specific landforms in later sections. However, it is appropriate to make some general points at this stage.

The surface form of an ice mass largely reflects its glaciological characteristics. In accumulation areas with gentle surface slopes, the ice surface tends to be fairly uniform. Once ice begins to move downslope in a valley the drag of the valley sides will create shear stresses that produce crevasses at an angle to the glacier margin and pointing up glacier (Fig. 8). In valley glaciers longitudinal compression and extension occur.

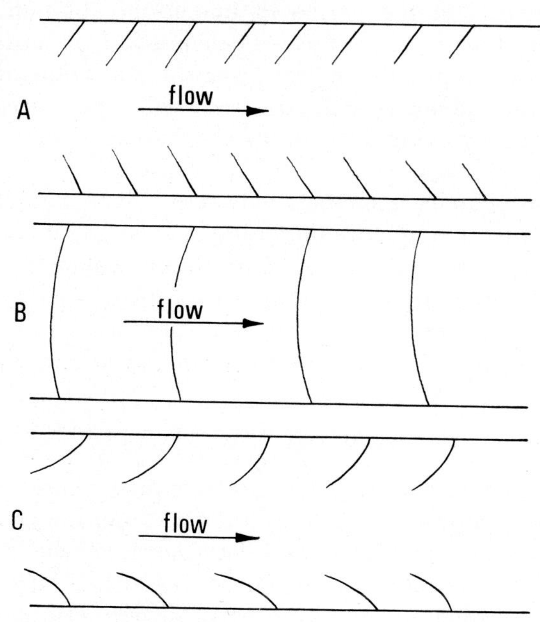

FIG. 8. Crevasse patterns in a valley glacier. *A*—Effect of shear stress of valley walls only; *B*—Shear stress and extending flow; *C*—Shear stress and compressive flow. (After Paterson, 1969 and Nye, 1952.)

Nye (1952, p. 88) states 'The result is that in the first case the forward velocity of the glacier decreases as one goes down glacier, because the ice is being compressed, and in the second case the velocity increases because the ice is being extended . . . one would expect transverse crevasses to form during extending flow but not during compressive flow.' Measured depths of crevasses rarely exceed 30 m, but crevasses exceeding 36 m in depth are known in the Antarctic and Greenland (Miller 1954). It is probable that deeper crevasses can occur in cold ice compared with temperate ice.

The occurrence of crevasses affects the penetration of meltwater and of debris into the ice mass. The development of crevasses on both the lateral and frontal margins of a glacier will of course affect the development of landforms and deposits. But it is in the form of the ice margin that glaciological characteristics play their most important part in affecting geomorphological processes. If an ice mass has a positive mass balance and this is reflected, possibly after a time lag, in high velocities, the ice margin is not only likely to be advancing but it will also be steep. With the advance of this steep vigorous ice front, that at least for part of the year will undergo ablation, there will be the release of large volumes of water. The energy of ice motion that results in glacial erosion, transport and deposition is thus supplemented by the energy of meltwater flow that also carries out erosion, transportation and deposition.

It has been stated above that, at least in temperate ice masses, maximum ice velocities are associated with the period of maximum ablation and therefore with periods of maximum meltwater discharge. The energy available for geomorphological work is therefore at a maximum during the advance of a temperate glacier during the ablation season. Actual advance of the ice front will take place during the ablation season if the amount of ice moving through the ablation zone is greater than the amount of ice lost by ablation.

During the wastage of an ice mass there may be less energy available for geomorphological work. The ice is moving more sluggishly or even not at all in the terminal area. However, if the climate has ameliorated much larger volumes of meltwater may be discharged during the ablation season than when the glacier was advancing. It is likely that fluvioglacial processes are much more significant during periods of glacier retreat than during periods of glacier advance, when glacial processes will be dominant.

GLACIAL MELTWATER

Those processes by which ice and snow are lost from an ice mass are collectively known as ablation. The most important results of ablation from the geomorphological point of view are the production of meltwater and the lowering of the ice surface.

The amount of ablation that takes place on an ice mass is largely determined by the number of hours of sunshine and the number of degree days above 0°C. However, ablation can result from several different combinations of weather conditions and correlations established in one area may not necessarily be true for others (Paterson, 1969, p. 228). Ablation involves heat exchange at the ice surface. If the temperature of the ice is below 0°C this heat exchange will simply change the temperature of the ice, but if the ice is at 0°C any heat received will cause melting.

Heat is supplied by solar (short-wave) radiation, long-wave radiation from carbon dioxide and water vapour, molecular and eddy conduction from the air, condensation of water vapour, conduction from the underlying ice and the freezing of rain. Heat is lost by outgoing long-wave radiation, molecular and eddy conduction, condensation and conduction from the underlying ice. Radiation is usually the single most important process (Paterson, 1969, p. 47).

If any ice surface receives more heat than it looses, the ice is warmed. After the temperature reaches 0°C the surplus heat melts and evaporates the ice. Evaporation usually constitutes less than 5 per cent of ablation so that although evaporation and condensation are significant terms in the energy balance because of their high latent energy, they are of little importance to mass balance because melting predominates.

Rates and total amounts of ablation over any given year will vary greatly from place to place because of variations in climatic and ice conditions. On the Antarctic ice sheet at low altitudes ablation from snow ranges between 20-80 cm of water equivalent per year; in Greenland rates of 100 cm per year are common. In the terminal areas of temperate glaciers at low altitudes daily ablation rates can be as high as 10 to 15 cm of water equivalent per day and as much as 12 m of ice can be lost from the terminus of some glaciers during one ablation season.

Ablation results in the lowering of the ice surface and the release of meltwater. Rapid lowering of the surface of ice masses undergoing rapid decay can result in distinctive landforms and deposits being produced in association with the fast-changing ice margins. The meltwater released by ablation occurs as sheet flow and channel flow on the ice surface and, in temperate ice masses, descends cracks, crevasses and moulins to flow within and beneath the ice. Some cavities within the ice are isolated and others are linked by channels. 'As long as the cavities and channels contain enough water the ice cannot flow in and close them.' (Paterson, 1969, p. 134).

Little is known about the behaviour of water within an ice mass. A temperate ice mass can be regarded as similar to a groundwater storage system. However, the behaviour of some streams fed by ice masses indicates that the analogy can be misleading. Floods on streams issuing from the Athabasca Glacier, Canada, have been studied by Mathews (1963). Over a period of thirteen summers, ten floods of 250 000 m^3 or more have been recorded. Since there were no ice-dammed lakes to contribute to those floods the water must have been stored in cavities within the ice. Storage could not have taken place in one large cavity because no subsidence of the ice surface was observed after the floods occurred. It is more likely that storage took place in a number of smaller cavities. Fisher (1963) has described a water-filled cavity estimated to be at least 1000 m^3 revealed by a tunnel on the Breithorn in Switzerland. The cavity was 20 m to 30 m below the ice surface.

The draining of marginal ice-dammed lakes and englacial cavities is not the only cause of fluctuations in the amount of meltwater issuing from an ice mass. The discharge of streams emerging at an ice front during the summer usually shows marked diurnal variations. The maximum discharge may be twice the minimum. 'The daily peak in discharge usually occurs around 17.00 or 18.00 hours, that is, a few hours after the daily peak in the rate of ice melt. Total daily discharge is usually greatest in late July or early August (in the northern hemisphere). Some waters continue to flow throughout the winter, the amount appears to be of the same order of magnitude as the amount of water which an average value of geothermal heat flux would melt from the base of the glacier.' (Paterson, 1969, p. 132).

The geomorphological work achieved by water flowing on and in glacial ice is primarily in the form of transportation of debris. However, some of these supraglacial and englacial streams do impinge on rock or drift deposits and cut channels into their surfaces (see Chap. V), and in other cases deposits are laid down on the floor of channels on and in the ice and are eventually let down on to the subglacial surface. (See Chap. VI).

The rate and periodicity of ice melt gives marginal and proglacial meltwater streams particular characteristics. They have been compared by Krigstrom (1962, p. 340) '... with water courses in arid and semi-arid environments. The similarity is principally one of great variations in discharge both in periglacial rivers and in rivers in climatic zones with marked fluctuations between dry and wet seasons. Due to heavy sediment load, areas of aggradation with braided river courses develop.'

Meltwater streams have the same mechanisms of erosion, transportation and deposition as streams in other environments. It is simply their association with ice that often leads to very unusual locations for stream channels and to a variable discharge and sediment load both over the short and long term. Variations in discharge are closely linked with the seasonal variations in ablation rates. On Breidamerkurjökull, Iceland, the mean annual loss of ice from the ablation area below 290 m over the period 1945 to 1965 has been calculated by Howarth (1968a) and Welch (1967) to be 150 million m^3 of ice. Assuming an ice density of 0·9 this is equivalent to 135 million m^3 of water. The total mean annual ablation from this glacier is likely to be at least 25 per cent greater than the calculated figure. However, just in terms of the measured ice loss over a part of the ablation area, the discharge of meltwater is equivalent to a single river with a discharge of 40 m^3 per second over a period of twelve months. A very high percentage of the flow occurs during the four to five months of the summer, and typical discharge rates for a single river could be expected to be of the order of 80 m^3 per second. These are relatively large discharges when it is borne in mind that the drainage area represented in the calculation is only approximately 55 km^2. Arnborg (1955a) has calculated that the discharge of the Hornafjardarfljot which drains Hoffellsjökull has an annual discharge of 44 m^3 per second, of which 40 per cent is contributed by precipitation and 60 per cent by ice melt. The drainage area involved in this case was 200 km^2. If similar relationships between area, ablation and precipitation occur on Breidamerkurjökull as occur on Hoffellsjökull it is likely that the total discharge from the Breidamerkurjökull drainage area amounts to 260 m^3 per second averaged over one year; this must represent peak discharges in excess of 500 m^3 per second.

The above calculations indicate that ablation is capable of at least doubling average annual discharge but is certainly capable of increasing summer discharge by a factor of four. The energy available for geomorphological work as a result of ablation is very considerable indeed.

Glacier ice and meltwater have been major geomorphological agents over 30 per cent of the earth's land area during the last two to three million years. The existence of ice sheets and glaciers at present has enabled glaciologists, hydrologists, physicists and climatologists to study the processes involved in the transformation of snow to ice, the build up and decay of ice masses, the nature of ice movement and the processes of ablation. It is in fact remarkable how much is known about glacier ice considering that the collection of data means the establishment of scientific programmes in isolated and difficult environments. Prior to 1950 much of the work was carried out by dedicated individuals willing to work in very unfavourable conditions. The advent of the International Geophysical Year (1957-1958) established research stations in many formerly inaccessible areas and the last 20 years have seen a growing number of persons involved in the study of glaciers and ice sheets in all their aspects. The results of the investigations of the last 20 years, combined with some of the excellent observations by geologists and geographers made at the turn of the century while exploring glacierized areas, now provide a basis for an interpretation of at least some of the processes which produce landforms and deposits in areas either presently or previously covered by glacier ice. The remainder of this book is concerned with those processes.

III GLACIAL EROSION AND TRANSPORTATION

THE ability of ice to modify the landforms over which it moves is widely accepted and there is a large volume of literature devoted to the description of landforms produced by glacial erosion. The actual mechanisms of ice erosion and transportation are poorly understood but this is not surprising for two reasons. First, the emphasis on describing and classifying the end products of glacial erosion limits the interpretation of the origins of the forms. It is extremely difficult to determine the extent of modification of landforms by ice erosion when the pre-glacial form is not known. In a similar way, multiple glaciation also complicates the interpretation of the final form. Secondly, it has proved exceedingly difficult to observe the mechanisms of glacial erosion at work because most of their work is done beneath considerable thicknesses of ice. The penetration of the basal layers of glaciers for the purpose of studying the mechanics of movement, erosion and transportation has only been achieved on a limited scale. However, the small sample of subglacial environments that has been studied has contributed proportionally far more to our understanding of the mechanisms of glacial erosion than have the volumes devoted to the description of the end products of glacial erosion.

The crux of the problem is what happens at the ice-rock (solid or drift) interface. Ice that simply sits on a rock surface without any movement will have little or no effect on that rock surface. Ice that moves across a rock surface can have a considerable effect in shaping the surface form of that rock. The ice must be regarded as the medium responsible for erosion although it is commonly asserted that clean ice moving across resistant, poorly-jointed rock is unlikely to affect that rock to any great extent. However, at the ice-rock interface beneath an active, temperate glacier, both the character of the rock and that of the ice is such that the movement of the ice over the rock surface results in fragments of rock being removed from the parent body and being transported by the moving ice to another location.

THE NATURE OF THE EVIDENCE

The mechanisms of glacial erosion have been deduced from observations made at the ice-rock interface, data concerning movement of the basal layers of ice in glaciers and ice sheets, and observations of rock debris in these basal layers. Deductions have also been made from theoretical interpretations of basal ice movement. As early as 1843, J. D. Forbes described basal ice at the Brenva Glacier (p. 204), '. . . set all over with sharp angular fragments . . . which were so firmly fixed in the ice as to demonstrate the impossibility of such a surface being forcibly urged forward without sawing and tearing any comparatively soft body which might be below it.' Forbes applied these observations near an existing glacier to his work on the Island of Skye, Scotland, (Forbes 1845) where he ascribed glacial erosion to two major processes, namely abrasion resulting from the incorporation of rock fragments in the basal layers of glacier ice, and plucking, which involved the removal of rock fragments from jointed or shattered bedrock.

Carol's (1947) observations of a *roche moutonnée* beneath the Ober Grindelwald Glacier supported the concept of plucking on the down-stream side of rock protuberances. He observed both the formation of meltwater as a result of local pressure melting, and the subsequent refreezing of the meltwater in a cavity on the down-stream side of the rock protrusion. The refreezing of the meltwater caused shattering of the bedrock and the shattered material was carried away by the moving ice. It must be emphasized that in this case the glacier ice itself is not responsible for the mechanical failure of the solid rock, the glacier simply is the transporting medium for rock fragments provided by freeze-thaw and other processes. Plucking is not achieved simply by the drag of moving ice. The tensile strength of solid rock is much greater than that of ice and it is unlikely that even simple adhesion of clean ice to rock by contact freezing is capable of removing anything but already loosened rock fragments. Plucking is probably facilitated by the protrusion of rock fragments already entrained in the ice at the ice-rock interface, and dragging other rock fragments along in the direction of ice movement. Boulders in the process of producing striations and in the act of quarrying the down-stream side of protuberances were observed at the side of Austerdalsbreen (Glen and Lewis, 1961).

Some very interested observations were made in a tunnel in the Blue Glacier, Washington, by Kamb and La Chapelle (1964). They observed basal sliding of the ice over the bedrock floor at the rate of 1·6 cm per day. They state (p. 162), 'When overburden pressure is removed from ice in contact with bedrock, by excavating away the ice above and around it, the basal ice freezes fast to the rock. When the basal ice is cut away quickly from the tunnel wall, without initial excavation and release of overburden pressure, it comes free from the sole and is not frozen to it. This observation demonstrates the presence of a thin layer of liquid water, at the pressure melting point, along the ice-rock interface.' This suggests that it is possible for local decreases in pressure to allow the ice to freeze to the bedrock thus providing a mechanism of rock removal. Kamb and La Chappelle also describe a regelation layer, ranging in thickness from 2·9 cm to zero where it pinches out against rock protuberances, but reforming on the down-stream side of the protuberances. They state (p. 164), 'The regelation layer is heavily loaded with debris in comparison with the ice above. The debris content varies markedly from place to place within the layer but is nowhere greater than about 10 per cent by volume of the layer. The debris consists of fine mud and of rock fragments up to 1 or 2 mm in size. . . . Rock debris accumulates on the up-stream side of bedrock protuberances. From the crest of such protuberances there extends down-stream a train of debris particles evidently derived from the accumulation on the up-stream side.' Although the authors do not state the source of the debris, it could be argued that it must originate from removal of rock fragments on the down-stream side of rock protuberances, as a result of shattering by freeze–thaw processes in the cavities known to exist in such locations and the subsequent entraining of the fragments in the regelation layer. Kamb and La Chapelle describe cavities beneath the ice: 'Where there are depressions below the general local level of the sole . . . the ice does not fill these in but instead bridges over them. This occurs for cavities ranging in width from 4 cm to at least 10 m, the corresponding separation of ice from bedrock ranges from about 1 cm to 20 cm.'

From the limited number of observations made of the ice-rock interface beneath ice at the pressure melting point, there can be little doubt that the four most important factors that determine the efficiency of ice as an agent of erosion are the presence of cavities between ice and the bedrock, the temperature of the ice, air and water in these cavities, the projection of rock fragments already entrained in the ice beneath the basal ice surface, and the rate of movement of the ice.

Since erosion implies the actual removal of rock fragments from their place of origin, the nature of ice movement in the basal layers of glaciers and ice sheets is of fundamental importance to an understanding of the mechanisms of glacial erosion. No erosion can take place under stationary ice. From direct observations in tunnels and by measurements of ice deformation in boreholes it is known that an important mechanism in the flow of temperate glaciers is the sliding of the ice over the bedrock beneath it. In the thicker parts of typical valley glaciers basal sliding normally accounts for about 50 per cent of their surface velocity, the remainder being due to internal deformation within the ice. In contrast to the basal sliding of temperate glaciers those glaciers that are below the pressure melting point at their base do not slide on their beds. In terms of the mechanisms of glacial erosion as they are presently understood it is difficult to envisage how erosion takes place under cold ice.

Kamb and LaChapelle (1964) measured the basal sliding at the side of the tunnel in the Blue Glacier and state (p. 162), '. . . essentially all the motion of 1·6 cm per day took place at the bedrock-ice interface . . . about 90 per cent of the total glacier motion at the observation site took place by basal sliding and only about 10 per cent by internal deformation.' There is no doubt that ice at the pressure melting point slides over the bedrock and this movement, along with the development of a regelation layer and of cavities as described above, is responsible for erosion at the ice-rock interface on the beds of glaciers and ice sheets.

Glen and Lewis (1961) measured sliding at the lateral margin of Austerdalsbreen. Measurements made near the base of an ice fall showed that side-slip of 26 cm per day amounted to 65 per cent of the maximum centre-line velocity. Lower down the glacier, side-slip amounted to 20 per cent of the maximum centre-line velocity. It appears, therefore, that the mechanisms of glacial erosion associated with basal sliding also occur along the sides of valley glaciers as a result of side-slip.

Although forward motion of temperate glacier ice over the bedrock has been observed and measured it is effective only in smoothing out irregularities at the ice-rock interface. It is only when these lines of movement occur at an angle to the rock surface that maximum erosion takes place. In 1952, Nye demonstrated the importance of extending and compressive flow in gaciers and in figures in that paper (Fig. 9) indicated slip-line fields that, theoretically, would develop in zones of compressive and extending flow. In a zone of extending flow the slip-lines descend towards the ice-rock interface, and in a zone of compressive flow the slip-lines rise from the ice-rock interface. Nye (1952, p. 88), states, 'A slip-line field is represented by two families of curves drawn so that their directions at any point give the two perpendicular directions of maximum shear stress (that is, the two directions in which the tendency to shear is greatest). In both the present cases they are parallel and perpendicular to the bed at the bottom but they turn so as to emerge at 45 degrees to the surface. The curves are parts of cycloids, and the field in compressive flow is the mirror image of the field in extending flow. The importance of the slip-lines is that they show the directions in which the ice has the

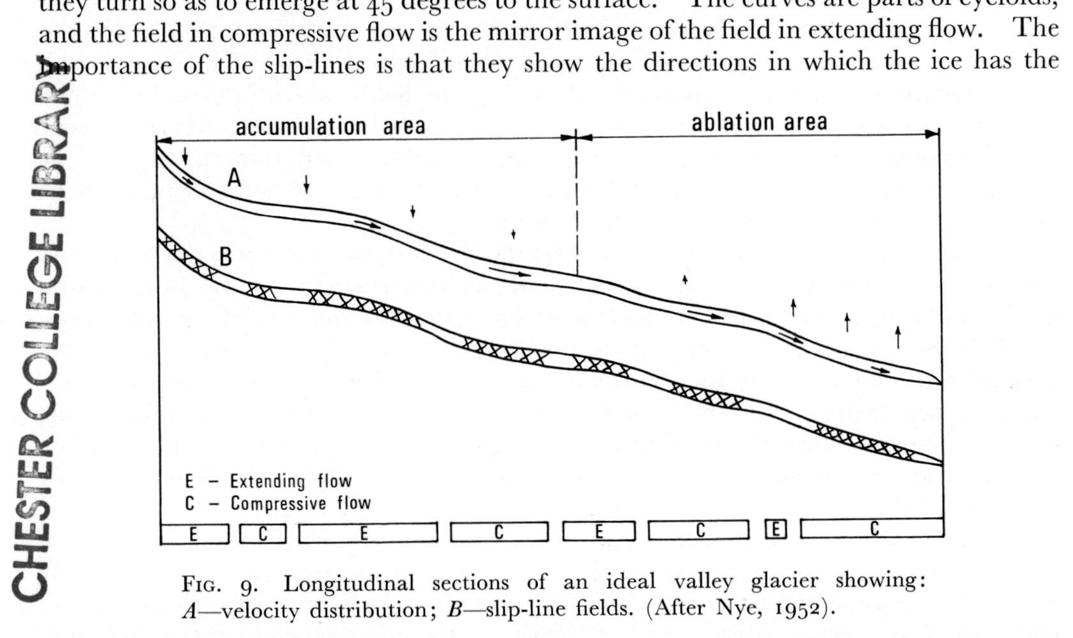

FIG. 9. Longitudinal sections of an ideal valley glacier showing: *A*—velocity distribution; *B*—slip-line fields. (After Nye, 1952).

greatest tendency to fracture by shear.' This concept of slip-line fields is related to the idea of rotational slip put forward by Forbes as early as 1848 (see Forbes, 1859). The irregularities of the ice-rock interface are therefore instrumental in developing slip-lines within the ice; these will develop curving up from the glacier bed at the base of steep slopes and near the snout of a glacier where compressive flow occurs. On the other hand, slip-lines curving down towards the glacier bed will occur over the crests of rises in the rock bed and in the accumulation zone of valley glaciers where extending flow tends to occur.

Nye and Martin (1967) subsequently extended the concept of slip-lines and dealt specifically with their effect on glacial erosion (p. 80): 'The glacier may be thought of as a giant sanding block. The long profile of the block is flexible, but whereas it can have any curvature in the concave sense it can only have a limited curvature in the convex sense. This block is covered by sandpaper of non-uniform roughness. It is then set to work on a rock surface of non-uniform hardness and is pressed down with a non-uniform distribution of pressure. The sandpaper is also endowed with a property of extensibility so that its velocity over the abraded material is non-uniform. The result would surely be to produce a smooth surface. . . . containing smoothly concave basins, with more sharply convex places where either the rock is harder, or where there were pre-existing convexities. The main point is that abrasion at a given place does not proceed at a rate given simply by the local conditions. . . . It depends also on the rates of abrasion occurring at other places.' The theoretical development of slip-fields as put forward by Nye is substantiated by field evidence, obtained by studying patterns of crevasses and shear planes on the surface and at the fronts of glaciers and ice sheets. Both the theory and the field evidence indicate the importance of structures which dip at an angle to the ice-rock interface and which allow the erosion mechanisms to proceed both on convex and in concave irregularities in the bedrock floor. The curvature of these irregularities affects the geometry of the slip-line fields, which in turn determines the shape of basins developed in the rock floor. Nye concludes that an observable rock bar need not be associated with a zone of relatively hard rock and states (p. 82), '. . . the eventual depth of the basins . . . as distinct from the time taken to make them, bears no relation whatever to the relative hardness of the rocks.'

In terms of observations made at the ice-rock interface the mechanics of erosion and the rates and methods of ice movement, and theoretical considerations of structural weaknesses and directions of ice movement, the actual mechanisms of ice erosion are poorly understood. Glacial erosion certainly results from the sliding of ice, at the pressure melting point, over the rock floor. However, the ice itself is so soft that without prior preparation of the rock material, so that it is easily incorporated within the ice, little erosion takes place. Mechanisms operating at the interface between ice below the pressure melting point and rock, if any, are unknown. They are certainly unrelated to sliding as the ice is frozen to the bed.

It has already been stated that erosion only takes place when rock fragments are actually removed from their parent body. There is much evidence of transport in the form of observations of rock fragments occurring within glacier ice in the basal layers of glaciers and ice sheets. Kamb and LaChapelle (1964) observed fragments up to 2 mm in size in the regelation layer in the Blue Glacier and similar observations have been recorded by Carol (1947), Haefeli (1951) and McCall (1952). Haefeli (1951) refers to inclusions of sand while McCall (1952) states that fragments up to 20 cm across were observed. More recently, Boulton (1970a, b) has described considerable amounts of debris derived from the glacier bed and contained in polar glaciers in Spitzbergen. The ice content of the debris bands ranges from 85 per cent to 10 per cent by volume. Boulton suggests that the debris was incorporated by basal freezing associated with regelation. Some distance up-glacier from the frontal margin, the englacial debris is concentrated in the basal layers, but nearer the frontal margin in a zone of compression both folding and thrusting occur and the englacial debris is transported to higher levels. The relatively heavy debris load described by Boulton is similar to that described by Koch and Wegner (1911) in East Greenland, Goldthwait (1951) in Baffin Island and Bishop (1957) in West Greenland. In contrast, the amount of included debris in temperate glaciers is usually much less, so Boulton suggests (p. 227), 'that many polar and sub-polar glaciers . . . carry very considerable amounts of englacial debris derived from the glacier bed, whereas temperate glaciers . . . carry very little derived englacial debris.' This statement seems to be in direct contradiction to the limited understanding that we have of the mechanics of glacial erosion. If basal sliding, regelation, and freeze-thaw processes are of prime importance in glacial erosion, temperate glaciers rather than polar glaciers should carry the most englacial debris. Boulton does suggest that, even though the Spitzbergen glaciers are polar at their margins, meltwater migrated from the interior and regelation took place when the meltwater entered the outer zone. He goes on to state (p. 228), 'This apparent contrast between the englacial debris load of temperate and polar glaciers supports the suggestion that temperature regime is the controlling factor in the englacial incorporation of large volumes of basally derived debris, and that only those glaciers in which basal freezing is an important process show large amounts of such debris. This carries the corollary that glaciers which are entirely frozen to their beds will contain very little englacial material.' It seems, therefore, that our understanding of the mechanics of incorporation and transport of rock fragments in situations where a large amount of debris is found in ice, which on the basis of other evidence is thought to be below the pressure melting point, is of a limited nature. On the one hand, the observations of cold glaciers indicate that they often contain more debris in their basal layers than do temperate ice masses. On the other hand, the mechanisms most likely to result in the incorporation of large amounts of debris appear to be related to ice that is at the pressure melting point. One of the problems in interpreting what appears to be contradictory evidence is to ascertain whether the large amount of debris observed in the basal layers of cold ice was included under the present temperature condition or at a time when the ice was at the pressure melting point.

From the above discussion it can be seen that the actual mechanics of glacial erosion are only poorly understood. However, there have been numerous attempts to estimate the total amount and rate of erosion accomplished by these mechanisms during a period of glaciation. A general idea of the lowering of an area by ice erosion can be obtained by measuring the load of streams draining a glacierized area. One of the earliest attempts was made by H. F. Reid (1892) in the area of the Muir Glacier, Glacier Bay, Alaska. He estimated that the silt content alone of a stream issuing from the glacier represented an annual lowering of the glacierized basin of 19 mm per year. Another study, again based on silt content of turbulent streams draining a glacierized area, in Iceland (Thorarinsson 1939a), produced a figure of 2·8 mm per year, and that represented a denudation rate five times greater than in an adjacent non-glacierized basin where the rate was calculated on the same basis. It is probable that the overall effects of glacierization over a given period of time are of the order of ten to twenty times greater than normal fluvial processes.

Another source of evidence about the mechanisms and the effectiveness of glacial erosion is revealed by examining those parts of the world where landforms resulting from glacial erosion are best developed. Some of the most dramatic forms of glacial erosion occur in the mountains of Norway, Scotland, British Columbia, south-eastern Alaska, and New Zealand. In these locations, thick, temperate glaciers descended rapidly from high mountains to sea-level over relatively short distances. It seems that steep gradients, large amounts of snow fall and generally temperate conditions were conducive to glacial erosion. On a small scale the repeated occurrence of cirques on the northern and north-eastern sides of mountains in the northern hemisphere points to the presence of controlling factors that influence the efficiency of the erosional mechanisms. In this case it seems likely that the length of time during which glacierization occurs is likely to be longest in these situations and therefore cirque development will be initiated and last longer in these locations.

Even a brief examination of topographic maps of areas known to have been glaciated quickly reveals evidence of the power of glacial ice to erode. Straight, steepsided troughs or low streamlined and grooved bedrock bear witness to ice erosion. Unfortunately, such evidence largely points only to the efficiency of the erosional mechanisms rather than to their mode of action.

FACTORS THAT INFLUENCE GLACIAL EROSION

The form of the surface upon which glaciers and ice sheets develop influences not only the character of the ice masses but also the mechanisms and extent of glacial erosion. One of the major problems in interpreting the development of landforms produced by glacial erosion is to establish the character of these landforms prior to the occurrence of erosion. In general terms, it is known that macro-features such as glacial troughs, cirques, horns and arêtes have been produced in mountainous areas where there was a considerable amount of relief prior to glaciation. Lowland areas, on the other hand, which prior to glaciation exhibited little relief, may contain grooves and troughs, but not of the same dimensions as those in mountainous areas. The relationship between amount of relief and the amount of glacial erosion is a function of the increased energy available for erosion in a glacier system which descends through a considerable range of altitude. Although increased ice velocity does not noticeably increase ice pressure, it does increase the rate at which the eroding and transporting medium moves across the surface being eroded. It seems likely that the rock particles entrained in and projecting below the ice mass play an important part in abrasion and plucking; high ice velocities mean that more particles pass over a given locality per unit of time, so more erosion can therefore be accomplished by fast-moving than by slow-moving ice. The correlation between the amplitude of the pre-glacial relief and the amount of glacial erosion accomplished is significant, but in practice it is often impossible to ascertain the nature of the pre-glacial relief after an area has been subjected to several periods of ice inundation. Apart from stating that the form of the pre-glacial relief is important, it is probably dangerous to try and estimate the amount of glacial erosion that has taken place in an area simply on the basis of comparative morphology.

The nature of the ice is of fundamental importance in determining the amount of glacial erosion. As well as its temperature, the velocity, thickness, surface slope and internal structures of the ice all influence glacial erosion. However, Nye and Martin (1967) point out (p. 80), '. . . abrasion at a given place does not proceed at a rate given simply by the local conditions (hardness of rock, amount of rock fragments carried by the ice, velocity of ice . . .). It depends also on the rates of abrasion occurring at other places . . .' The theory of slip-line fields as applied to relative rates of erosion beneath an ice mass emphasizes the curvature of the bed, which in turn places a restriction on the relative rates of erosion occurring at different points on the bed. It appears, therefore, that there is a complex series of interactions between the surface slope, velocity distribution, structural planes of weakness and the shape of the ice-rock interface. It is not possible to regard these factors independently because each affects the other. As pointed out by Nye and Martin (1967, p. 82), '. . . The shape of basins is determined not only by the relative hardness of the rocks but also by the large-scale geometry of the slip-line field.' But surely the geometry of the slip-line field reflects the pre-glacial shape of the ice-rock interface, which in turn may reflect relative resistance to erosion by processes other than ice. It seems likely that glacial erosion may well accentuate irregularities in the surface which existed prior to glaciation. However, it is possible for new basins to be eroded and their location need not necessarily be related to the occurrence of relatively softer rocks. Similarly, rock bars may develop that represent convexities in the ice-rock interface which are the result of the development of slip-line fields within the ice and are not related to the occurrence of more resistant rock at that location.

One of the most important controlling factors that influences the ability of glaciers and ice sheets to erode is the nature of the preparation of the rock that is subjected to erosion. The importance of rock fragments embedded in the basal layers of the ice to the mechanisms of erosion has already been discussed. Probably of even greater significance is the preparation of the rock surface by various forms of weathering prior to glaciation, and structural failure, both before and after the initiation of the ice cover. The occurrence of joint planes or vertical or near vertical bedding planes facilitates glacial erosion. If these lines of weakness are also attacked prior to the arrival of the glacier by means of frost-shattering then the extent of glacial erosion is much greater. Once the initial cover of ice has removed the weathered material, however, the rate of erosion may well be slowed unless the structural weaknesses themselves are exploited by the mechanisms of ice erosion, or further freeze–thaw cycles are possible in the subglacial environment.

The removal of rock material by ice erosion can produce structural weaknesses in the form of joints, developed roughly parallel to the ice-rock interface. The continued removal by glacial erosion of solid rock, which is usually at least twice as dense as the glacier ice, results in expansion of the rock when the overburden pressure is removed. (Lewis 1954, Linton 1963). Such relationships in the development of structural weaknesses as a direct result of glacial erosion could assist the erosional mechanisms in their continual attack on the subglacial surface.

One specific rock type which is believed to be very susceptible to erosion by ice is massive limestone. Sweeting (1966) has stated that the original scar and ledge relief of some limestone areas is believed to have been produced by glacial erosion, the most striking areas of limestone pavement occurring where scouring affected the most massively bedded limestones. Evidence of the importance of glacial erosion in producing limestone pavements is found in the occurrence of striae beneath erratic blocks perched on the pavements.

It is generally accepted that during the Pleistocene several glaciations occurred. In areas that have been covered on more than one occasion by an ice sheet or valley glacier, the effects of glacial erosion are likely to be more severe. This may be the result simply of the longer time period during which glacial mechanisms have operated. On the other hand, the preparation of rock surfaces by periglacial processes during the interglacials could also have facilitated glacial erosion during the subsequent glaciation.

MECHANISMS OF GLACIAL EROSION

In 1888, Chamberlin stated (p. 208), 'The ice of a glacier has of itself but little abrading and practically no striating power. It can neither groove nor scratch . . . The little effect that it may produce is of the nature of polishing.' This statement is perhaps a little strong but it is in essence true. Clean ice passing over unweathered and non-jointed rock accomplishes very little erosion. However, glacier ice is often far from clean in its basal layers and the rock over which it moves is often weathered and jointed.

The size of the debris incorporated in the basal layers of a glacier largely determines the type of abrasion which takes place. If the debris is rock-flour, silt or sand, rock surfaces become polished. If the debris is of gravel or boulder size, the rock surface will be striated (scratched) or grooved. These striations and grooves can be used as indicators of ice movement but it must be remembered that directions of basal ice movement, which is what striations reflect, can change over short distances and even at the same location over a period of time. A large number of striations must be studied before a generalization about the regional direction of ice movement can be made.

When a boulder is embedded in the basal ice with only a small area projecting, it possible that it is held sufficiently firmly for the projection to scratch and to groove the underlying rock surface. If the block rotates within the ice a new cutting edge will be exposed and possibly a second scratch or groove may be cut parallel with, but not continuous with, the first scratch or groove (Edelman 1951). Striations usually occur in large numbers rather than in isolation and they therefore reflect a relatively high concentration of debris in the basal ice. They can range in length from 1 cm to 1 m and have been observed on level, sloping, vertical and curved rock surfaces. They are most common on gently sloping surfaces up which the ice has ascended. They can begin and end quite abruptly. It is difficult to determine, with certainty, which is correct of the two possible directions that alignment of striations suggests for the direction of movement of the ice. However, the choice can usually be made on the basis of other evidence. Since striae are often only of the order of millimetres in depth their development is strongly affected by the lithology and structure of the rock in which they are cut and by the rate of weathering of the surface after the ice has left the surface.

The dividing line between abrasion and the other major mechanism of glacial erosion, plucking, is not easy to define. Abrasion involves erosion at the ice-rock interface without movement of the parent rock, and with removal of only small parts of the parent rock. Plucking, on the other hand, involves removal of larger fragments of the parent rock. The actual mechanisms involved are only poorly understood but are certainly related to ice at the pressure melting point. The incorporation of rock fragments into the basal ice may result from simple adhesion by freezing of ice on to loosened bedrock. This can only occur when the thin layer of meltwater that usually occurs at the ice-rock interface freezes, as a result of a local drop in pressure. Kamb and LaChapelle (1964) have observed the development of a regelation layer which involves melting on the up-glacier side of rock protuberances as a result of increase in pressure and the subsequent refreezing of the meltwater on the down-glacier side of the protuberance. Rock fragments in the regelation layer demonstrate the feasibility of this process.

Another form of plucking occurs because the projecting portions of rock fragments already included in the ice, drag other rock fragments along at the ice-rock interface. This process removes rock fragments from their parent outcrop and transports them along the ice-rock interface, but it does not necessarily involve their entrainment in the ice. Entrainment may subsequently result from regelation or pressure melting of the rock fragment upwards into the ice. Similarly, the development of shear planes associated with slip-line fields may produce differential movement along the plane of shear and this may allow the migration of rock fragments, loosened by the plucking mechanism, to travel along the shear planes up into the ice.

The plucking mechanism seems likely to be most effective on the down-stream side of well-jointed or weathered rock protuberances. It can also function in locations where flow-lines come into contact with the rock floor and allow transport of the plucked material up into the ice.

The strict definition of glacial erosion necessitates that glacier ice be directly involved in the entrainment and removal of the rock fragments. However, it has already been pointed out that various weathering processes and the development and subsequent refreezing of meltwater at the ice-rock interface are often very important factors in glacial erosion. Rarely is there a simple relationship between moving ice and solid rock. If the definition of glacial erosion be extended to include all processes that result in erosion beneath glacier ice, then it is necessary to examine the work carried out by the movement of till, mixtures of till and water, mixtures of ice and water, of water with some debris load and water without a debris load. There is a continuum ranging from ice at the one end to water at the other, with each intermediate stage representing various combinations of ice and debris and water and debris. The erosive work of water, and water containing some debris, will be discussed in Chapter V, since there is a sufficiently large number of landforms produced by fluvioglacial erosion to justify their separate treatment.

Numerous examples of cavities, sinuous courses, grooves or trenches are described in the literature (Demorest, 1938; Edelman, 1949; Ebers, 1961; Dahl, 1965; Gjessing, 1967). Gjessing concludes, because there is a difference between the direction of movement responsible for these forms and the main direction of movement of the ice sheet in Norway, that either these features were produced by a plastic ice mass, underneath the more rigid ice of the ice sheet, or they were produced by the scouring action of water-soaked ground moraine. He states (p. 27), 'Such a soaked ground moraine could appear as an independent substance underneath the inland ice, and with rock material and water in different proportions it could have the property of a plastic substance or a liquid of a viscosity which kept the flow laminar throughout the whole system of sinuous and bending courses.' Gjessing implies that the movement of the water-soaked till is initiated by the movement of basal layers of ice, but because the pressure in the soaked till is hydrostatic the pressure will be transmitted evenly in all directions through the till. This leads to a movement in the till, in the general direction of ice movement but capable of deflection around rock protuberances and into troughs and channels. Although Gjessing's suggestion seems plausible there is little or no evidence available to substantiate the effectiveness of water-soaked ground moraine as an effective erosional agent.

Johnsson (1956) suggests that forms similar to those described by Gjessing (1967) could be produced by a mixture of rock-flour, ice and water. Other workers (e.g. Hjulström, 1935; Ebers, 1961; Dahl, 1965) have suggested that these forms are mainly produced by meltwater under high pressure and involving cavitation.

The combination of ice, rock debris and water in varying proportions is certainly an effective agent of erosion. The details of the mechanisms involved may be poorly understood but the results of their work are widely distributed in areas that have suffered glaciation.

LANDFORMS PRODUCED BY GLACIAL EROSION

Prior to 1950 little was known about the conditions that actually prevailed beneath glaciers and ice sheets and nearly all the statements about the process of glacial erosion were deduced from the forms. Some of these statements have had to be modified as our understanding of glacier physics and the nature of the ice-rock interface has increased.

In areas of considerable local relief the results of glacial erosion are easily recognized. The straight, steepsided troughs, the cirques, horns and arêtes all bear witness to the erosion carried out by the passage of ice over the pre-glacial relief.

GLACIAL TROUGHS

The importance of the existence of a valley system and of the preparation of the rocks of this valley system by jointing and weathering before the initiation of an ice cover is widely stressed in the literature. The subsequent occupation of the valley system by a glacier system then modifies the pre-glacial relief and the characteristic straight troughs of Alpine areas are developed. The prime difference between fluvial and glacial erosion is the size of the channel occupied by the water on the one hand and the ice on the other. Whereas the river is in direct contact with only a very small proportion of the cross profile of the valley, the glacier is in direct contact with a very high proportion of the cross profile (Fig. 10). Various mechanisms of glacial erosion are at work on the valley floor and the valley sides, and the net result is that protuberances in both a lateral and a vertical direction tend to be removed. The typical glacial trough therefore, tends to be relatively straight, with steep sides, and in cross profile often approximates to a parabola. Svensson (1959) measured the cross profile of the Lapparten Valley in Sweden and defined it by the formula $y=0{\cdot}000402x^{2{\cdot}046}$, where y is the height and x is the distance from the valley centre. Because the exponent is almost 2, the formula closely approximates to that of a true parabola. This parabolic cross profile can be achieved by the widening of the pre-glacial V-shaped cross profile without over-deepening, or by both widening and deepening of the pre-glacial valley by glacial erosion. Graf (1970) has suggested that, as the intensity of glacial processes increase, valleys become relatively narrower and deeper.

If a valley system consisting of a trunk valley with numerous tributaries is occupied

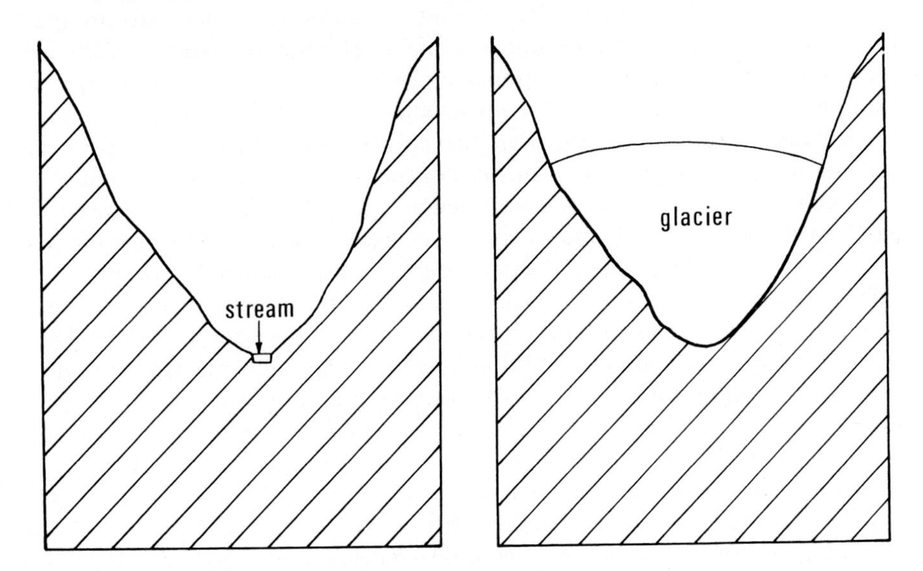

FIG. 10. Comparison of the proportion of a valley cross profile occupied by a water channel (stream)and an ice channel (glacier).

by a system of valley glaciers, the mere fact that the original trunk valley was deeper and wider than its tributaries usually means that the glacier occupying the trunk valley is wider and deeper than its tributary glaciers. Consequently, the ice-rock interface is of greater extent in the trunk valley than in the tributary valleys and more glacial erosion is achieved in the trunk valley than in the tributary valleys. This often results in the trunk valley being more deeply eroded, and the tributary valleys are left hanging above the floor of the trunk valley.

The long profiles of glaciated valleys are often steplike and frequently contain over-deepened sections in the form of rock basins. The step form can be compared with large-scale *roche moutonnées*, whereas the rock basins are probably related to lines of shear within the ice. Nye and Martin (1967) state (p. 82), '. . . the eventual depth of the basins . . . as distinct from the time taken to make them, bears no relation whatever to the relative hardness of the rock. Could this confusing situation possibly account for some of the confusing field evidence on the question of what determines the location of basins and overdeepened places in glacier valleys?'

Fiords are very similar to normal rock basins except for their coastal location and inundation by seawater. The rock bars that often occur at the mouths of fiords simply represent the down-valley lip of the normal rock basins. Some rock basins can be hundreds of metres deep, the deepest in Scotland being Loch Morar (1017 feet). According to Sissons (1967), there is no regular pattern of the deepest part of the rock basins in Scotland and he believes that this is the result of the local geology and relief affecting rates of glacial erosion.

Not only are glaciers capable of over-deepening their valley floors but they are also capable of breaching water-sheds. If a valley becomes full of ice any pre-glacial col along the watershed between that valley and its neighbour will tend to carry ice across the pre-glacial watershed. These diffluence channels can participate in the development of a complex system of valley glaciers which eventually may obliterate much of the pre-glacial divide system as the ice mass takes on the form of an ice cap or ice sheet. The most significant type of erosion in this context involves the destruction of a major pre-glacial watershed to produce a transfluence channel. The development of diffluence and transfluence channels may involve reversals in directions of ice movement as the initial system of valley glaciers is superseded by an ice cap or ice sheet.

Sugden (1968) discusses the selectivity of glacial erosion within an ice sheet and the implication that ice moves in streams within ice sheets. He quotes work by Dansgaard (1961) and Fristrup (1964) in Greenland and by Gow (1965) in Antarctica that revealed streaming of ice within the ice sheets of those areas. Sugden concludes (p. 90), 'It appears that streaming within ice sheets is a reasonably common occurrence and that under certain conditions such streams are capable of eroding troughs. . . . In the Cairngorms . . . in some places ice accomplished virtually no erosion, whereas in immediately adjacent areas it was highly effective as an agent of erosion.' Similarly Clayton (1965) writing about the Finger Lakes Region, states (p. 54), 'In all cases the erosion is concentrated in the valleys; the plateau may be increasingly segmented, but there seems no reason, even in the complex through valley zone, to regard it as seriously lowered by the passage of ice.' It seems therefore that even during the ice sheet stage pre-glacial valleys are modified into glacial troughs, divides may be breached and cols developed into diffluence channels with perhaps little or no modification of intervening plateau areas or ridges. In this context glacial erosion is certainly selective.

Linton (1963) has classified glacial troughs in terms of their relationship to the pre-glacial topography. Although such a classification presupposes that it is possible to distinguish the pre-glacial elements in a glaciated landscape, it does form a useful system of classification. In situations in which glacial troughs simply result from partial modification of the pre-glacial valley system and considerable areas of high ground overlook their accumulation areas the troughs are classified as the *Alpine type* (Fig. 11*a*). The *Icelandic type* (Fig. 11*b*) is related to accumulation of ice on plateau surfaces, generally above the altitude of the troughs, the troughs being cut by the discharge of ice from the plateau to the lower ground. In other circumstances, when an ice cover found the pre-glacial valley system inadequate in coping with the discharge of a complex valley glacier system, new troughs were cut either as diffluence or transfluence channels. The *Composite* troughs (Fig. 11*c*) which form in such a situation result from the amalgamation of segments of pre-glacial valleys, with new segments developing from the partial or complete destruction of pre-glacial divides.

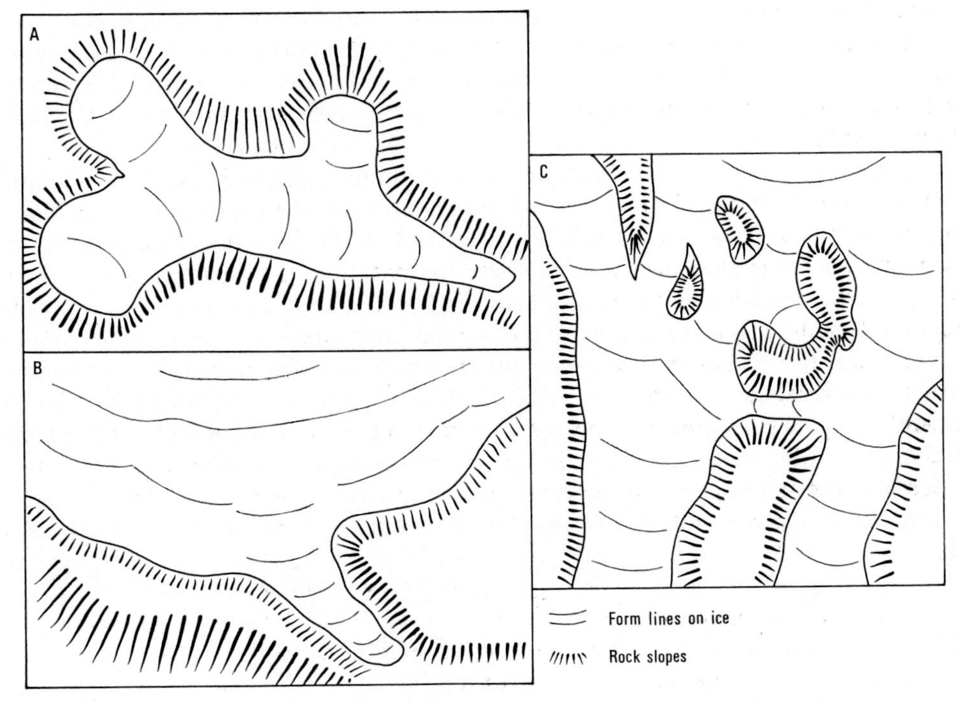

FIG. 11. Alpine (*A*), Icelandic (*B*) and Composite (*C*) glacial troughs.

CIRQUES

The term *cirque* was introduced in 1823 by Jean de Charpentier in a study of the Pyrénées. It is now widely used in connection with steep-sided depressions that are semicircular in plan, the floors of which are sometimes over-deepened to form rock basins. Cirques range in size from small shallow depressions tens of metres across to the very large feature on Mount Lister near McMurdo Sound, known as Walcott cirque; it is 16 km wide and has a back wall of 3000 m.

Cirques are most frequently found in areas of considerable relief where the processes of glacial erosion have eaten into a pre-glacial mountain system. They are not entirely limited to pre-glacial valley heads because they can also develop on smooth slopes subjected to nivation processes. Cirques tend to develop on north-west, north and north-east facing slopes in the northern hemisphere and south-west, south and south-east facing slopes in the southern hemisphere. (Lewis, 1938; Seddon, 1957; Schytt, 1959; Gage, 1961; Andrews, 1965; Temple, 1965; Sissons, 1967). Factors which influence the location of cirques include the occurrence of shade, the direction of the prevailing winds and the general alignment of the high ground on which they occur. These factors affect the amount and rate of snow accumulation and therefore the intensity of, and length of time during which, processes of glacial erosion are active.

The facts that cirques have a fairly consistent plan shape, that their lengths are usually about 3 times their heights (Manley, 1959) and that they often occur cutting across different rock types, suggest that the processes responsible for their development are similar from place to place and in general are little affected by local variations in rock type. Cirque development is affected by the length and number of glaciations and by the nature of the pre-glacial relief. Although rock type may not be an important controlling factor in cirque development, rock structures such as joints, faults, foliation and shatter planes have been shown to significantly aid the mechanics of erosion on cirque walls. (McCabe, 1939; Battey, 1960). The distribution of cirques in areas known to have been glaciated, and in close association with other features known to have been produced by glacial erosion, leave little doubt that they, too, are the products of severe glacial erosion. However, the actual mechanics of the erosion processes involved are still the subject of considerable speculation. The importance of headwall shattering by freeze–thaw processes certainly plays an important part in cirque development, but it is unlikely that such processes occur on any significant scale at the base of *bergschrunds* (Battle, 1960). Fisher (1963) has suggested that if the boundary between cold ice and ice at the pressure melting point migrates up or down a cirque headwall this may produce a zone where freeze–thaw processes are very active and allow erosion

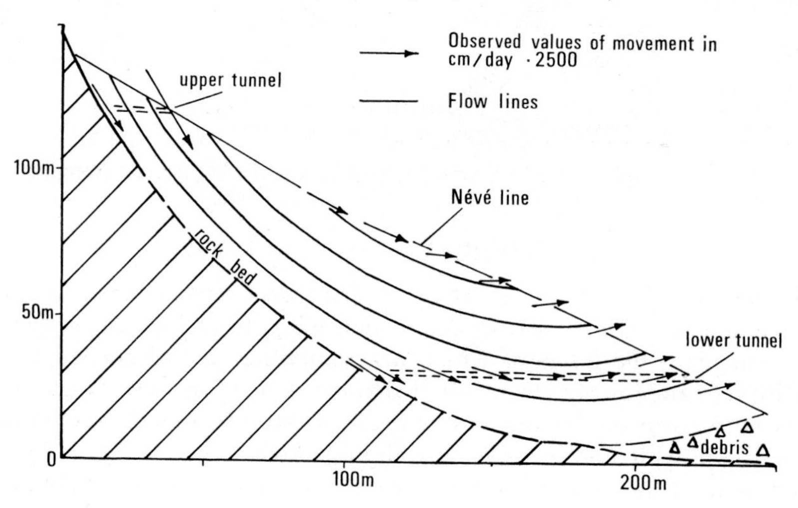

FIG. 12. Long section through Vesl-Skautbreen showing flow-lines and velocities. (After McCall in Lewis, 1960.)

of the headwall. Attack on the headwall and sides, by whatever mechanism which may be active, allows for the overall enlargement of cirques, but since many cirques have over-deepened floors, another mechanism must be sought to explain the development of these rock basins.

The work carried out by W. V. Lewis and his associates in Norway during the period 1950-1953 (Lewis, 1960) provided the first detailed information about the internal structure and movement of cirque glaciers. McCall's work on Vesl-Skautbreen demonstrated that the rotational movement of the ice (Fig. 12) favoured the formation of a rock basin. The backwall of the cirque was attacked by sapping, this process being greatly assisted by well-developed joints in the headwall. The material removed from the headwall was included within the basal layers of the glacier, which were seen to have a high concentration of debris. This debris acted as the tools for abrasion by the ice as it moved across the cirque floor. McCall points out (in Lewis, 1960, p. 57), 'The depth of ice required to give the vertical load of 2 kg/cm^2 is 22 m. Any depth greater than this would cause the ice to flow around and under the block and the bearing of the block on the bed would not be increased. Thus the amount of abrasion is directly proportional to the effective area of a fragment and is independent of the glacier depth at depths greater than 22 m. In other words, given the same bed conditions and ice temperatures and the same type of cutting tools, an extremely deep glacier would not cause any more abrasion than would a thin glacier whose depth was greater than

FIG. 13. Severe glacial erosion: cirque development with associated arêtes and horns, Mineral King, California. (After U.S.G.S.)

22 m.' However, the tendency to abrade the basin floor is related to the slip-line fields in the cirque glacier so that maximum abrasion occurs some distance back from the ice front and least abrasion occurs at the ice front. Little or no ice movement occurs at the ice front, material is piled up beneath the ice and this tends to accentuate the basin form of cirques.

Cirques rarely develop in isolation. As they cut back into the pre-glacial relief steep-sided ridges develop, separating one or more cirques (Fig. 13). These steep-sided ridges are known as arêtes. Glacial erosion by cirque development is so significant that mountain masses are largely destroyed, their remnants forming spectacular horn peaks.

GLACIAL EROSION IN AREAS OF LOW RELIEF

The location of landforms produced by glacial erosion is not limited to areas of considerable relief. Rock basins are known to occur in lowland areas. Linton (1963) describes the troughs on the south-eastern side of the Kilpatrick, Campsie and Ochil Hills in central Scotland where the local relief rarely exceeds 300 m. A trough in the lower Devon Valley contains basins which descend over 100 m below present sea-level. The reasons for the development of such troughs are presumably linked to the development of slip-line fields within the glaciers responsible for them.

The more common features of glacial erosion in areas of low relief are parallel groovings, irregular knobs and depressions, generally streamlined features, *roche moutonnée* and crag and tail features. It must be stressed, however, that similar forms can also occur in areas of considerable local relief.

In an area of some 80 km^2 in north-west Canada, H. T. U. Smith (1948) has described grooves that range from 2 to 40 m deep, 6 to 100 m wide and 70 to 10000 m long. In cross profile they have both V- and U-shaped forms, with crests varying from flat to rounded or sharp. They are generally straight and parallel and appear to be indifferent to the structure of the country-rock. From other evidence it is known that these grooves parallel the general direction of ice movement in the area. Such continuous grooving is rather unusual, but nearly all low-lying areas where solid rock has been overrun by glacier ice exhibit general lineations in the surface form that parallel the former direction of ice movement. In some locations erosion appears to have been more selective and an irregular surface of depressions and rock knobs has developed. If a particularly resistant rock protuberance occurs, a crag and tail may develop, with the crag exhibiting some evidence of glacial erosion in the form of striae and fluting on the stoss end and sides. The tail may be formed by less resistant solid rock that has not been eroded, on the lee side of the more resistant rock knob, or it may consist of glacial or fluvioglacial deposits laid down in the lee of the rock knob.

The term *roche moutonnée* has been applied to a wide range of solid rock outcrops that exhibit glacial polishing, striations and general smoothing on the stoss side, and broken, shattered irregular surfaces on the lee side. The mechanisms involved in this contrasting erosion of protruding rock knobs have already been discussed above. They can range from features that are of the order of a few metres in height to major units a hundred or so metres high and kilometres in length. They may occur on valley floors or valley sides, on interfluves or plateau surfaces or on lowland plains.

GLACIAL EROSION OF DRIFT DEPOSITS

Throughout the above discussion of glacial erosion it has been assumed that the ice-rock interface has been solid rock as opposed to glacial or fluvioglacial deposits. Although it is convenient to separate zones of glacial erosion from zones of glacial and fluvioglacial deposition it is rather unrealistic. Both sets of processes can occur together or at least glacial erosion can occur on glacial and fluvioglacial deposits laid down during an earlier phase of glaciation. The fact that glacial and fluvioglacial deposits are unconsolidated and often contain water in between rock fragments means that they are relatively easily incorporated into the basal ice by regelation and can be dragged along by rock fragments embedded in the basal layers of the ice.

The problem of differentiating between streamlined and fluted landforms produced by glacial erosion and those produced by glacial deposition is a complex one. Extentsive areas of fluted drift deposits are quite common but whether they are the product of erosional or depositional processes beneath the ice is hotly debated. Very often, if the drift deposits are characterized by a large percentage of clay or silt on the one hand or of coarse gravel on the other, they may not be easily incorporated in the basal ice. This is particularly true of clay-rich deposits with no joints or of open gravel that is unfrozen and tends to act like a mass of ball bearings, allowing the ice to pass over the deposits without disturbing them greatly. The most positive evidence for erosion of a drift area by glacial ice is the truncation of distinct beds (i.e., till on gravel or gravel on till), indicating that the surface form post-dates the deposition of the material. Similarly, if a fluted area consisting of till has lineations of the surface form that have a different orientation to the dominant till fabric of the moulded material, it is likely that the moulding is the product of a period of erosion post-dating the deposition of the till cover. A more detailed account of fluted drift surfaces will be given in chapter 4.

TRANSPORTATION BY GLACIER ICE

Erosion and transportation must be considered together. The actual mechanism involved in the detachment of rock particles from their original location only becomes significant when they are subsequently transported by the eroding medium.

There is a considerable amount of evidence that glaciers and ice sheets are capable of transporting large volumes of material over considerable distances. However, in the discussion that follows it must be remembered that not all of the material transported by glacier ice has been derived by glacial erosion. Large amounts of material that occur on and within glacier ice have been supplied by mechanical weathering on rock surfaces above the ice, by landslides, debris slides, avalanches, mud flows, and extraglacial streams that carry material on to the ice surface. Volcanic activity associated with wind transport can produce both coarse and fine deposits on the ice surface.

Material can be transported by glacier ice in three distinct environments: supraglacially, englacially and subglacially. Supraglacial debris is more extensive on valley glaciers than on ice sheets, simply because the sources of the debris are more extensive where more rock surfaces are exposed along the sides and at the head of valley glaciers than in association with ice sheets, where most rock surfaces are beneath the ice surface.

A comprehensive description of supraglacial debris on a stagnant valley glacier has been provided by Sharp (1949). In his discussion of the debris on the surface of the Wolf Creek Glacier in the Canadian Yukon, he points out that there are two possible sources from which the debris is derived (p. 292), '(1) Material from extraglacial areas dumped directly on to the ice below the firn line and (2) englacial material brought to the supraglacial position through lowering of the ice surface by melting. This englacial material, in turn, may be (a) extraglacial debris dumped on to the ice above the firn line, (b) subglacial material brought to an englacial position by movement along shear planes, or (c) the ground moraine of inset or superposed ice streams.' He concludes that 90 per cent of the debris on Wolf Creek Glacier is derived from an englacial source. In contrast, Speight (1940) states that most of the debris on the Tasman Glacier in New Zealand has always been supraglacial. Sharp supports his statement by pointing out that if the debris on the Wolf Creek Glacier was evenly distributed over its surface it would only form a layer 0·6 m thick, and since over 150 m of ice has melted from the ablation zone only a fraction of 1 per cent of the melted ice consisted of debris.

Garwood and Gregory (1898), Ogilvie (1904), Tarr and Martin (1914) and Ray (1935) have all emphasized the coarse, angular, poorly-sorted character of ablation moraine. This may be true of such moraine on actively moving glaciers but on stagnant ice masses, melt water activity produces sub-rounded and rounded fragments and undertakes a certain amount of sorting. Whatever the character of supraglacial debris it can occur either in relatively narrow (tens of metres) bands as lateral or medial moraines, or as quite extensive spreads in the terminal zone, where the material of medial moraines has been spread out as a result of differential ablation rates, and where that material has been supplemented by subglacial debris brought to the surface along flow-lines. Extensive areas of supraglacial debris have been described in Alaska by Tarr (1908) and Tarr and Martin (1914). The material is rarely more than a couple of metres thick although Tarr (1908) suggests it may reach 6 metres in thickness near the glacier margin. However, during its transport on the ice surface in either medial or lateral positions it is rarely more than 'one boulder' thick. The presence of any material that protects the ice surface from melting results in the ice beneath that material forming a ridge standing above the surrounding ice surface. In ablation zones both medial and lateral moraines may stand above the glacier surface by as much as 10-60 metres and have slopes of 30°-45°. Such a situation results in the redistribution of the supraglacial debris across the ice surface by sliding down these steep slopes. Transport is achieved not only by the moving ice but by redistribution across an irregularly melting ablation surface.

The distance travelled by rock debris in lateral and medial moraines is a function of the length of these features along the side of or in the central portions of the glacier. It may amount to tens or even hundreds of kilometres.

The englacial source of supraglacial debris involves transport, up through the glacier ice, of debris derived from the ice-rock interface. The transport takes place along

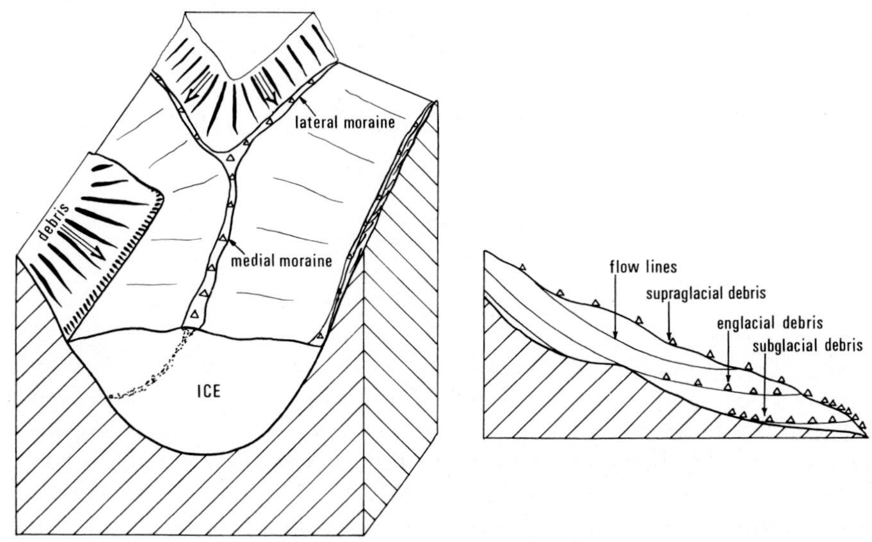

FIG. 14. Sources of debris associated with a valley glacier.

flow-lines and the material may be derived from the valley bottom or from rock knobs projecting up into the ice (Fig. 14). Debris bands outcropping on the surface of glaciers and ice sheets have been studied in Baffin Island (Goldthwait 1951), Greenland (Weertman 1961b) and Spitzbergen (Boulton 1970a, b) and the author has observed debris arriving at the surface of temperate glaciers in Alaska, Switzerland and Iceland, along structural planes which dip at a steep angle towards the bed of these glaciers in the frontal zone. Whatever the mechanism that entraps this material at the ice-rock interface, there can be no doubt that debris is transported from the subglacial environment up to the surface of the glaciers.

The fact that glaciers are very effective transporting agents can clearly be demonstrated by the study of erratics. These blocks that now occur sitting on rocks of entirely different lithology, and which can be traced back to their source outcrops, give an indication of the size of material that can be carried by glacier ice and the distance over which this transport is effected. Some erratic blocks are up to 800 m long. The largest in Germany is 4 km by 2 km by 120 m thick. A quartzite erratic in Alberta is 24·5 m × 12 m × 9 m and weighs 18000 tons.

Apart from sheer size of material carried, ice is a distinctive transporting agent because it is capable of lifting rock fragments high above their original location. Differences in altitude between original outcrop and final resting place of 500 m have been recorded in Scotland and 1500 m in the United States.

In terms of distance of transport, erratics have been traced from their source outcrops for distances of several hundred kilometres. Presumably the only limiting factors are the distance from the point of entrainment to the margin of the ice mass that is acting as the transporting agent and the ability of the erratics to survive weathering and erosion during transport.

IV GLACIAL DEPOSITION

INTRODUCTION

Something of the order of ten per cent of the earth's land area is covered by deposits laid down by glacier ice and its associated meltwater. Prior to the acceptance of the Glacial Theory in the middle of the nineteenth century, most of these surface deposits had been interpreted as being the result of the Biblical flood or as having been deposited by icebergs at a time of relatively high sea-level. The term used to classify these deposits was 'Drift'; with the development of the Glacial Theory the adjective 'glacial' was added and the term glacial drift was applied to all deposits associated with glaciation.

Glacial drift includes glacial, fluvioglacial, glacio-lacustrine and glacio-marine sediments. The thickness of glacial drift accumulated during the Pleistocene glaciations was very great. The maximum thickness of glacial drift that can be observed in any location at present probably represents only a fraction of the total drift accumulated at that location during the Pleistocene, as large amounts of drift were removed by both glacial and fluvioglacial erosion.

Flint (1971, p. 150) quotes the following maximum drift thicknesses in North America: Central Ohio 231 m, New Hampshire 121 m, and the Spokane Valley 400 m. In England, 143 m of drift has been recorded in East Anglia and 175 m in the Isle of Man.

The problem of the classification of glacial drifts is a complex one. It is very easy to get involved in circular reasoning in an attempt to classify the different types of deposits included in the term glacial drift. Most classifications are based on one or more of the following: sedimentary character, relationship to ice and form. The most satisfactory classification for geomorphologists would be based on the genesis of the deposits, but unfortunately in many instances this is not clearly understood. However, it is possible to define the two central members of a two-fold subdivision of glacial drifts. Some drift deposits are laid down by glacier ice while others are laid down by meltwater. This provides a two-fold subdivision on the basis of the medium of deposition, and in very general terms allows the adjectives 'unstratified' and 'stratified' to be applied. The main characteristics of each of these subdivisions of glacial drift can be summarized as follows:

GLACIAL DRIFT

Unstratified	Stratified
Ice deposited	Water deposited
Unsorted (particle size)	Sorted
Angular unwashed rock particles	Washed
Striated particles	Rounded rock particles
Wide variety of rock type	Wide variety of rock type
Till	Sand and Gravel

It is fairly easy to define the central members of each of these subdivisions but in reality the two subdivisions are part of a continuum, and it is often very difficult to establish the medium of deposition solely from the sedimentary characteristics of the end product. It has been pointed out in the previous chapter that the erosion and transportation of rock debris often involves the presence of some water. Similarly the deposition of rock debris by the ice is often accompanied by the release of some meltwater. It will be demonstrated later in this chapter that the deposition of till containing relatively large amounts of meltwater plays an important part in the development of certain glacial landforms. In the same way sands and gravels laid down by meltwater frequently contain lenses of unstratified till. The two methods of deposition are by definition closely associated in a glacierized area. Tarr (1909, p. 98) has clearly described the conditions that often prevail along the margins of existing glaciers: 'Till and stratified sand, gravel and clay are often inextricably mixed, and the variations, both of texture and stratification, appear according to no known law.'

The medium of deposition is, of course, closely related to the general environment of deposition. On the basis of general environment it is possible to further subdivide the unstratified and stratified deposits:

Unstratified	Stratified
Subglacial (e.g. ground moraine)	Ice-contact:
Supraglacial (e.g. ablation moraine)	Subglacial
	Supraglacial
Marginal (e.g. moraine ridges)	Englacial
	Marginal
	Proglacial:
	Fluvial (e.g. sandar)
	Lacustrine
	Marine

Any classification of glacial drift which emphasizes form is liable to create more problems than it solves. Drift deposits with similar forms can have different origins, and so it appears that the most satisfactory means of classification employs both sedimentary character and the general environment of deposition, the latter including both the medium of deposition and position in terms of the ice mass. A further factor has often been included, namely an evaluation of the condition of the ice mass as well as its positional relationship with the deposit. Drift deposits have been further subdivided on the basis of their formation in association with active or stagnant ice. However, it is often a dubious activity to attempt to determine glaciological conditions from sedimentary or morphological characteristics. Also, it is now known that certain forms which in the past have been ascribed to stagnant ice can also be formed by at least slowly moving ice. This then introduces the problem of defining stagnation.

Any classification system is only as strong as the criteria used. By far the most information about glacial drift has been derived from the study of deposits after they have been laid down. Only very limited amounts of data have been derived from areas in which glaciers still exist and where glacial and fluvioglacial deposits are currently developing. Until the mechanisms of deposition are better understood, subdivision of complex deposits will not have a sound basis.

However, the basic subdivision of unstratified and stratified drift remains as a useful classification of complex deposits. There are major differences between deposits laid down by and from the ice itself, with only relatively small amounts of assistance from meltwater, and those drift deposits laid down after transport, even if only over short distances, by meltwater. The remainder of this chapter will be concerned with unstratified or glacial deposits.

Some problems in the terminology associated with glacial deposits must be clarified. The most common deposit is glacial till, which has sometimes been referred to as boulder clay. However, not all glacial tills contain a mixture of boulders and clay and throughout this book the term 'till' will be used. To complicate matters even further, some workers use the term 'moraine' as the equivalent of 'till'. Moraine is a morphological term whereas 'till' is sedimentological and they are therefore not interchangeable. Throughout this book 'till' will be used to refer to the sediment whereas 'moraine' will be used to refer to the morphology of the sediment.

MECHANISMS OF DEPOSITION

During the period 1890-1932 work carried out in Alaska, Spitzbergen, Greenland and Iceland (Garwood and Gregory, 1898; Tarr and Martin, 1906; Tarr, 1909; Koch and Wegner, 1911; Lamplugh, 1911; Tarr and Butler, 1909; Tarr and Martin, 1914; Gripp, 1929; Todtmann, 1932) provided a great deal of descriptive data about glacial sediments, on, under, and at the margins of existing glaciers. This literature was the source of many of the interpretations of glacial sediments in areas of former glaciation. Early workers established the complexity of the ice marginal sedimentary environment. They described the distribution of debris along lateral margins, in medial moraines and in the ablation zone in general. Descriptions of debris bands outcropping parallel to frontal margins led to suggestions of the movement of debris along flow-lines and the squeezing of material into crevasses and other cavities. The development of zones of debris concentration both on and in the ice, in relationship to the structural properties of the ice mass, were also established. Much of this work depended on the observation of deposits and forms in the process of development, and although attempts were made to hypothesize about the actual mechanisms involved, no studies of the mechanisms themselves were made. This has also been true of much of the literature since the 1930's (e.g. Goldthwait, 1951; Okko, 1955).

In 1957, Harrison studied shale-loaded debris planes on the edge of the Greenland ice sheet, near Thule. These debris-rich zones contained 24 per cent glacier ice by weight, and dipped 37° up ice. Harrison concluded (p. 287), '... the dominant orientation in glacier ice of disc and blade-shaped particles, from fine sand to boulder sizes, is one of parallelism of the maximum projection planes of the particles with the megascopic ice foliation planes.' Although this was more a study of debris in transport than in the processes of deposition it did demonstrate the relationship between the attitude of individual particles in a glacier till after deposition to their mode of transport in the ice. The method of deposition was such that, when the ice of the debris-rich zone melted away, the debris became a till which inherited a fabric closely related to structures previously occurring in the ice.

Direct observations of till deposition at the ice-rock interface are relatively rare. Boulton's work (1968, 1970a and 1970b) in Spitzbergen is an excellent contribution to our understanding of glacial deposits. He argues (1970a) that the major part of the subglacially derived debris is transported in an englacial position in the Spitzbergen glaciers. Incorporation of the debris into the glacier ice is believed to be by a process of basal freezing. As a result of ablation much of this debris appears on the ice surface as a flow-till (Boulton, 1968), but the debris can also be deposited as a result of the melting of debris-rich stagnant ice buried beneath sediments or beneath active ice under the glacier.

At Dunerbreen, Boulton (1968) observed stagnant ice beneath a till cover that contained sufficient debris to produce a till 5 m thick after all the ice had melted. Tills can be produced by the top melting of debris-rich buried ice masses, so long as the top surface of the buried ice mass lies within the depth of maximum summer thaw. Tills produced in this manner sometimes move down slope and thus structures inherited from the parent ice will be destroyed and a flow-till will result. On the other hand, if the debris melts out into a stable position the till preserves structures derived from the parent ice, and Boulton (1970b) calls these 'melt-out tills.'

Boulton describes a sequence of upper till, laminated silts and sands, sand and gravel, and lower till, all of which rest on debris-rich glacier ice at the margin of Erikabreen in Oscar II Land (Fig. 15). This sequence was frozen to within 2 m of the surface in early August. He states (Boulton, 1970b, p. 233), 'There is good evidence to show that the laminated silts and sands were laid down in a proglacial lake into which, at a later stage, the overlying till flowed, and the gravels presumably represent supraglacial fluvial activity before the lake formed. The glacier-ice surface is irregular but the overlying lower till maintains an almost uniform thickness of about 0·4 m. If this till were a flow till, its maximum thickness would tend to be distributed at the low points of the ice surface. Also, had the gravels been deposited directly upon this till, the high velocity depositing current would have tended to scour into this pre-existing silt-clay till, but of this there is no sign. It is therefore suggested that the gravels were deposited directly on to an ice surface and that the till was subsequently released at this interface by melting, because of the existence of a supraglacial lake of relatively high thermal capacity. The freezing of the sediment column then occurred after the disappearance of the lake.'

With this example, although demanding rather special ground temperature conditions in association with the margin of a polar glacier, Boulton has demonstrated how tills can be produced by melting of a debris-rich ice mass from the surface downwards. Bottom melting of buried, debris-rich, stagnant ice also occurs but is likely to take place at a very much slower rate than top melting.

Boulton also describes till deposition beneath Nordenskioldbreen after material was incorporated by basal freezing into the sole of the glacier. A layer of debris-rich, (and

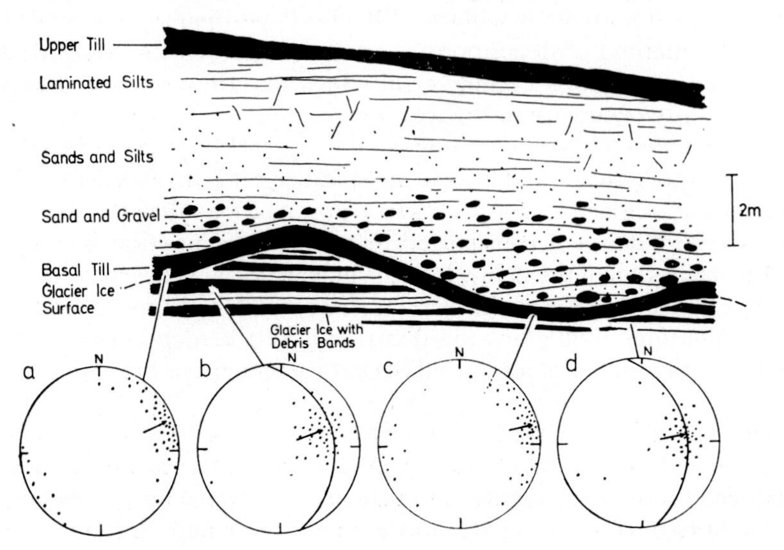

FIG. 15. A supraglacial section of the margin of Erikabreen in Oscar II Land. (*a*), (*b*), (*c*) and (*d*) are equal area stereographic plots of long axes of blade-shaped stones. (*a*) and (*c*) are from stones in the lower till, and (*b*) and (*d*) are of englacial stones. In (*b*) and (*d*) cyclographic traces show the dip of the glacier ice foliation, which is approximately parallel to present day glacier movement. The arrows show the direction of the computed mean vector. (After Boulton, 1970b.)

therefore less plastic), ice existed, over which the relatively debris-free ice, which was relatively more plastic, could move more quickly. This situation allowed masses of basal, debris-rich ice to develop around the flanks of bedrock obstacles where the rate of flow of the debris-free ice increased and overrode the debris-rich ice. The geothermal heat flux melted the ice in the debris-rich stagnant mass and produced an unfrozen till. Meltwater was expelled from the thawing till by the pressure of over-riding ice and the till became more compact. Masses of stagnant, debris-rich, basal ice need not only occur in association with bedrock obstacles. They can also occur as a result of the decrease in plasticity of the basal debris-rich ice as it , and the relatively clean ice above it moves more quickly towards the margin. Mickelson (1971, p. 43), working on the Burroughs Glacier in south-eastern Alaska, has calculated that basal melt-out tills accumulated at rates ranging from 0·5 to 2·8 cm/yr. He states that '. . . till deposition began quite late in Neoglacial time and had ceased or slowed down when the ice was

70-100 m thick. Since the tills can be differentiated as units, deposition of the till may not have been continuous. . . . Depending on the amount of water present in the till at the time of deposition, some reorientation of the till which had melted out might be expected.'

Observations on and under existing glaciers indicate that melting is the main method of accumulating glacial till. Whatever the mechanism of the incorporation of debris into glacier ice, if a mixture of ice and rock debris exists melting must occur before till is deposited. Melting can take place at the bottom surface of debris-rich, stagnant, basal ice and at both the bottom and top of buried, debris-rich, stagnant ice. Boulton has demonstrated that these processes can occur in association with polar glaciers but there seems no reason why they should not occur in association with temperate ice, except that some mechanism other than basal freezing has to be envisaged for the incorporation of debris into the ice.

Debris which does not contain interstitial ice arrives at the surface of temperate glaciers (Fig. 16) along the lines of old crevasses and flow-lines. When the material is deposited on to the subglacial surface, it, too, will have all the characteristics of a till except, perhaps, that its fabric will not have a very strong orientation. Another product of the melting of ice from beneath a debris cover, rather than within it, is ablation moraine. This can take the form of a narrow medial moraine or a complete debris cover over a valley glacier where redistribution of several medial moraines by differential melting (Sharp, 1949) has produced an extensive cover of debris.

Glacial deposits, in terms of conditions observed under, in, on and in front of existing glaciers, are produced either by the melting of interstitial ice, that ice having a significant debris content, or by the melting of ice upon which there is a debris cover. Further information about mechanisms of glacial deposition has been derived from the study of the form and internal sedimentary characteristics of glacial deposits after the ice responsible for them has left the area.

Glacial deposits can be subdivided into two groups on the basis of their morphology. The first group is characterized by very little local relief and can best be described as a *till-plain* or *till-sheet*. These deposits tend to mask even major irregularities in the bedrock relief so that they are thicker along depressions in the bedrock and thinner over ridges or plateaus. The second group has a very irregular surface form and can consist of simple linear ridges or complexes of ridges, hollows and mounds. The simple term *hummocky topography* has been used to describe these deposits.

The internal character of these deposits, the patterns of surface form and their general disposition have been the subject of detailed investigations in many areas. Data have been obtained from which it has been possible to hypothesize about the direction of movements in the ice mass responsible for the deposits, and about the mechanism of deposition and the evolution of the resultant landforms. By examining the rock types present in a glacial till and referring to the local bedrock geology, directions and amounts of transport by the glacier ice can be established. Direction of ice movement can be determined by striations and other erosional and streamlined features. However, in recent years a great deal of attention has been paid to the orientation and dip of the long axes and maximum projection planes of particles in till.

Fig. 16. Debris emerging at the ice surface, Breidamerkurjökull, Iceland.

This work has yielded some information relating to the mechanics of till deposition and about secondary transport after the till has been deposited.

The tendency for pebbles in glacial deposits to have a preferred orientation was known last century (Hind, 1859; Miller, 1884). However, it was not until 1936, when Richter established quantitatively the relationship between the preferred orientation of pebbles in till and the direction of ice movement, that the full significance of till fabric was widely realised.

Holmes (1941), working in New York State, set the pattern for many of the till fabric studies to be undertaken in the following thirty years. He concluded that the majority of stones tend to have their long axes aligned parallel to the former direction of ice movement but that in a few localities the stones were aligned at right angles to that direction. Fabric patterns from successive layers of till recorded occasional shifts in the direction of glacier flow. Holmes believed that the ground moraine, or till, accumulated gradually beneath the moving glacier and that it moved as a plastic solid and he states (p. 1350), 'Stones carried above the floor were subject to rotation about an axis lying in a plane of uniform shear and normal to the direction of glacier flow. Ordinarily the longest axis was the axis of rotation, though under some conditions, probably following temporary orientation by sliding, rotation about the intermediate or the shortest axis probably occurred. Stones rotating on or near the floor were subject to shoving into the deposited till by thrust from higher and faster-moving debris. Many of the stones thus imbedded remained permanently in that position.' Perhaps one of Holmes' most significant conclusions was that the form, roundness, size and relative axial lengths of the stones themselves affected the orientation they achieved in the till. This latter point has been substantiated by other workers (e.g. Andrews and King, 1968) but unfortunately a large body of till fabric work does not take account of this important factor. Holmes implied a slow accumulation of till in the subglacial environment. Although he indicated that stones became oriented by sliding and rotation in the ice and beneath it, he did not specifically explain how the material in the ice was finally deposited.

Harrison (1957), working on data collected from ground and end-moraine areas near Chicago and from the Greenland ice sheet near Thule, did suggest a mechanism for the preferred orientations he observed and a mechanism for final emplacement of the till. The orientation and dip of disc, blade and rod-shaped particles of shale, 4-8 mm in size, were measured. In samples from ground moraine and end moraine, both disc and blade-shaped particles tended to be preferentially imbricated up-stream. Harrison then examined debris within ice at the margin of the Greenland ice sheet and concluded (p. 275) '. . . that the bulk of the till fabric has been inherited from the transportational environment i.e. it represents little altered, remnant, englacial debris fabric. The fabric pattern maxima of disc, and blade-shaped particles are believed to define the vanished slip planes of the debris-charged basal zones of the glacier which carried the till. . . . It is suggested that the bulk of the till was deposited by slow melting out from the stagnant basal zones of the various glaciers which transported it. Deposition seems to have been confined in time almost entirely to the deglaciation phase of glaciation.' These two statements about the method and time of emplacement of glacial till are extremely interesting. Melting is suggested as the mechanism for the release of the debris from the ice.

Harrison also studied the variability of till fabric between individual till sheets and in vertical section. He concluded that lateral variations over distances of 225 m are considerable but that over very short distances (p. 281) '. . . the horizontal variability of the fabric at a given horizon is less than the vertical variability between horizons.' The opposite conclusion was reached by Young (1969) who established that, over distances of 60 cm, vertical variations in long axis orientation are greater than in the horizontal direction. Working on one section in which three tills (1·5 m, 4·3 m and 2·4 m thick) were separated from one another and capped by various thicknesses of sand and gravel he studied till fabric at seventeen sites over a vertical distance of 11 m. The extreme case of vertical variation in preferred orientation amounted to 130° through a distance of 60 cm. Young concludes (p. 2350), 'Variation in preferred orientation within the same till unit was often greater than variation between units, and the occurrence and maintenance of any trend in orientation within the vertical plane appeared to be quite unpredictable.' He therefore argues that it is unlikely that the till unit was produced by a plastering-on mechanism or that the rapid changes in preferred orientation reflect inherited fabrics from changes in direction of movement in the parent ice mass. Quoting Hoppe (1963), Young stresses the necessity of melting-point temperatures in the subglacial environment, which permit glacial deposition to take place. Local environmental conditions therefore will affect the amount of water present. These local conditions will change throughout the period of deposition and sudden flushes of water, mud flows or the collapse of basal ice would upset the fabric of the newly deposited till.

Whereas Harrison (1957) saw the preferred orientation of stones as originating in the transportational environment, and surviving the melting-out processes from basal stagnant ice, Young (1969) stresses the importance of the disturbance of any inherited fabric, by meltwater allowing post emplacement movement of the till, or by impact from falling ice blocks or large boulders. Okko (1955), in his description of glacial drift in Iceland also stresses the importance of basal melting but he claims that in general the till remains undisturbed, preserving the fabric it developed during transport, but that in some places the ice squeezes the till into crevasses and in front of the margin.

The conflicting evidence that has been presented regarding the preservation of fabrics derived from transport in the ice is hardly surprising. Data available on the topic are very limited in terms of the wide range of conditions likely in the subglacial environment in which till is deposited. When pebbles in till do have a preferred orientation it is likely that they reflect movement of the transporting medium. However, it is possible that such preferred orientations may be weakened or destroyed as the debris melts out of the ice and is affected either by injections of meltwater allowing flowage in the till or by post-emplacement disturbance by falling blocks of ice or large boulders.

The fact that till forming moraine ridges often has a distinctive fabric unrelated to ice movement directions has provided information about post-emplacement till movement and mechanisms for the development of certain types of moraine ridges, hummocks and plateaus. Hoppe (1952), on the basis of strong preferred orientations at right angles to moraine ridge crests in Sweden, suggested that the ridges were formed by the movement of water-soaked till into cavities and crevasses beneath the ice. Similar deductions were made by Andrews and Smithson (1966), in Baffin Island, and by Price (1970) in Iceland, in regard to moraine ridges developed at the edge of a polar ice cap and temperate valley glacier respectively. Andrews and Smithson (1966) emphasized the seasonal supply of meltwater to the subglacial till that allowed it to be squeezed up at the ice front to form moraine ridges, with fabrics being derived from the movement of the till by internal flowage. Price (1970), also demonstrated strong preferred orientation in moraine ridges but in the Icelandic situation the seasonal supply of meltwater is less significant than in the Baffin Island situation, as the base of the glacier is at the pressure melting point.

Another mechanism which tends to destroy till fabrics inherited from debris transport in ice, is pushing by an advancing ice front. Okko (1955) has described push moraines around the margin of Vatnajökull, and those associated with the Pleistocene ice sheets have been described in Germany, Holland, North Wales, Canada and the United States. Suitable material, usually in the form of glacial drift from an earlier glaciation, has to be present during an advance of an ice margin. This material, if it is frozen, can be faulted, folded and transported by the pressures induced by the advancing ice margin. The fabrics of this disturbed drift bear little relationship to the movement direction of the ice mass responsible for the tectonics of the newly formed moraines.

The study of transport and deposition of debris by existing glaciers and the till fabric of ground moraine and moraine ridges in areas of former ice cover has provided information about the mechanics of glacial deposition. There are really only three possibilities. The material may either be released from the ice in the basal zone or at the ice surface by melting, or it may be transported on the ice surface and dumped by the melting of ice beneath it, or it may be transported between the basal ice and the solid rock or other glacial drift beneath it. The last possibility seems most unlikely and there is little evidence from existing glaciers or from the internal character of glacial deposits to suggest that this mechanism, called 'plastering', actually takes place. Plastering involves a receiving surface, the material to be plastered, in this case till, and the plastering medium, in this case glacier ice. If tills were accumulated over wide areas by a plastering mechanism it is hard to see how a void was created, between the ice and the underlying rock or drift, to allow the plastering to take place.

The occurrence of debris-laden basal ice in both polar and temperate ice masses has been established by observation of existing glaciers. It is not surprising that melting of the basal ice occurs beneath temperate ice masses and it has also been established that preferred alignment of stones in till derived from such debris-laden ice can reflect former directions of ice movement. However, it has also been established that similar deposits with strong fabrics can be produced by the melting of debris-rich ice at the margin of the base of the Greenland ice sheet (Harrison, 1957) and beneath glaciers in Spitzbergen (Boulton, 1970). It is interesting to note that in both situations structures in the glacier ice, local stagnation and basal melting are all important factors affecting the nature of the glacial till produced. Basal till seems to be intimately associated with melting and, just as in the case of glacial erosion, it seems difficult to envisage glacial deposition taking place beneath ice masses that are frozen to their bases. From the limited number of observations of till being laid down at the ice-rock interface it also seems likely that local stagnation is involved, although the deposits themselves can inherit fabrics developed during previous transportation in moving ice. It would be dangerous to make wide generalizations from the limited data available but it does seem increasingly certain that till deposition is most likely to occur when the basal ice is at the pressure melting point and when little or no movement is taking place in the ice in the zone of deposition. If this generalization has any validity subglacial tills are most likely to accumulate during the wastage of an ice mass and the shrinkage of its limits, as opposed to during its build up and the extension of its limits.

Debris can also be released by melting of debris-rich ice at the upper surface of a glacier. This occurs when debris-rich bands outcrop in the ablation zone. Such tills may have an inherited fabric but their location is such that they are likely to be highly disturbed by the melting out of clean ice from beneath them. Alternatively, they may develop new fabrics as they flow across slopes developed by differential ablation of the ice surface.

Rock fragments can also be transported on the ice surface without actually being incorporated in the ice. Such material occurs in medial and lateral moraines and is brought to the surface of the ablation area along flow-lines or crevasses. When the material on the surface of the ice becomes unstable, due to the steepness of the ice surface, it is dumped along the ice margin and can either accumulate as a ridge, if the ice margin is stationary, or can be spread out as a sheet if the ice margin is retreating. Such accumulations of glacial deposits are characterized by their lack of internal structure and the small percentage of fine particles, the latter having been removed by meltwater activity on the ice surface prior to their deposition. This dumping mechanism is unlikely to produce tills more than a few metres in thickness, e.g., 0·3-6·0 m on the Martin River Glacier (Reid and Clayton, 1963).

The above outline of possible mechanisms of glacial deposition suggests that melting of interstitial ice from debris-rich layers either at the base, or on the surface of an ice mass, and melting of ice from beneath debris carried on the surface of an ice mass, are the two most important aspects of glacial deposition. It would be useful if terminology could be established to distinguish the two principal methods of debris accretion. The term 'lodgement till' is widely used in the literature to refer to till deposited subglacially, as opposed to 'ablation till' resulting from the disappearance of ice from beneath a debris cover. Unfortunately, similar deposits to those generally referred to as lodgement till can be produced by the melting of debris-rich ice from the upper surface downwards (Boulton, 1970). It would also be confusing to use the term 'melt-out till' for subglacial till unless the ablation till was referred to as the 'melt-under till' because it, too, is related to the melting process even though the debris is not actually in the ice.

In the present state of our knowledge about tills it seems necessary to employ descriptive terminology wherever possible. A till that is deposited in the subglacial environment should simply be referred to as a subglacial till in contrast to a supraglacial till (ablation till). It seems necessary to distinguish between different types of supraglacial till in that some have developed distinctive characteristics by dumping or flowage. Each of these can be qualified by the required adjective.

Throughout the previous discussion the term till has been used solely to refer to a particular sedimentary deposit. The term moraine is only used with reference to the morphology of the deposit. If the till is a simple sheet with little or no local relief the term ground-moraine can be used. If it is necessary to distinguish between a ground moraine consisting of subglacial till and one consisting of supraglacial till then the appropriate adjective can be applied.

SEDIMENTOLOGY OF GLACIAL DEPOSITS

In a useful guide to the description of till, Scott and St. Onge (1969) stress the lithologic and physical heterogeneity of the deposit. The mechanisms of deposition described above appear to be capable of producing tills having widely different particle size distributions, particle shapes, rock types, colour, porosity, permeability and compaction. Although tills from different locations can vary greatly in their physical and lithological characteristics, it does appear that a given till unit is relatively homogeneous over considerable areas. Contrasts between till units have been revealed by close study of their various lithological and physical characteristics, and some understanding has been obtained of the factors that control these characteristics.

An interesting study by Frye, Glass, Kempton and Willman (1969) of the tills of northwestern Illinois can be used to demonstrate the usefulness of a study of the lithologic and physical characteristics of till sheets. The study of grain size distribution in tills has been widely used for identifying and correlating till units in Illinois by other workers (Kempton, 1963: Kempton and Hackett, 1968a, b; Johnson, 1964), and they have demonstrated the usefulness of studying the fraction smaller than 2 mm. It was also established that it was important to define the range of grain size as well as the medians. Over eight hundred determinations of the sand-silt-clay percentages were made by sieving and hydrometer methods, as well as determinations of the pebble and cobble content. Some till units were found to be quite homogeneous throughout their extent; one unit did not vary more than ± 5 per cent from 37 per cent sand, 38 per cent silt and 25 per cent clay. Other units differed quite markedly from one location to another or included several distinctly different grain size compositions in stratigraphic superposition. Frye *et al* (1969, p. 11) state, 'Because the grain size of a till is the result of both the regimen of the glacier and the source rocks over which the glacier moved, the sand-silt-clay percentages are not an infallible identification of any individual till. Rather, this is only one of the many properties that can be used to categorize the deposits.' Kelly and Baker (1966), working on tills in North Dakota, were able to distinguish between two tills on the basis of mean size and standard deviation of particles in the till, even though the range of these values for the two tills overlapped. They also concluded that skewness and kurtosis values for the two tills were not diagnostic. Järnefors (1952), working on the tills from northern Sweden, and Schneider (1961), working on tills from Minnesota, found textural similarity in tills that were different in terms of their colour, rock type, structure and the size of larger boulders. It seems, therefore, that particle size distribution can sometimes be used to differentiate between two till units; there are occasions, however, when two units have similar particle size distributions but are very different on the basis of other characteristics.

A useful characteristic of till for descriptive and differentiation purposes is colour. Various shades of grey and brown are most common and these can be accurately described with the use of a Munsell colour chart. Care needs to be taken when assessing till colour to ensure that local variations are not the product of weathering or high moisture content. Kelly and Baker (1966) analysed two tills of different colour and deduced that the colour difference was the product of different lithologies in each till unit.

According to Anderson (1957) the relative amounts of a given lithology in the sand and pebble grade sizes are dependent upon the following factors: (a) lithological properties, especially those affecting durability in the sand and pebble sizes; (b) distance from source; (c) rate of comminution; (d) dominant type of bedrock erosion. There is certainly no single factor that is dominant in determining the lithology of tills. It used to be widely accepted that local bedrock material predominated in tills. However, Dreimanis and Vagners (1969) have shown that only the higher concentration of local bedrock amongst the larger fragments was used to substantiate the statement. If the finer materials are examined the generalization is shown to be false. However, in areas of soft bedrock, and in situations where new material is added to the glacial drift after only short distances of transport and has therefore suffered little comminution, local material may predominate in tills. On the other hand, if material moves up into the ice, where debris particles are subjected to less comminution, the resultant tills will be richer in far-travelled rock types. Holmes (1960) has demonstrated the decrease in larger particle sizes in favour of an increase in finer particles with increasing distance from the parent outcrop.

Particles in tills tend to be angular and subangular rather than round. There are exceptions to this generalization, particularly in tills that consist of reworked fluvioglacial material (Price, 1969) that has not been transported very far by the ice. Samples of granite gneiss from deltas, eskers, kames and moraines have been studied (King and Buckley, 1968) in terms of their mean roundness values, which were generally twice as large for the fluvioglacial deposits as for the glacial deposits.

Holmes (1960) examined over 3000 pebbles and cobbles from 27 till localities in New York State. All the pebbles were of either sandstone, shale or limestone. The greatest distance of transport of these particles was 32 km. The shape and roundness of particles changed with increasing distances of glacial transport, there being a relative decrease of about 25 per cent in wedge-form and in rhombohedroid stones and a relative increase of more than 100 per cent in ovoids. Holmes concluded that, although wedge forms tended to be more conspicuously striated than others, the process of rounding predominated over faulting, and that prolonged transportation by the middle latitude ice sheets tended to produce stones of ovoid rather than wedge shapes.

Easterbrook (1964) undertook a study of void ratios and bulk densities of glacial tills and till-like sediments. He demonstrated that by studying these characteristics it was possible to determine whether or not a till-like sediment was deposited beneath glacier ice. However, since both void ratio and bulk density depend on particle size distribution as well as on degree of compaction, differences between till and other till-like sediments are significant only if the samples have similar particle size distribution. Various workers have undertaken consolidation tests on tills in an effort to estimate the thickness of the ice beneath which the tills were accumulated (e.g., Harrison, 1957). It seems likely though, that numerous factors are involved in consolidation of tills and that to base calculations of ice thickness solely on a direct relationship between ice pressure and post depositional consolidation values is a gross oversimplification of a complex relationship.

Clay mineral analysis can also be used (Frye, Glass and Willman, 1968) to differentiate tills. To use this method meaningfully it is necessary to consider the type, rate and depth of mineral alteration caused by weathering. X-ray diffraction and differential thermal analysis have been used to differentiate till units (Goodyear 1962) and to determine the source areas from which materials constituting specific till units were derived (Glentworth *et al*, 1964). Along with clay minerals, the petrology of the coarse fraction of tills is extremely important, both in terms of unit differentiation and in giving the overall character in terms of particle size and shape and colour to till units.

The sedimentary characteristics of tills not only provide information about the direction of movement of the transporting ice and assist in differentiating various till units, but also provide information about the environments of till deposition. There is little doubt that the greatest percentage of till deposition in the form of ground moraine takes place in the subglacial environment. The amount of water present may be reflected by the presence or absence of lenses of fluvioglacial material within till units, or by the destruction of strong preferred orientation of particles as a result of flowage of the till after its emplacement and water saturation. Carey and Ahmad (1961) have distinguished between wet-base and dry-base glaciers, i.e. whether the basal ice is at or below the pressure melting point. This seems to be a useful classification in terms of sedimentary environments. They then go on to distinguish between subglacial environments beneath wet and dry-based glaciers when they enter the sea (p. 883), 'The grounded shelf zone of a dry glacier does not differ in any significant way from the terrestrial zone. The full load of dry ice loaded with debris grinds the pavement. There can be little sedimentation, if any.' In contrast, sedimentation beneath a wet-based glacier which terminates in the sea will be of considerable volume and involves rapidly changing lithological characteristics. The interbedding of glacial and fluvio-glacial sediments and structures associated with flow tills and turbidity currents are common.

Further study of glacial sediments by specialists in sedimentology will add much to our understanding of glacial processes.* The landforms consisting of till cannot be studied without reference to their internal characteristics. Unfortunately the wide variations, in terms of petrology, lithology and sedimentology that occur in glacial sediments make interpretation and generalization very difficult, suggesting that glacial environments are very complicated and produce highly variable sediments.

GROUND MORAINE

The morphological expression of a till sheet or sheets is known as *ground moraine*. From the point of view of the processes involved in the emplacement of the till it may consist entirely of basal till (subglacial till) or it may have a capping of supraglacial till (ablation till or flow till). The thickness of the till sheet is usually of the order of tens of metres but if more than one till sheet is present the combined thickness of the till sheets may be in excess of one hundred metres.

Ground moraine is characterized by having low local relief values, usually less than ten metres. A thick cover of till tends to mask the underlying bedrock topography and pre-glacial valley systems may be completely obliterated, the till being thickest in bedrock depressions and thinnest over bedrock protuberances. In some instances

* See 'Till: a Symposium', edited by Goldthwait *et al*, published by Ohio State University Press. This volume was not available until the present book was at the press.

bedrock ridges may stand above the general level of the ground moraine. Some authors prefer to use the term *till sheet* to describe a ground moraine which is so thin that it reflects the underlying bedrock. However, it is more satisfactory to restrict the term till sheet to its sedimentological usage and the term ground moraine to the morphological expression of the till cover, even if it reflects the underlying bedrock form.

Ground moraine covers very large areas in the plains areas of North America and Europe. Superimposed or projecting through ground moraine, moraine ridges break up what would otherwise be vast areas of ground having no transverse linear elements.

In these areas of limited local relief, across which the large Pleistocene ice sheets advanced and retreated on several occasions, evidence in the form of multiple till units can be examined which forms the basis of Pleistocene stratigraphy and chronology. It is necessary to understand the mechanisms of ground moraine formation before the evidence of multiple till units can be properly interpreted.

The actual mechanisms of glacial deposition have already been discussed but one further consideration in the form of the stratigraphic relationships which occur within ground moraines must be examined. The problem can be illustrated with reference to a study of a drift section in Midlothian, Scotland (Kirby, 1969). In a 14 m section a basal till (2·8 m) is overlain by sand and gravel (2·8 m) which in turn is overlain by a sequence of four thin beds, each between 10 and 50 cm in thickness, of alternating sand and till (Fig. 17). The top half of the section consists of 6 m of till overlain by 1·5 m of sand. This is a relatively simple sequence and except that in this example the upper till is overlain by fluvioglacial deposits, it represents much of the stratigraphy that tends to underlie areas of ground moraine. The interpretation of stratigraphic sequences such as these is the basis of much of Pleistocene chronology and correlation. The sediments represent alternating environments in which dominantly glacial processes were superseded by dominantly fluvioglacial processes. Since each set of processes is generally characterized by subglacial and proglacial environments respectively, many workers would argue that tills interbedded with fluvioglacial deposits indicate separate periods of ice cover to account for each till layer, superseded by deglaciation to account for the fluvioglacial deposits.

Kirby interprets the Midlothian section in the following manner. On the basis of similar till fabrics in the three lowest tills, he concludes that deposition occurred (Kirby, 1969, p. 52), '. . . by the same agency moving in the same direction with similar momentum. It is extremely improbable that the original ice sheet depositing the bottom layer of till would have been followed by further separate ice sheets or tongues readvancing across open ground in exactly the same direction or with the depositing mechanism developed to the same extent to deposit the higher layers of till. The three lowest layers of till are therefore all regarded as part of a single till; the interstratified sand representing merely breaks within a single process of glacial deposition.' Kirby also points out that all the layers of till can be associated on stone count evidence and that there is gradation between the successive layers of till and sand excluding the topmost till-gravel junction. He therefore concludes (p. 52), '. . . it is thought that such transition in grain-size and therefore in the amount of water washing, particularly from glacio-fluvial deposits into till, is exceptional where the successive deposits relate to

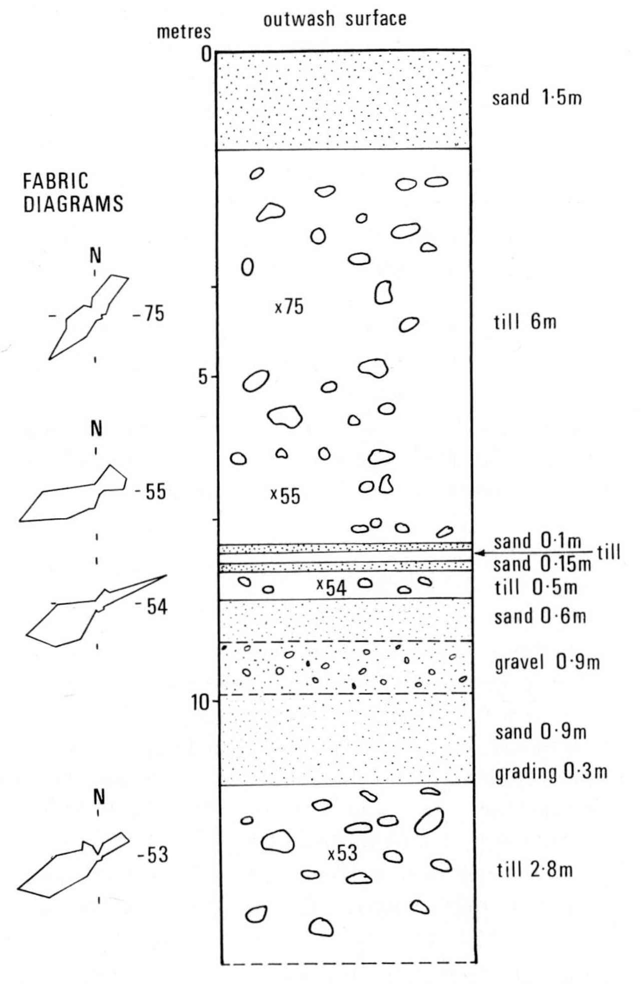

FIG. 17. Section at Park Burn, Midlothian, Scotland, showing the alternation of till and fluvioglacial deposits. (After Kirby, 1969.)

different ice sheets. The gradation is a further point suggesting that the total deposit, excepting the topmost sand and gravel, represents a single glacier load.'

Kirby's interpretation of the Midlothian section has much in common with the very controversial ideas put forward by Carruthers (1947-1948, 1953). Carruthers interpreted sequences of stratified and unstratified sediments involving a basal till, shear clays, suspension structures, sand layers with current bedding, gravel and an upper boulder clay, as resulting from bottom melting of heavily debris-laden stagnant ice. There can be little doubt that changes in the subglacial environment during the melting of a heavily debris-laden ice mass could account for the sedimentary characteristics described by Carruthers and Kirby. Whether or not such conditions could occur over a wide area to allow such sequences to develop on a regional scale is difficult to decide. Such rapid changes through considerable thicknesses of deposits must reflect the development in the ice of caves of considerable size. The maximum size of such features in a stagnant ice sheet is unknown. If, however, sequences of stratified and unstratified sediments occur over wide areas it is highly likely that they represent fluctuations in an ice margin. Certain other lines of evidence can sometimes be investigated to indicate whether successive till sheets represent more than one period of ice cover. If cryoturbation structures occur in a till sheet or if it shows evidence of weathering or soil development then it is highly likely that the ice left the area before deposits higher in the sequence were laid down.

Although ground moraine may not be a very dramatic morphological feature, the information provided by it, and the deposits beneath it, often play a vital part in the interpretation of processes associated with a former ice cover. Not until a more thorough understanding is obtained of both glacial and fluvioglacial processes of deposition will it be possible to interpret much of the evidence revealed by sections and bore-holes in the materials beneath the surface of ground moraine.

FLUTED GROUND MORAINE

In some areas ground moraine has a distinct lineation approximately parallel to the former direction of movement of the glacier or ice sheet by which it was deposited. The ground moraine surface in such a situation may have a local relief of only a few metres, with maximum amplitudes reaching tens of metres. In Quebec, Henderson (1958) has described ridges that are 100 m apart and 3 km long and with a height of only a few metres. Morrison (1966) has described ridges that range between 40 and 160 m apart, are 2 m high and are 400-1200 m long. In North Dakota, Lemke (1958) recorded ridges 50-70 m apart, 2-5 m high and 200-600 m long.

Similar features have been observed in front of existing glaciers (Dyson, 1952; Hoppe and Schytt, 1953; Lemke, 1958). Lemke observed that (p. 278), 'small transverse recessional moraines and deposits . . . are superimposed on the linear ridges.' Hoppe and Schytt (1953) proved that the flutes extended beneath the glacier and were associated with till accumulation on the lee side of boulders. The importance of boulders in the mechanism involved in the development of fluted ground moraine was also pointed out by Dyson (1952).

The author has observed fluted ground moraine in Alaska with much bigger amplitudes and wave lengths than the features discussed above. However, when fluting of the order of tens of metres in local relief is observed it is very difficult to establish whether the ground moraine was laid down in the form of flutes, or whether it and underlying fluvioglacial sediments or solid rock have been eroded by an ice advance after the till had been deposited. It is difficult to decide whether small or large flutes are the product of depositional or erosional processes. In general small flutes with amplitudes of a few metres are believed to be produced by the squeezing of water-soaked till either into cavities developed on the lee side of boulders or into basal crevasses (Dyson, 1952; Hoppe and Schytt, 1953). Some workers favour a combination of processes with glacial erosion initially shaping the ridges in the solid rock or drift beneath the glacier, followed by the deposition of a till sheet which takes on the form of the ribbed material beneath it. If erosion is involved, then the basal ice surface must have a corrugated form and there are no data available about the detailed shape of basal ice surfaces over extensive areas.

It is interesting that some workers have noted similarities of long profile shape between fluted ground moraine and drumlins. Lemke (1958) actually describes a fluted surface as 'narrow linear drumlins' and both he and Morrison (1966) state that the ridges forming the flutes are higher at their proximal than at their distal ends. Gravenor (1953) referring to a fluted surface in southern Ontario suggests that these flutings may result from the same process as that which formed the drumlins.

DRUMLINS

In terms of increments in the amount of local relief with trends approximately parallel to the direction of ice movement, there is a progression from ground moraine with little or no local relief, through fluted ground moraine, to drumlins. However, drumlins do not always consist simply of streamlined ground moraine, and any explanation of their origin must include a consideration of drumlins consisting of a wide variety of unconsolidated sediments and solid rock.

Gravenor (1953) provides a summary of the literature on drumlins in which the available data, up to that date, on the distribution, size, shape and internal characteristics of drumlins are presented. He concludes that any theory of drumlin formation should satisfy the following conditions (p. 678): '1. Drumlins may consist of (a) clay till, (b) sandy or loamy till, (c) rock, (d) pre-existing drift. 2. They frequently have lenses or layers of stratified materials which sometimes are faulted or folded. 3. Rock drumlins are found side by side with other varieties and have the same shape. 4. Many glaciated areas do not support drumlins. 5. They exist in fields wider than most moraines and rarely occur singly. 6. They have a streamlined shape with the stoss end usually pointing upstream. 7. Lamination may or may not be present. 8. Some drumlins have cores but most do not. 9. They are found behind terminal moraines which mark approximately the outer limit of the ice advance. 10. Their long axes parallel the directions of ice movement.'

More recent work on drumlins has tended to concentrate on three aspects: drumlin shape and distribution, stone orientations within drumlins and theoretical considerations of pressure distributions within the ice and the till. Drumlins usually occur in *en echelon* groups. Vernon (1966), in a study of drumlins in County Down, Ireland, noted that they are concentrated in bands both perpendicular and parallel to ice flow but that their spacing is variable. Doornkamp and King (1971) point out that where drumlin density is high, drumlins tend to be small. This is not as obvious as it seems, because small drumlins, by being spread sparsely, could result in low density values. It seems likely, therefore that the conditions conducive to the development of regularly spaced, but few, drumlins are also conducive to the development of large drumlins. In contrast, drumlins are more numerous in those areas where conditions were marginal for their formation and the ice moved less consistently and slowly. Wright (1957), working on the Wadena drumlin field in Minnesota, showed that at nearly all the sites he investigated the orientation of long axes of stones in the till forming the drumlin tended to be parallel to the crest of the drumlins and parallel to the former direction of ice movement. Some 64 per cent of the stones also plunged up-glacier at an average angle of 23°, and on this basis Wright believed that the till deposit was related to shear zones in the ice and therefore the drumlins were deposited by active ice. He concluded that the drumlins were formed by accretion of basal material from the lowest layers of the ice.

Chorley (1959) was concerned with the streamlined form of drumlins and their formation by a moving medium. Drumlins are not symmetrical as in an ellipse, they are symmetrical only about the long axis, with a rounded end facing into the flow and the tapering end down-stream. The long profile is not symmetrical either, with a steeper up-stream face. Chorley compared the shape of drumlins with that of aircraft wings, in which the maximum width is situated about three tenths from the leading edge. The greater the air speed the wing has to withstand the greater is the length relative to breadth; that is, the form is more elongated. Chorley therefore argues that long drumlins were associated with a powerful ice flow. There is not a simple relationship between drumlin size, elongation or symmetry and ice flow because other factors, such as supply of debris, must also come into play.

Reed *et. al.* (1962) compared drumlins with ellipsoids. They found that at the lower contour levels on drumlins there was good agreement with the ellipsoid shape but there was less agreement at the higher contour levels. They also found that there was a variety of spacing with a multiple modal distribution.

Hoppe (1959), working on the drumlins in northern Sweden, showed that stone orientations were generally parallel to the long axis of the drumlins. In contrast, a section through a drumlin in Wensley Dale, Yorkshire, (Andrews and King, 1968, p. 456) showed preferred orientations of stones to be at a considerable angle to the long axis of the drumlin, with a gradual increase in divergence between elongation of the drumlin and the direction of the mean vector from the bottom, upwards. The authors suggest that ice was pressing against the side of the drumlin to a greater extent in the lower than in the higher parts.

The formation of drumlins in terms of the dilatancy of the material of which they are composed has been considered by Smalley and Unwin (1968). Dilatancy is a property of a granular mass and dilatant materials are more resistant to shear stresses. The authors contend that it is the high resistance to initial deformation of till which leads to the formation of drumlins, (p. 378); 'In the suggested drumlin-forming mechanism the glacial till is being continuously deformed by the movement of the glacier. . . . Within the thin deformed layer of till there is a certain variation in stress level.' If the stresses drop below a certain level then the expanding material collapses into the static stable form and there are no stresses of sufficient magnitude available to cause sufficient dilation to get the compacted material moving again, so the till flows around it, shaping it so that it causes the minimum of disturbance in the flowing stream of till. In some situations the stress level is either too high and the ice mass sweeps everything before it, or too low, resulting in continuous deformation of till and no drumlin formation.

Drumlins therefore form (Smalley and Unwin 1968, p. 379), 'Towards the periphery of the ice sheets (where) the stress levels drop . . . and further towards the periphery the stress levels drop too low to allow drumlin formation and end-moraine structures form.' This theory of drumlin formation depends heavily on the presence of concentrations of relatively large rock fragments in the till allowing dilatancy to operate, and a direct relationship between ice thickness and pressure distributions in a continuously deforming till layer beneath the ice. The authors go on to consider erosion at the ice-drift interface, in which less erosion would take place where high boulder concentrations in the drift occurred with consequent moulding of these particular drift deposits into streamlined forms. They can therefore explain drumlins formed from pre-existing stratified drift as well as drumlins formed from more recent till.

Throughout the considerable amount of literature which deals with the origin of drumlins, two theories dominate. Drumlins are either explained in terms of erosion of a pre-existing drift cover, or in terms of subglacial accumulation of till into streamlined forms. The erosional theory is certainly necessary when there is evidence of the drumlins having been formed from previously existing material. Shaler (1889) thought that the drumlins of New England were formed by two glaciations. The first glaciation provided an irregular till surface and the second scoured this surface leaving the drumlins. Tarr (1894) found that rock drumlins and till drumlins have the same shape, and consequently he concluded that the same erosive processes produced both types. However, as Thwaites (1956) pointed out, one of the problems about the erosional theory is the shape of drumlins. If the dominant process is erosion then surely they should have a shape similar to *roche moutonnée*, with their blunt and steep ends pointing down-stream rather than up-stream as is common with the majority of drumlins.

By far the most common explanations of drumlin formation depend in some way on deposition (Chamberlin, 1883; Upham, 1894; Russell, 1895; Millis, 1911; Alden, 1918; Goldthwait, 1924; Fairchild, 1929; Flint, 1947; Vernon, 1966; Hill, 1971). The complex internal character of drumlins, however, does not suggest a simple depositional mechanism. It seems highly likely that the drumlin with its fascinating streamlined form is polygenetic. In this case the detailed studies of shape, size, distribution, internal constituents and fabric that have been made, point to both erosional and depositional processes that produce very similar forms. Although the term drumlin is usually only applied to streamlined forms consisting of glacial till, the existence of similar forms consisting of other types of glacial drift and partly or wholly of solid rock means that the term has meaning only in a morphological sense and not in a genetic sense. Even for drumlins consisting entirely of till there may well be more than one process of development involved.

MORAINE RIDGES

Glaciers and ice sheets are capable of depositing till in the form of ridges which can be classified in several ways. Prest (1968) has adopted a classification based upon the relationship between the moraine ridge and the directions of ice movement in the ice mass responsible for it. Other workers have stressed the dynamic character of the ice responsible for the formation of moraines and have separated ridges formed by active ice from those deposited by stagnant ice (e.g. Gravenor and Kupsch, 1959). Some workers have stressed the overall morphology or location in terms of other relief features in their classifications (Hoppe, 1952; Andrews, 1963). However, in line with the general theme of this book I will emphasize the process of formation as a means of classifying moraine ridges.

There are two basic approaches to the study of moraine ridge formation. Moraine ridges can either be studied in the process of formation in association with existing glaciers or they can be studied long after they have been formed by examining their morphology, distribution and internal characteristics, and by developing hypotheses of formation based on the available data. Both approaches will be used in the following treatment.

There are probably at least three processes capable of producing moraine ridges: dumping, pushing and squeezing. Some moraine ridges have been created by more than one of these processes and in such circumstances it is often difficult to determine which was the most significant. It is also possible that the external form and internal character will not reveal the process of formation. In some cases factors other than the actual formative process are more important. For example, the environment of deposition may be of paramount importance in determining the form of a moraine ridge so that ridges formed in an ice marginal terrestrial environment may be different from those formed in a subaqueous environment. In the same way structures in the ice may be more important than the depositional process in determining the pattern of moraine ridge development.

MORAINE RIDGES PRODUCED BY DUMPING

In the frontal zones of glaciers and ice sheets debris occurs on the surface. This debris may have been transported on the surface of the ice mass as medial moraine or it may have been brought to the surface along shear planes or crevasses. Goldthwait (1951) described the accumulation of debris in a zone 135 m wide on the margin of the Barnes ice cap (Fig. 18). The dirt layer was usually not more than 2·5 mm thick and rested on a slope of 15°. He observed that dirty ice melted more quickly than clean ice above it, so that there was a gradual steepening of the dirty ice slope and material moved down the slope to accumulate at its base, developing a deposit 0·5-1·2 m thick. A trough then developed at the junction of the dirty ice slope and the thicker till layer on the outer margin. The trough separates the developing ice-cored moraine from the dirty ice and in time the ice core melts and a low moraine ridge 1-5 m high and 15-150 m wide is formed. There has been considerable debate as to the mechanism by which debris is actually brought to the surface along the margins of a cold ice mass (Weertman, 1961b), such as the Barnes ice cap, but once the material does arrive on the ice surface the stages involved in the development of moraine ridges by the dumping of the material have been clearly outlined by Goldthwait. Similar debris accumulations have been described by Souchez (1966) in the Antarctic, and by Hooke (1970) in Greenland.

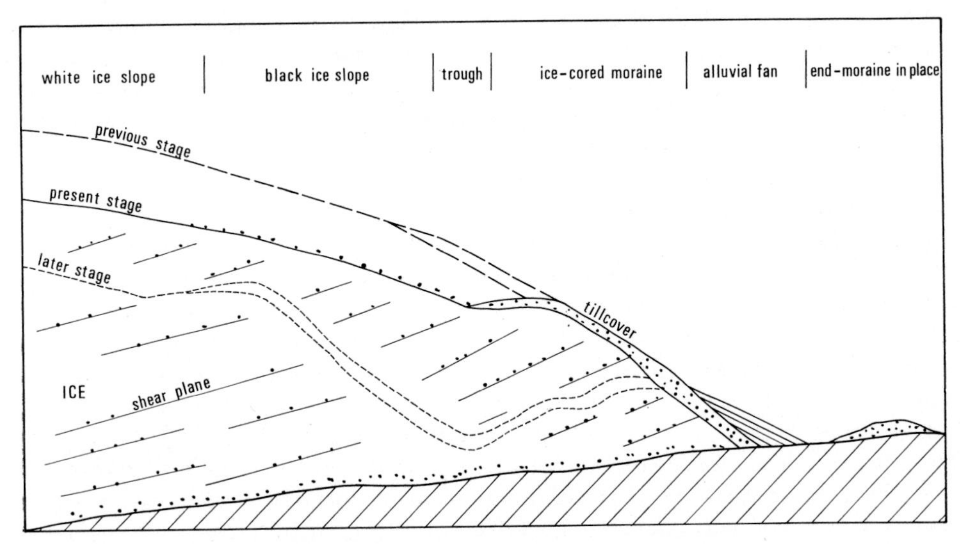

FIG. 18. The retreating margin of the Barnes Ice Cap. Dots indicate dirt in transport up inclined shear planes or gathering on the surface as ablation moraine, and finally deposited as end moraine. (After Goldthwait, 1951.)

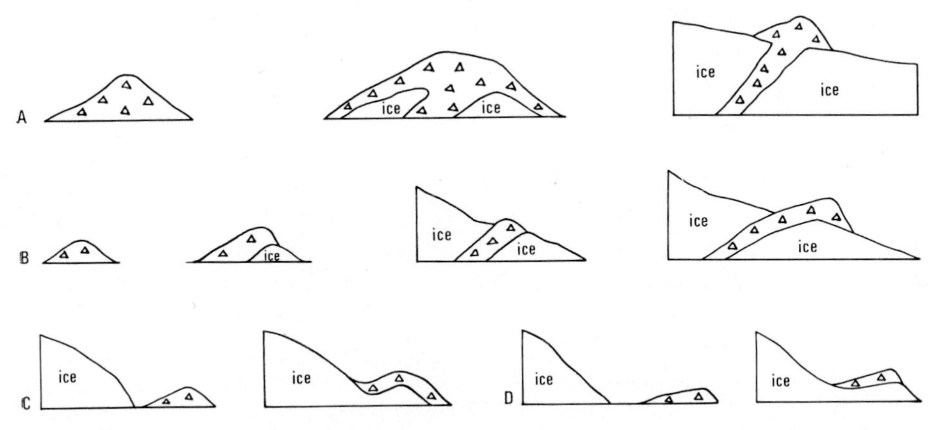

FIG. 19. Subglacial and englacial debris arriving at the ice surface due to ice wastage to produce moraine ridges. (Successive stages: right to left.)

The writer has observed moraine ridges in the process of formation in the frontal zones of various glaciers in Alaska and Iceland. Outcrops of debris trending roughly parallel to the glacier margin are not uncommon. Depending on the nature of the material, the slope of the ice surface, the structures in the ice, and the presence or absence of meltwater streams on the ice surface, the type of moraine that is produced varies widely (Fig. 19). On some occasions a ridge of debris stands up above a wasting ice surface (Figs 19*a*, *b*). This material can protect the ice beneath from melting as fast as the clean ice. If this ice-cored moraine ridge survives the process of downwastage, eventually the ice core will melt out and a moraine with concave slopes will result. There are several variations of this mechanism (Figs 19*b*, *c*, *d*), depending upon the dip of the bed of debris and the slope of the ice surface. Since all these ridges involve ice melting out from beneath the debris it is unlikely that the moraine ridges so produced will have any distinct internal structure or fabric. It is also unlikely that ridges larger than a few metres can be produced in this way.

Dumping of material along the lateral margins of valley glaciers produces lateral moraines. However, there is considerable debate about the percentage of material in lateral moraines that is derived from weathering and mass movement on the rock walls above the glacier and material derived from the glacier itself. The increased steepness of the surface of the ice towards the valley wall is a common feature on valley glaciers and there is a natural zone of accumulation in the trough formed between the glacier and the valley wall. Dumping of material that has been transported on and in the glacier certainly plays a part in the construction of lateral moraines.

Østrem (1963, 1964), has studied ice-cored moraines in Norway and Baffin Island. Prior to this work it was generally assumed that the ice forming the ice cores of moraines near existing glaciers was glacier ice. Østrem discovered that although in some cases glacier ice did form the core, in other instances the ice had originated from a snow

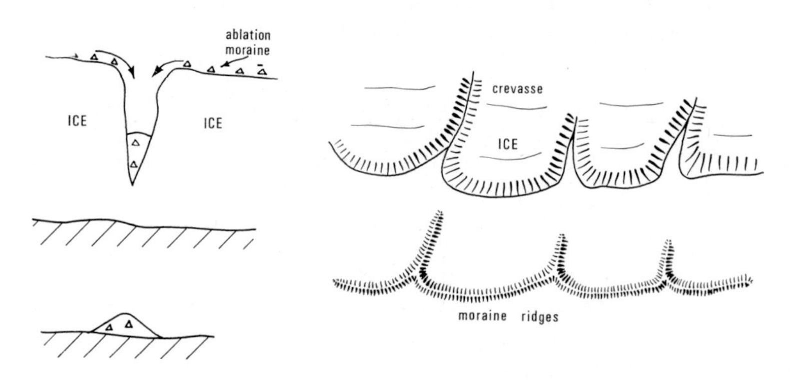

FIG. 20. Dumping of material in surface and marginal crevasses to produce moraine ridges.

bank that has been piled up against the ice margin. It appears that conditions which favour the formation of ice-cored moraines are those where accumulation and ablation are small.

Dumping of material from the ice surface may occur in open crevasses or other re-entrants into the ice margin (Fig. 20). The sliding of material into the ice-walled channels allows a ridge to develop, which, when the supporting ice walls and the ice beneath it melts, will result in the formation of a moraine ridge.

Boulton (1967) has described the development of both single ridges and complex ridges (hummocky moraine) from debris accumulated on the surface of an ice mass. He demonstrated how a series of debris bands result in an irregular ablation surface; when the debris is let down by the melting out of the ice from below it, an area of hummocky moraine can be produced. He states, (p. 733), 'It is now possible to envisage another mode of origin for many "controlled disintegration features," i.e. they are derived from debris bands which crop out on the glacier surface and have survived melting of the underlying ice. Complex series of "controlled" ridges could then be derived from closely packed debris bands, and single, widely spaced ridges could be derived from widely spaced debris bands. The composition of the ridges would depend on the composition of the debris bands from which they are derived, and they might vary from till to silt, sand and clay.'

The process of dumping includes the deposition of material that simply slides off the ice surface into the proglacial environment, and material that is let down by the melting of ice from beneath or within it. In the latter case the material may have inherited a fabric produced by its emplacement in the ice so long as it is not disturbed by flowage during deposition. In most of the other ridges produced by dumping there is little or no preferred orientation of the rock fragments.

MORAINE RIDGES PRODUCED BY PUSHING

As the margin of an ice mass advances it is possible for it to act as a bulldozer and push material into a ridge form (Gwynne, 1942; Dyson, 1952; Okko, 1955; Hewitt, 1967). The moraines produced in this way tend to have more convex slopes of which the distal slope is the steepest. However, it is by their internal character that they are most easily recognized. The deposits forming push moraines show signs of tectonic disturbances in the form of faulting and thrusting. There is a tendency for imbricate structures to develop. If distinct faults are present it is likely that the material was in a frozen state when the push moraine was formed.

In order for push moraines to develop there must be suitable material and an advance of the front over not too great a distance. It seems unlikely that individual moraine ridges would survive a prolonged period of pushing.

MORAINES PRODUCED BY SQUEEZING

Moraine ridges that are produced by the movement of water-soaked till as a result of pressures induced from the overburden of ice or through pressure differences within the till sheet are of wide occurrence. The squeezing of till into cavities or from beneath the ice front has been mainly interpreted from the dominant orientation of pebbles in the till ridges at right angles to the ridge crests.

Hoppe (1952, 1957, 1959), working in Sweden, and Gravenor and Kupsch (1959), working in Canada, have provided a great deal of data relating to the formation of moraine ridges. Hoppe (1952) was dealing with a topography consisting of moraine plateaus, rim ridges and dead ice depressions. The ridges ranged in height from 5·35 m and were 50-200 m wide at the base. Stone orientations and dips were measured and showed a strong orientation at right angles to ridge crests with generally low angle even on the outsides of ridges. Hoppe (1952, p. 24) concluded, 'Rim ridges, terraces and terrace ridges (as well as isolated ridges) have been proven to be subglacial formations shaped through the transport of till from dead ice depressions towards cavities

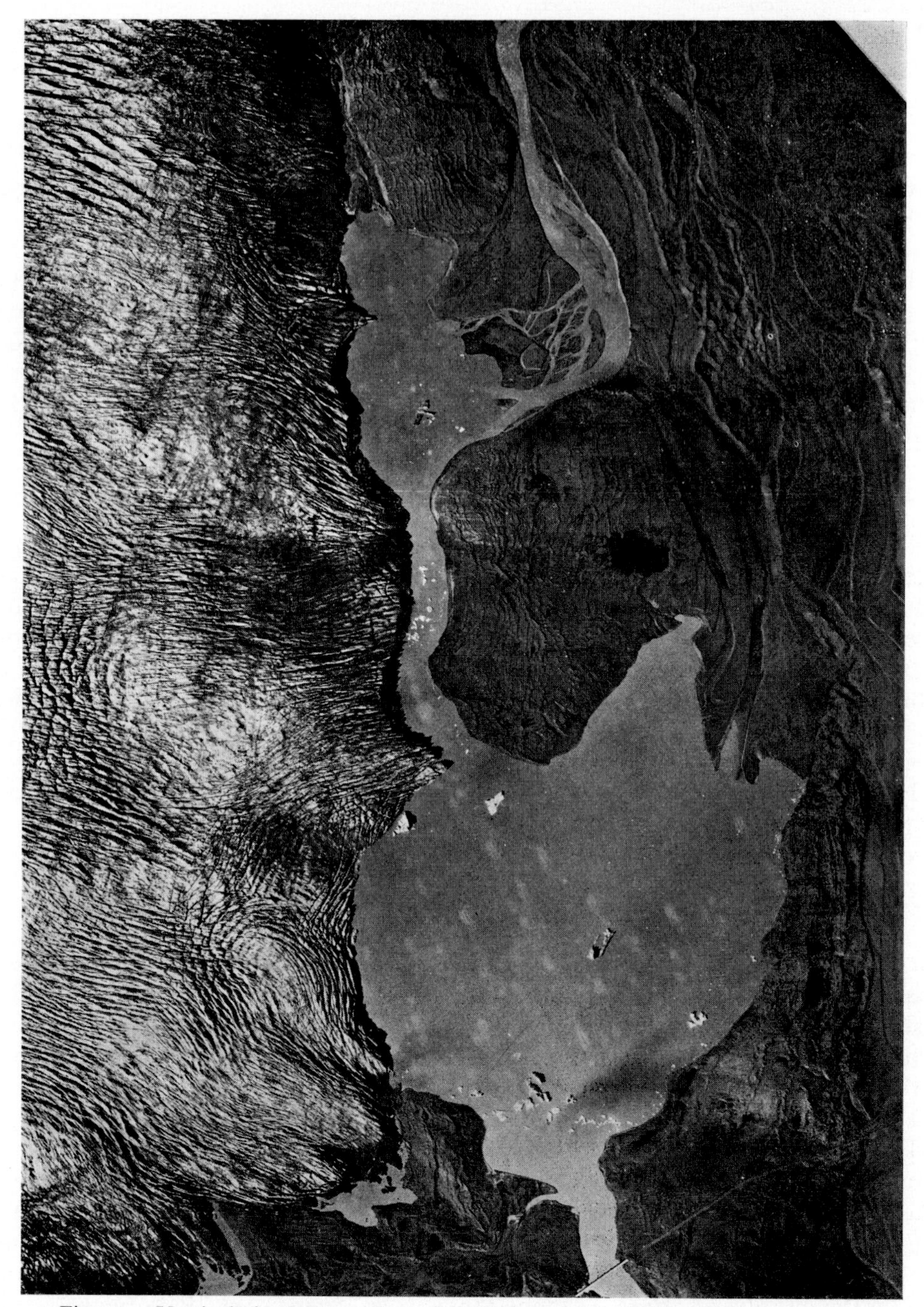

Fig. 21. Vertical air photograph (1965) of the margin of Fjallsjökull, Iceland and its proglacial area (see Figs 22 and 23). (Photography by Landmaelingar Islands for University of Glasgow.)

FIG. 22. Photogrammetric map (1965 photography) of moraine ridges along the eastern margin of Fjallsjökull, Iceland. (After Price, 1970.)

in or below the ice. These formations are heaped on the moraine plateaus which consequently must be older and therefore also subglacial.' He also states (p. 54), 'The moraine ridges . . . were formed underneath the ice through squeezing into basal cavities (crevasses, cavities caused by meltwater etc.) of moraine material which was soaked with water and therefore in a plastic state. The weight of the ice itself exerted the necessary pressure.'

Gravenor and Kupsch (1959), discussing a wide variety of features associated with the disintegration of an ice sheet in Alberta, Canada, also suggested a squeezing mechanism to produce moraine ridges, although they believe that similar forms can also be produced by the letting down of till that has slumped into surface crevasses on to the subglacial floor. They state (p. 56), 'The features resulted from the letting down of till due to ablation, from the squeezing-up of till into openings at the base of the ice, or from a combination of both causes.'

Stalker (1960), also working in Alberta, accounted for moraine ridges by (p. 33), '. . . pressing or squeezing of sub-ice material into nearby tunnels, holes and crevasses.' It should be noted that Hoppe, Gravenor and Kupsch, and Stalker all emphasized the importance of subglacial deposition by the squeezing process.

As a result of a detailed study of the till fabrics of the cross-valley moraines of north-central Baffin Island, Andrews and Smithson (1966) concluded that moraine ridge formation was related to the influx of summer meltwater which, when it reached the base of the ice cap, would cause a rapid loss of strength in the till. The weight of the overlying ice mass would initiate a flow of semi-liquid till towards the ice front where it would be concentrated along the foot of the ice cliff as a moraine ridge. They also concluded that those moraine ridges aligned perpendicular to the ice cliff probably represent the injection of liquid till into subglacial cavities. Andrews and Smithson were dealing with a cold ice mass so the arrival of summer meltwater was an important triggering mechanism in their squeezing process.

Price (1970), working on moraine ridges very similar to those described by Andrews and Smithson (1966) but closely associated with the temperate glacier Fjallsjökull, Iceland (Figs. 21, 22, 23, 24), discovered that the pebbles within the moraine ridges were either at right angles or at a high angle with the ridge crests no matter whether the ridges were straight or curved. All the moraine ridges had been formed during the last 30 years and were one to four metres high and four to ten metres wide at the base. Many of the distal slopes had angles of between 25° and 35° compared with 20° and 25° for proximal slopes. The material forming the ridges was a gray till with subrounded and subangular particles, mainly gravel and cobbles in a sandy matrix. Less than 5 per cent of the material was silt or clay. From the shape and distribution of the ridges it was concluded that each moraine ridge had paralleled the ice front at the time of its development. There was no evidence of push structures in the older ridges or in those currently being formed (Figs. 25, 26). The dominant orientation of pebbles in those moraine ridges produced parallel to the former ice margin and on slopes perpendicular to the former ice margin was at right angles to the ridge crests (Fig. 27). In the majority of the ridges the dip of the pebbles was towards the former ice margin on both the proximal and distal sides of the ridges.

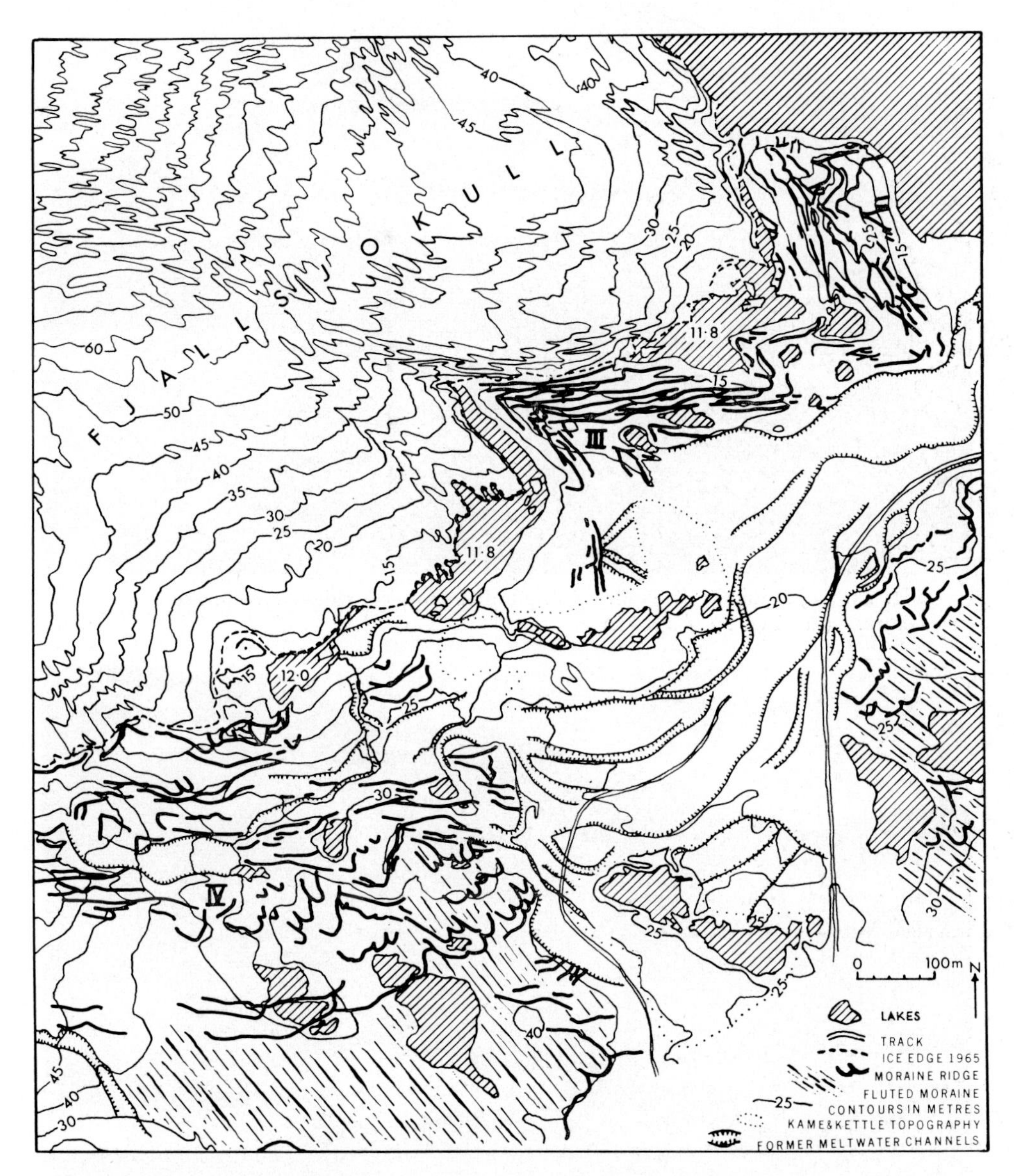

FIG. 23. Photogrammetric map (1965 photography) of moraine ridges along the southeastern margin of Fjallsjökull, Iceland. (After Price, 1970.)

Fig. 24. Fjallsjökull, Iceland. Sub-parallel moraine ridges crossing mound in centre. (See Fig. 22, Area II.)

FIG. 25. Fjallsjökull, Iceland. Arcuate moraine ridges 2-3 m high along ice margin.

Fig. 26. Fjallsjökull, Iceland. Moraine ridge, 2 m high at ice margin.

The observations made of those moraine ridges in the process of formation and the morphological and sedimentological characteristics of the older moraine ridges suggest that they were both formed by the same mechanism. That mechanism is believed to involve squeezing of water-soaked till (Fig. 28) from beneath the glacier to produce ridges which are parallel to the ice margin and reflect any crenulations it may possess. The semi-liquid till builds into a ridge with a steeper distal and less steep proximal slope because of the drag effect on the basal layers. The pressure that produced the movement in the semi-liquid till is derived from above as the glacier ice sinks into the water-soaked till. This may be the result of the lower bearing capacity of water-soaked till compared with dry till, or it may result from fracture of ice blocks along crevasse lines causing settling of the ice blocks into the till. Both mechanisms could operate simultaneously. It is tempting to suggest an annual occurrence of whatever mechanism-produced the moraine ridges when fifteen ridges are known to have been produced in sixteen years. The mechanism could be triggered off by the early summer meltwater descending to the base of the glacier as suggested by Andrews and Smithson (1966). It is possible that more than one ridge could be produced in one year and perhaps none in other years. It is therefore impossible to prove annual formation of

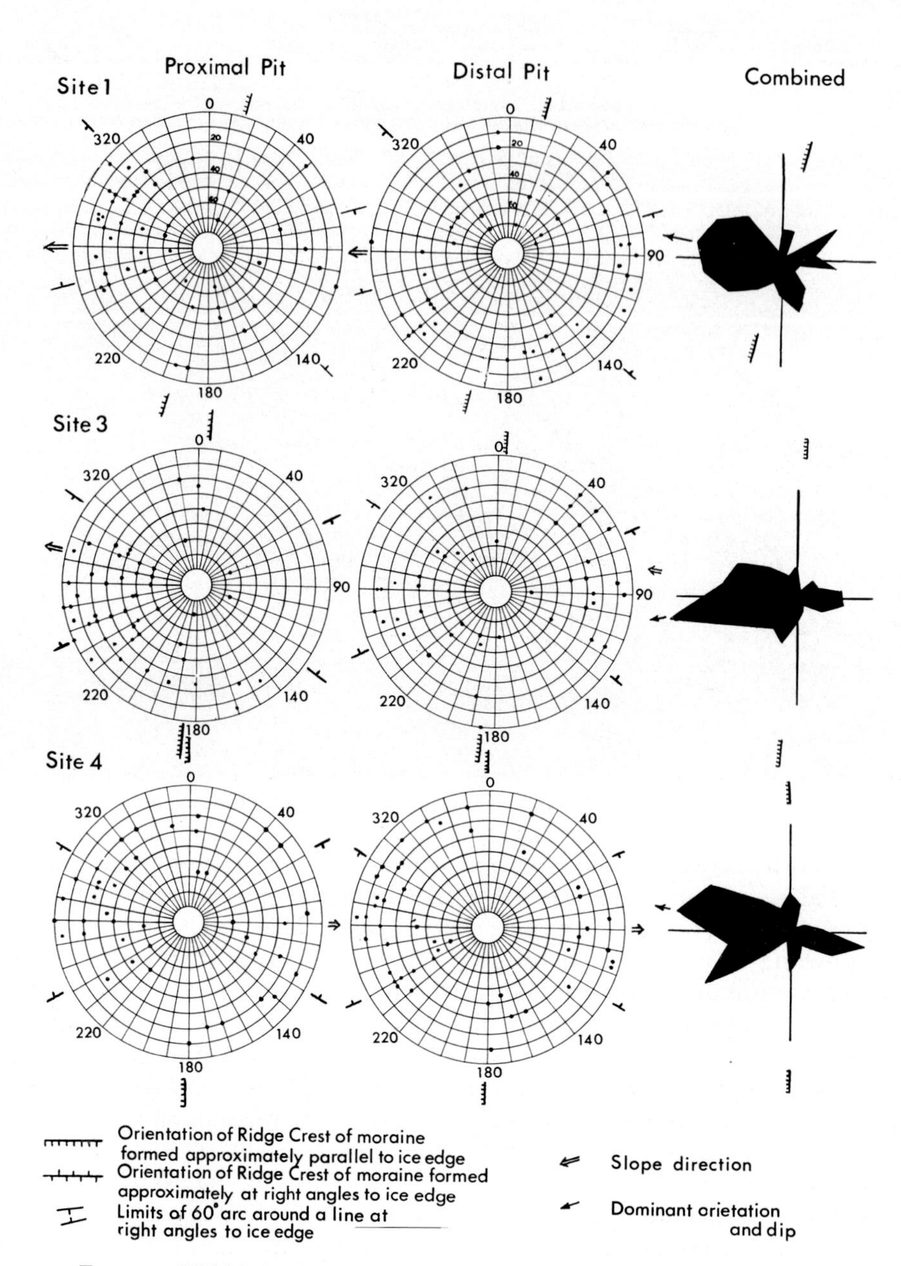

FIG. 27. Till fabrics from moraine ridges in front of Fjallsjökull, Iceland. (After Price, 1970.)

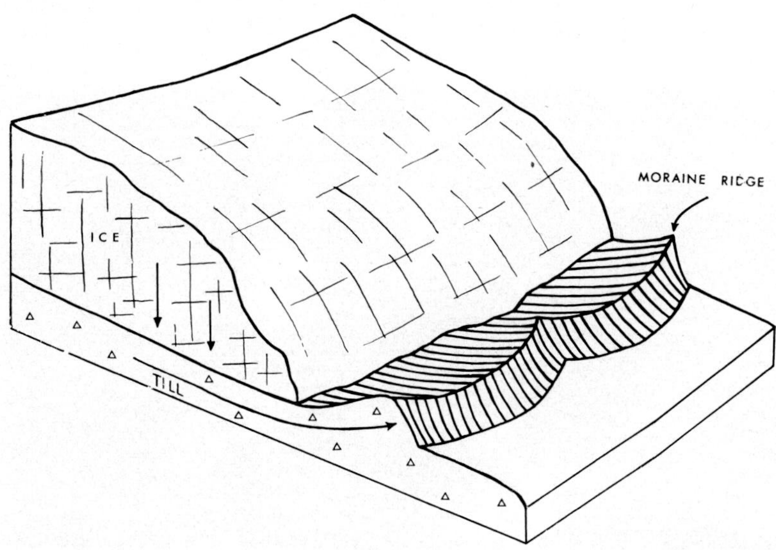

Fig. 28. Diagrammatic representation of the squeezing of till from beneath an ice front to produce proglacial moraine ridges. (After Price, 1970.)

ridges simply on the basis of the correlation between number of ridges and number of years.

The development of moraine ridges both in the subglacial and proglacial environment is only poorly understood. Undoubtedly some moraine ridges are produced by a combination of dumping, pushing and squeezing. The environment in which glacial deposition takes place can change rapidly over short time periods and short distances. Moraine formation may or may not be strongly influenced by structures in the ice but it will certainly be influenced by whether or not the ice is stagnant. The development of moraine ridges may result from direct deposition or from post-depositional transport of the material to produce ridges.

Frequently when the moraines of an area that was glaciated many thousands of years ago are being studied, only a very small part of the initial moraine system has survived to provide evidence of the process of formation. Fluvioglacial erosion tends to destroy large parts of proglacial and subglacial moraine ridges soon after they have been formed. Fluvioglacial deposits can also bury all or part of moraine ridges. In many ways the complete arcs of moraine, 50-300 m high which surround Convict Lake in the Sierra Nevada (Sharp 1969) and Kviarjökull (Fig. 29) in Iceland are remarkable on two accounts: first their completeness which is so unusual, and secondly their size. It is hard to imagine the processes of dumping, pushing and squeezing producing such big features except over extremely long periods of time. Not only do depositional environments and processes play an important role in moraine ridge formation but also the time period over which these environments and processes function in a particular locality are of prime importance.

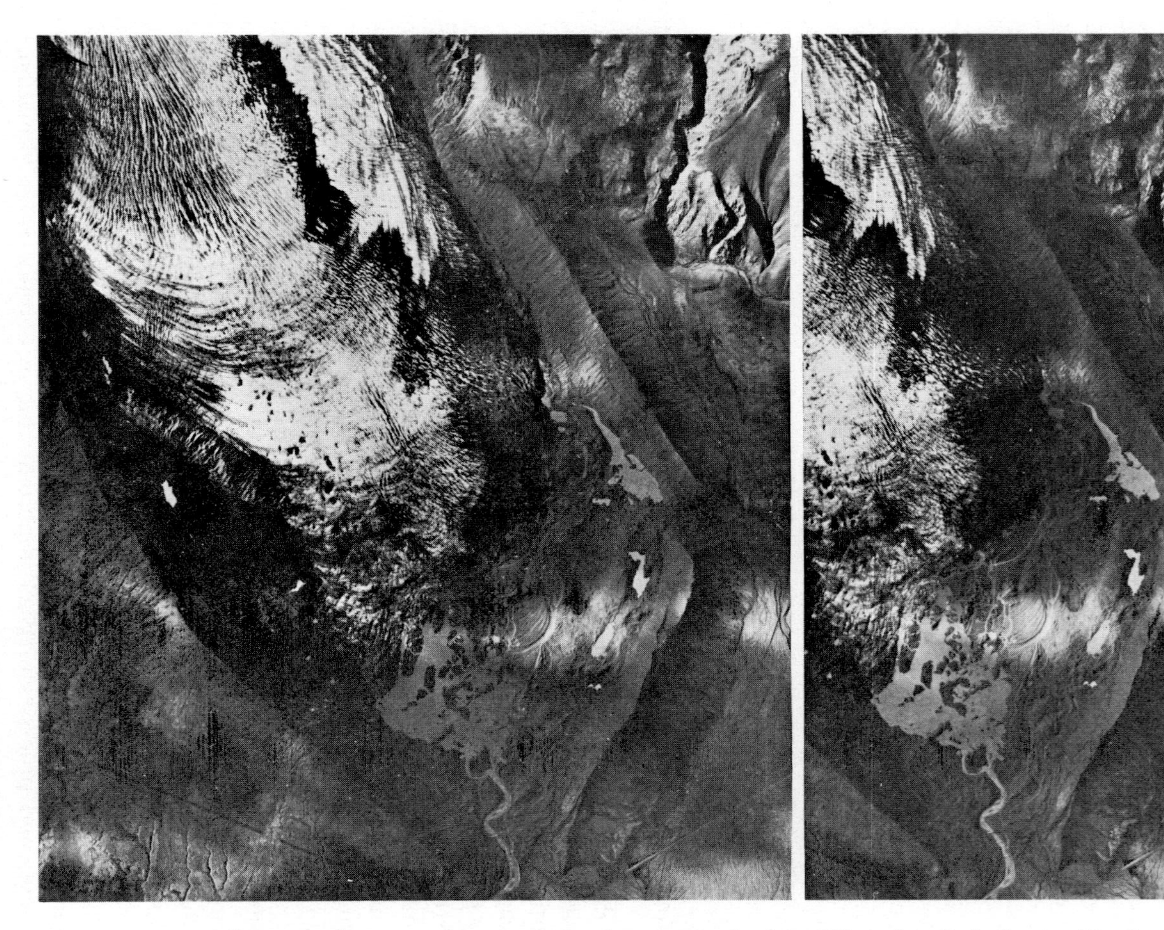

Fig. 29. Stereoscopic pair of air photographs (1965) of Kviarjökull, Iceland. The large lateral and frontal moraines are up to 100 m high. (Air photography by Landmaelingar Islands for University of Glasgow.)

V FLUVIOGLACIAL EROSION

INTRODUCTION

In Chapter 2 the significance of the change from solid to liquid state of glacier ice and snow, as a result of ablation, was discussed. The large volumes of meltwater released when an ice sheet wastes away are capable of a great deal of erosion and transportation in a distinctive environment, so producing landforms that are different from normal fluvial landforms. The significance of the channelled flow of meltwater on, in, and under glacier ice was overshadowed for many years by the emphasis on the importance of drainage associated with ice-dammed lakes.

The deglaciation of an area has often been studied in terms of the development of ice-dammed lakes and their associated drainage channels. The recognition of sites of former ice-dammed lakes, on the basis of the occurrence of glacial drainage channels (meltwater/overflow channels), and associated deposits, led various authors to reconstruct successive glacier margin positions to account for the development of the lakes and the channels. Such reconstructions often ignored not only glaciological principles but also the lack of any evidence, apart from the channels themselves, for the former existence of the ice-dammed lakes. Kendall's (1902) interpretation of the glacial drainage channels and ice-dammed lakes of the Cleveland Hills has long been the model for studies of fluvioglacial erosion and deposition associated with the deglaciation of areas in Britain. Charlesworth (1926b, 1955) for example, indicated on maps of Scotland hundreds of former ice-dammed lakes without shorelines to justify the existence of so-called overflow channels and to delimit ice margin positions. Stimulated by the publications of Scandinavian workers on the processes involved in the deglaciation of the Norwegian and Swedish Highlands, a reappraisal of the processes involved in the origin and significance of meltwater channels and their associated deposits was commenced in Britain. Since 1958 a series of papers by Sissons (1958a, b, c, 1960a, b, 1961a, b, 1963), Price (1960, 1963a), Derbyshire (1961), Embleton (1964), Bowen (1964-66), Bowen and Gregory (1965), Gregory (1965), and Clapperton (1968), applying the concepts so strongly put forward by the Scandinavian geomorphologists (e.g. Mannerfelt, 1945, 1949; Hoppe, 1950; Gjessing, 1960), has resulted in a new approach to the problems of deglaciation and the old ideas, primarily based on Kendall's work, have had to be largely rejected.

MELTWATER CHANNELS ASSOCIATED WITH EXISTING GLACIERS

When melting takes place on a glacier or ice sheet the behaviour of the meltwater is determined by the temperature and structural characteristics of the ice, the slope of the ice surface and the character of the ice-rock interface. The ice mass, whether it is a valley glacier or an ice sheet, covers a pre-glacial land surface and if the ice buries all of that land surface above sea level the ice-rock interface is entirely subglacial. When an ice mass is of more limited extent and terminates on land, the junction between the ice and the ice-free slopes is defined as the ice margin. The nature of ice margins is extremely variable depending upon the temperature of the ice, its structure and velocity, and the steepness and detailed form of the slopes against which the ice abuts. A major concern in the discussion that follows is whether or not the junction between the ice and the slope on which it rests represents an interface across which water cannot move. An ice mass that is below the pressure melting point and is frozen to its bed will not allow the penetration of meltwater along this interface unless there are suitable structural features, such as crevasses, present. When an ice mass is at the pressure melting point the interface between ice and rock will allow meltwater penetration but it is possible for the junction to be water-tight and therefore for water to move along the line of the ice margin rather than penetrate beneath it. However, it is unlikely that water flowing along the margin of an ice mass at the pressure melting point will remain at the surface for long distances because of the common occurrence of marginal crevasses and tunnels which will carry the meltwater below the ice surface.

The form of the ice surface on a glacier or ice sheet is a function of the mass balance. In an accumulation area meltwater may occur for short periods during the summer and may modify the surface form and therefore the character of the ice margin for relatively short periods. In the ablation area the summer meltwater is of great significance as all the ice surface suffers ablation for long periods. Where an ice mass terminates against a rock slope, the reflected radiation from the rock slope produces more rapid ablation near the ice margin than over the rest of the ice surface. This phenomenon is responsible for the common convexity of cross profiles of valley glaciers and for the form of the ice surface near a rock ridge emerging from beneath a downwasting ice sheet.

Terminology relating to ice margins can be confusing. In the discussion that follows two types of ice margin will be distinguished. Those that are approximately parallel to the general direction of ice movement will be defined as lateral margins while those at right angles to ice movement will be defined as frontal. The variety of ice surface gradients, of slope angle, morphology and material that can be encountered along ice margins is so great that further classification would be absurd.

Of particular relevance to any interpretation of the origin and significance of meltwater channels in areas of former glaciation is the information available in the literature on meltwater streams and meltwater channels associated with existing glaciers. Writing in the Quarterly Journal of the Geological Society, London, E. J. Garwood (1899) described meltwater streams that he had observed on glaciers in Spitzbergen (Figs. 30, 31). He paid particular attention to meltwater streams flowing in tunnels

FIG. 30. Supraglacial stream, Highway Glacier, Spitzbergen. (After Garwood, 1899.)

near the surface of glaciers, in the vicinity of nunataks, and to voluminous streams that had cut channels into the surface of ice in King James Land.

The work done during the period 1890 to 1912 on the glaciers of Yakutat Bay by Russell, Tarr, Butler and von Engeln included many observations on meltwater drainage. Tarr and Butler (1909) described a valley on the eastern margin of Hayden Glacier, through part of which a stream flowed. The valley had ice on one side and gravel bluffs 200-400 feet high, or a steep rock face, on the other. They stated (p. 142), that '. . . in one place the stream has cut a gorge across a rock spur. When Hayden Glacier has disappeared, this gorge will form a gulch on the mountain slopes high above the valley bottom. The course of the valley elsewhere will be marked by a precipitous, trimmed slope and by a complex of stream and marginal lake-deposits with irregular topography, but the other valley wall will be missing. Other glaciers present similar phenomena, and short gorge sections are not uncommon where the glacier, by crowding against rock spurs, has forced the stream across them.' The authors described one of these gorges, half a mile long, that was cut in granite to a depth of 200 feet.

Writing in 1909, R. S. Tarr described marginal drainage channels of several different types. Some were benches formed by streams flowing between a rock wall and the ice margin; others were well-developed channels with two rock walls while others were deep gorges. Tarr pointed out that many marginal channels were characterized by lack of continuity as the meltwater streams that cut them either flowed beneath the ice

FIG. 31. Englacial stream, Kings Glacier, Spitzbergen. (After Garwood, 1899.)

or left the ice margin and cut across rock spurs. He also stated (p. 100); 'Where engaged in gorge cutting these streams work with great rapidity, for the volume is great, the sediment load heavy, and therefore, with sufficient grade to prevent deposition, they are active agents of erosion.'

Von Engeln (1912) concluded that large volume marginal streams only occurred after a rapid advance of valley glaciers. He believed that when a glacier had existed for a long time without experiencing any spasmodic disturbance, its lateral drainage would normally be submarginal, that is, '. . . under the ice but near its lateral edge.' (p. 109). His conclusions were based on observations made of supraglacial streams becoming confluent and descending moulins, and of other streams draining the valley sides, being relatively warmer, cutting courses under the ice margin. Von Engeln described meltwater streams emerging at the front of Hidden Glacier, near each lateral margin. He took temperatures in the waters draining off the valley sides towards the ice, and since they were as high as 6·7°C in July and early August, it was not surprising that they cut submarginal courses.

The description of meltwater drainage associated with glaciers in Spitzbergen and Alaska can leave no doubt as to the extent and efficiency of such systems. The marginal channels occupied by fast-flowing, heavily laden streams indicate the ability of such streams to erode deep channels in solid rock. The observations of Von Engeln also established the probability of submarginal drainage and the development of englacial and subglacial drainage in stagnant ice. The existence of large marginal streams only in association with glaciers known to have been recently active has not yet been denied.

Sharp (1947) described meltwater streams flowing on, and at the margin of, the Wolf Creek Glacier, Alaska. The gradients of the supraglacial streams were relatively high, they were turbid with suspended matter and, due to the low co-efficient of friction in the ice channels, even the small streams were transporting relatively large fragments by traction. The marginal streams of the Wolf Creek Glacier were fed by direct melting of ice walls, supraglacial and englacial streams and by land streams. The courses of

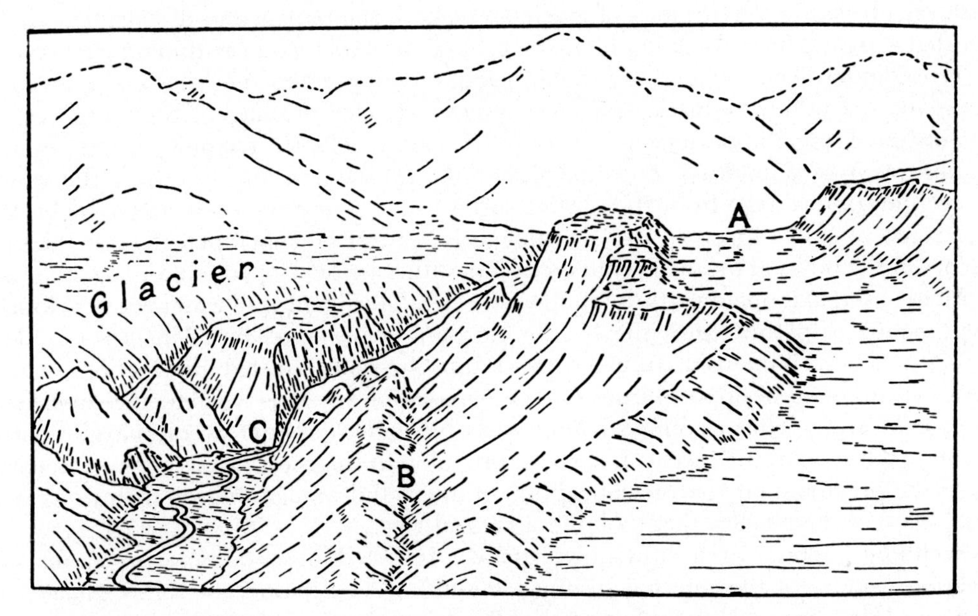

FIG. 32. Meltwater features near the Wolf Creek Glacier, Alaska. (After Sharp, 1947.)

the marginal streams were not long as they soon followed courses into the ice. Some abandoned bedrock channels, above the present ice margin, were 6-45 m deep, narrow, steep-sided and partly choked with boulders (Fig. 32). Sharp (1947, p. 51) states that, 'Some are short, aligned gulches crossing successive spurs between tributary valleys; others are clearly cut-off loops of entrenched meanders left hanging at both ends. Longer and straighter bedrock channels are continuous with remnants of narrow terraces representing the parts of the channels along which the glacier formed one wall. The position of some bedrock channels is determined from a morainal cover, but others are located along the lines of overflow from ponded drainage.'

Work on the Moreno Glacier, Patagonia, by Nichols and Miller (1952) supports the concept of subglacial and englacial drainage of marginal lakes. Further work in Norway by Liestøl (1955) also deals with the drainage of ice-dammed lakes and again the subglacial and englacial routes are proved.

Although a great deal of information is available about meltwater streams on the surface and along the lateral margins of glaciers and ice sheets, very little is known about the englacial and subglacial hydrology of ice masses. A conference held in Cambridge, England, in September 1969 and sponsored by the Glaciological Society, considered some 40 papers on various aspects of glacier hydrology. The papers were primarily concerned with observations of discharge in glacierized drainage basins, the development of theoretical models of the internal movement of water in glaciers, the effect of meltwater on ice movement, and catastrophic floods produced by the drainage of ice-dammed lakes. The symposium indicated a great deal of interest on the part of earth scientists in the hydrology of glaciers but both the papers and discussion sessions revealed a great paucity of data about the behaviour of meltwater within an ice mass.

Work undertaken by Stenborg (1968, 1969) on temperate valley glaciers in Sweden represents the only systematic study on internal glacier drainage. Stenborg (1968) demonstrated the importance of vents or cracks of 'glacier tectonic' origin in the establishment of a meltwater drainage system. He stated (p. 52), '. . . the down penetration of drainage from the glacier surface only develops when prepared by the occurrence of cracks or crevasses. Analogously it must be concluded that the further reaches of the internal drainage courses cannot either be developed without making use of cracks or singular openings through the ice, at least until bodies of water are developed, deep enough to permit emptying by the mechanism proposed by Glen (1954). The further reaches of internal drainage must thus be concentrated in positions where cracks or any kind of opportunities for the percolation of water beyond the superficial cracks can exist. The positions are most probably found near the lateral margins of the glacier. The occurrence seems much less probable in the central parts, where according to the predominant routine conception of internal drainage the water finds its way to the base of the glacier, there joining a central river.'

Stenborg (1969), in a subsequent paper, reported observations, based on salt injection studies, of the internal drainage system of two temperate valley glaciers in Sweden. His observations confirmed that the internal drainage occurred as semilateral systems fed by meltwater streams that followed oblique lateral crevasses from those points where the meltwater descended below the glacier surface. The semilateral systems are believed to be relatively narrow both vertically and horizontally.

Although Stenborg's observations confirm the importance of structural control in determining the location of the englacial drainage systems they also raise problems about the mechanics of the development of internal drainage channels. When meltwater moved through cracks and crevasses in the upper part of the ice mass and developed discrete channels, these channels were widened by melting, due to the heat supplied by friction and eddy viscosity energy, by corrasion by rock fragments in transport or by melting due to circulating air. None of these processes is possible unless water is flowing and therefore none of them is capable of initiating new channels. Stenborg (1969, p. 27) concludes: 'For the initiation of drainage courses we should therefore consider either the possibility of penetration through vents made by some other force than water or the development of the drainage courses by standing water.'

The development of vents by some other force than water is most likely to be associated with glacier-tectonic activity resulting from differential ice movement. In temperate glaciers crevasses may be 20-30 m deep, but whether minor cracks sufficient to allow initial water movement can penetrate to greater depths is not known. In a valley glacier vents of glacier-tectonic origin are commonly located in lateral positions where relatively high shear-strain rates occur.

Glen (1954) suggested an explanation involving standing water as the cause of the development of drainage courses. Ice flow occurs as a result of the pressure differences between a water-filled hole and the surrounding ice. However, this mechanism requires a deep water-filled hole (under certain conditions it must be 150 m deep), and with the exception of ice-dammed lakes this presupposes the existence of tectonic vents down to the depth required for the initiation of the 'Glen mechanism'. The actual mechanisms whereby meltwater develops a channel system at depths greater than about 30 m in temperate glaciers are poorly understood. The geomorphological evidence of meltwater penetration into ice masses suggests maximum penetration through 100 m to 150 m (Clapperton 1968, p. 209). However, this evidence is based on the assumption that a subglacial chute or a subglacially engorged esker was developed throughout its length from top to bottom over a relatively short period of time. The fact that ice structures play an important part in the development of internal drainage systems is not surprising. Until more is known about the structures that develop within glaciers and ice sheets, as a result of differential rates of movement and irregularities at the ice-rock interface, it will be extremely difficult to discover the behaviour of meltwater streams within glaciers. However, the existence of meltwater streams at, beneath the margin of, on the surface of, within, and beneath glaciers and ice sheets, all capable of eroding channels in ice, solid rock and drift deposits has been clearly established.

MELTWATER CHANNELS IN AREAS OF FORMER GLACIATION

Waters overflowing from ice-dammed lakes have long been postulated as the cause of many meltwater channels. As early as 1863, Sir Archibald Geikie (1863, p. 28), suggested that some channels in the Tweed Valley near Drumelzier, and others in the Manor Valley, Peeblesshire, Scotland, '. . . seem as if they have been formed when the lateral glens were dammed up so as to form lakes, and the pent-up waters escaped by cutting out channels for themselves by which they escaped into the main river.' J. Geikie (1869, p. 18) writing about the same channels suggested that, '. . . they should be referred to a time when the drainage of the district was greatly modified by large accumulations of snow and ice.'

The great controversy over the origin of the parallel roads of Glen Roy, Scotland, that started with the publication of Jamieson's classic paper (1863) embodied the two major problems associated with the existence of large bodies of water dammed by ice, namely, the efficiency of ice as an impenetrable barrier to meltwaters, and the development of shoreline and lake-bottom deposits. Jamieson (1863, 1892) postulated ice-dammed lakes overflowing through various cols in the Lochaber area, to account for the parallel roads as the beaches or shorelines of the lakes, and he used the example of the Marjelen See as an illustration of the strength of ice barriers. Prestwich (1879), however, pointed out that the Marjelen See was drained almost every year through crevasses in the glacier. Prestwich cited other examples of englacial drainage and came to the conclusion that glacier ice at the end of Glen Roy could not have withstood the hydrostatic pressure of lakes of the size suggested by Jamieson.

Some 23 years before Kendall published his work on the Cleveland Hills, therefore, there was doubt as to the efficiency of glacier ice as an impenetrable dam and Prestwich had suggested englacial and subglacial drainage. Between 1890 and 1912 much evidence of meltwater drainage on, into and under glaciers was made available by studies of Alaskan glaciers by Tarr (1909), Von Engeln (1912), and Russell (1892), and from other areas by Tarr (1908), Garwood (1899) and Rabot (1905). In the introduction to his work on the Cleveland Hills, Kendall mentioned the theoretical possibility of englacial and subglacial drainage. Although aware of the work on existing glaciers he did not apply it to the Cleveland Hills and nowhere does he refer to drainage into or beneath the ice. The record of the discussion that followed the reading of Kendall's (1902) paper indicates that it was very well received. Referring to Kendall's work on the Cleveland Hills, Harmer (1907, p. 470), wrote, 'It is clear moreover, that such cases (of ice-dammed lakes) must be typical and not anomalous.' This general acceptance of the development of ice-dammed lakes and the cutting of overflow channels continued for many years.

Sissons (1960b, 1961a) has discussed in detail the limitations of Kendall's work and the effect it has had on studies of deglaciation in Britain. He has shown that the existence of glacial drainage channels does not necessarily indicate the former existence of ice-dammed lakes. The rarity of shoreline features, the rarity of definite deltas and the absence of proven lake deposits all suggest that ice-dammed lakes of considerable extent were the exception rather than the rule. The problems that are encountered in explaining channels with up-down long profiles, the absence of channels in localities where they would be expected on the basis of the lake hypothesis and the occurrence of channels on the reverse slopes and minor summits of spurs, are very difficult to explain if lake overflows are held responsible for the formation of the channels. Sissons also pointed out that although Kendall's work has been widely accepted and applied, a few workers have expressed doubts as to the validity of Kendall's views and suggested that the supposed lake sites were occupied by stagnant ice. Hollingworth (1952), believed that at the stage of maximum glaciation, areas such as Kendall's Lake Pickering would be more likely to be occupied by masses of stagnant ice rather than by bodies of open water.

The difficulties encountered by workers attempting to apply Kendall's hypothesis to other areas began to be appreciated following the publication of papers by Mannerfelt (1945, 1949) and other Scandinavian workers. Publications by Sissons (1958a, b, c, 1960b, 1961a, b, 1963) related the concepts developed by the Scandinavians to the deglaciation of parts of the British Isles and showed clearly the weakness inherent in Kendall's hypothesis. The strength of the Scandinavians' approach was that they drew both on their own personal experience and that of other workers, obtained in areas of existing glaciers, and applied their knowledge of present-day processes to areas of former glaciation. No one would seriously suggest that ice-dammed lakes never existed in association with the Pleistocene ice sheets and valley glaciers, and they certainly exist in presently glacierized areas. However, it cannot be denied that Kendall's paper on the Cleveland Hills had a profound effect on the interpretation of meltwater channels in Britain for some fifty years and, until origins other than association with ice-dammed lakes were considered, only a very limited understanding of the origins of meltwater channels and their significance had been obtained. There can be few workers who have affected so profoundly the development of ideas in glacial geomorphology as did Kendall.

Two papers by Mannerfelt (1945, 1949) were the forerunners of many others that applied the observations made in Alaska and Spitzbergen to the channels mapped in the Swedish–Norwegian mountains. The ice sheet covering the mountain mass began to downwaste, the higher peaks and ridges emerging above the ice surface. The meltwater channels occurring in cols, that were observed by Mannerfelt, originated between the highest peaks as they emerged above the ice surface. The channels were not generally the result of outlet drainage from ice-dammed lakes, but were formed (1945, p. 224), '. . . by streams which flowed along through the mountain passes at the bipartition of the shrinking ice mass.' Channels were also cut between the ice and the rock of the emerging spurs and ridges but Mannerfelt pointed out that (p. 23), 'A typical feature is that the lateral meltwater tries to take a sublateral or subglacial course under the ice margin.' He also realized that the ice remained longest in the valleys and that along the margins of these downwasting ice masses marginal channels were cut that in places were traversed by subglacial chutes. Mannerfelt suggested that the occurrence of flights of marginal channels separated by fairly constant vertical distances, indicated that the ice surface was lowered by ablation of between 3 and 5 m each year. Since these channels were formed at the ice margin they would also indicate the slope of the ice surface. After measuring the gradient of many marginal channels, Mannerfelt concluded that the gradient of the ice surface was between 2 and 3 m per 100 m.

Both of these ideas put forward by Mannerfelt have been subjected to criticism. To determine the gradient of the ice surface and the rate of downwastage (i.e. annual ablation) by using data derived from meltwater channels, there must be no doubt about the marginal origin of the channels. Mannerfelt himself stated (1949, p. 197), 'There is always a natural tendency for the lateral meltwaters to undermine the margin of the ice and find their way into subglacial chutes. Sublateral gullies are the commonest.' It has proved almost impossible to attribute with certainty a marginal position to any channel during its formation as even channels that are parallel to the present contour pattern could be submarginal. Therefore, it is usually impossible to use meltwater channels to determine the slope of the ice surface and the rate of downwastage.

Publications by Gillberg (1956) and Holdar (1957), continued the work of Mannerfelt and added more detail to our knowledge of fluvioglacial erosion. Gillberg pointed out that for marginal drainage to develop, crevasses and other hollows at the ice edge must have been closed and for this to happen the ice must have been moving a little. If the ice was completely stagnant, as was implied by Mannerfelt (1945, 1949) and Hoppe (1950), much of the meltwater drainage would be subglacial. Gillberg described three types of marginal drainage channel: strictly lateral, extra lateral and sublateral, and he stated that most marginal forms are sublateral. Channels that Gillberg called 'sublateral' are usually termed 'submarginal'. He also stated that (p. 446), 'Subglacial chutes . . . always begin at terraces of glacial lakes (in this area)—they have probably all been subglacial outlets.' The absence of many strictly marginal features made it impossible to determine the gradient of the ice surface.

Holdar (1957), stressed the importance of the deglaciated land areas as a source of water that eventually drained into the marginal channels. He described the development of marginal and subglacial drainage as well as the drainage of ice-dammed lakes, over the ice, under the ice and over land passes. Since annual ablation varies considerably, Holdar pointed out that the similar vertical interval between successive marginal channels, stated by Mannerfelt to be caused by annual formation, was open to question. Holdar expected the vertical interval to be irregular.

The study of fluvioglacial erosion received stimulus in Britain from Peel's detailed work (1951) on two large Northumbrian meltwater channels, in which he posed problems that required further study. The two channels, interpreted by Dwerryhouse (1902) as lake overflows, have marked up-down profiles difficult to reconcile with that interpretation. Peel tentatively suggested a reversal of flow of waters from two lakes to explain the up-down profiles and did not consider (p. 87), 'the attractive—though here quite untenable—conception that the channels were cut in their entirety by subglacial waters under hydrostatic pressure.'

Twidale (1956a, b), answered Peel's request for more detailed studies of similar forms in a paper on some channels in north Lincolnshire. He described four channels with up-down long profiles and suggested that they were formed in two ways: (1) as the result of the pre-glacial form of the ridge that was cut through by the overflow channel, and (2) as a result of the reversal of flow of waters from two ice-dammed lakes. A third process was also discussed but it was only concerned with explaining changes in the gradient of the long profile of an overflow channel and not with up-down profiles. Similarly, the first process discussed by Twidale, that of the reversal of pre-glacial drainage, is not concerned with erosional up-down profiles produced by meltwaters.

In 1956, Peel commented on Twidale's papers, pointing out several weaknesses in the hypothesis put forward to explain up-down profiles by reversal of flow. Peel referred to the work of Mannerfelt (1945) and suggested (p. 486) '. . . some at least of the channels which exhibited an anomalous up-and-down profile may have been excavated in part at least, subglacially.' The broader implications of down-wasting ice occupying the proposed lake sites were not discussed and it was not until 1958 that papers were published which discussed marginal and subglacial drainage channels formed in association with downwasting ice in Britain.

Sissons' contribution to our understanding of the part played by, and the significance of, fluvioglacial erosion during the deglaciation of certain parts of Britain, has been considerable. He was the first worker to apply the concepts developed by Mannerfelt and other Scandinavian workers to the deglaciation of upland areas in Britain. During 1958, three papers were published by Sissons that included evidence of the former existence of masses of stagnant, downwasting ice, around and beneath which meltwater streams cut channels in positions that were determined by the presence of ice. Sissons located the successive positions of the ice margin, in the three areas he mapped, by means of the position of marginal channels and ice contact deposits. Recent work, including some done by Sissons himself, has shown the limitations of this technique. Similarly, the calculations of the gradient and rate of downwastage of the ice surfaces in the Eddleston Valley and East Lothian based on the gradient of, and vertical distance between, marginal channels is also unreliable because the strictly marginal position of the channels cannot be proved.

Gjessing (1960), in his work on the drainage of the deglaciation period in Northern Atnedalen, Norway, was unable to obtain evidence of the gradient of the ice surface from the form and position of meltwater channels. However, he did conclude that all the channels, even those formed below the ice surface, indicated the direction of slope of the ice surface.

Two papers by Sissons (1960b, 1961a), summarize the development of the concepts of meltwater drainage associated with downwasting ice masses. He pointed out the limitations of applying Kendall's hypothesis of the formation of meltwater channels to every area regardless of the absence of definite evidence of the former presence of ice-dammed lakes. The significance of submarginal and subglacial drainage in the development of individual channels and channel systems has been clearly established. (Sissons, 1960a, 1961b; Price, 1960, 1963a).

Literature dealing with the processes of deglaciation in general and the formation of meltwater channels in particular, is extensive. From the sound basis of observations made by Von Engeln, Tarr, Martin, Sharp and others in areas of existing glaciers, meltwaters flowing on, at the margin of, within and beneath the ice are known to be capable of eroding deep channels when they come into contact with solid rock or drift. The development of marginal benches, marginal channels and submarginal channels has been observed in areas presently glacierized. The hypothesis of subglacial erosion has been clearly demonstrated in Scandinavia, Britain and North America, the simple model put forward by Kendall in 1902 to explain glacial meltwater channels having been superseded by a much fuller and more comprehensive model. The existence of ice-dammed lakes is not denied but depositional evidence in the form of shorelines, bottom deposits and deltas must be added to the presence of channels before the origin of the channels can be attributed to the overflow of water from such lakes. The channel segments in drift deposits and solid rock, in areas of former glaciation, represent parts of a meltwater drainage system associated with the waste of the former ice cover. Some channels may represent outlets of ice-dammed lakes but the majority are the products of drainage systems along the margin of, on, in and under the ice. The direction of meltwater flow is generally similar to the direction of slope of the ice surface and therefore roughly parallel to the direction of ice movement. Detailed examination of the meltwater channel systems suggests that certain channel segments are probably cut under hydrostatic pressure. In some areas fluvioglacial erosion is superseded by fluvioglacial deposition at approximately the same altitude over considerable areas. Such a relationship is suggestive of an englacial water-table below the level of which fluvioglacial erosion has not occurred.

THE IDENTIFICATION AND CLASSIFICATION OF MELTWATER CHANNELS

The existence of channel forms, in areas of former glaciation, unrelated to the present drainage system, requires explanation, and their detailed description and mapping have suggested possible modes of formation. The channels may be small features only 5-10 m wide, 2-5 m deep and 20-30 m long, or they may be very large features over 100 m deep and kilometres in length. They may be cut in solid rock or drift deposits. Characteristics that make meltwater channels distinctive from other fluvial channels are as follows:

1. They begin and end in areas that under normal fluvial activity would not be likely locations for the initiation and termination of channelled flow.
2. They frequently cut across present drainage divides.
3. They sometimes have open in-take ends that bear no similarities to other types of water-eroded channels.
4. They frequently have very steep sides and are either unusually straight over considerable distances or develop tight meanders over short distances.
5. They do not usually widen appreciably in the down-stream direction.
6. Tributary channels are often ungraded at their junction with the main channels.
7. They either lack a present stream compatible with the size of the channel or are completely dry.
8. Shoreline remnants or lake-floor deposits, deltas, kame terraces or eskers can occur near channel in-takes.
9. Large outwash fans, kames or eskers can occur at the distal end of channels.

Any individual channel may exhibit only a few of the above characteristics. It is often difficult to be certain that an individual feature is the product of meltwater activity but when examined in the broader context of other similar features in an area, a stronger case can often be made. The two extreme cases, on the one hand an isolated channel, and on the other a major valley trending at right angles to a former ice margin, often present the geomorphologist with a difficult task in ascertaining the extent to which meltwaters were responsible for their form. In the case of the isolated channel it must be remembered that the drainage system that may have been responsible for it may have been, over much of its length, either on or within the ice. Meltwater channels cut in solid rock or in drift deposits only represent those segments of the meltwater drainage system that came into contact with the ice-rock interface. Such channels as were cut may have also suffered modification by post-glacial streams, slumping, solifluction, soil and peat development. It is often difficult to determine the significance of structural control or the form of the pre-glacial relief. In areas that have experienced multiple glaciation it is probable that channels have been occupied more than once, while others which developed during an earlier glaciation have been obscured by the deposits of subsequent glaciations. In the case of major valleys trending away from a former ice margin the contribution of meltwater in the development of their present form is virtually impossible to judge. It could be argued that the Mississippi and its tributaries represent one of the world's largest meltwater channel systems as a resul of the vast quantities of meltwater that passed through that system during the Wisconsin glaciation.

Having established the criteria for recognizing meltwater channels it is necessary to discuss the problem of their classification. In 1961a, Sissons stated (p. 15), '. . . owing to our present inadequate knowledge, . . . to attempt a complete classification of channels or to discuss all types of channels, . . . is not advisable or practicable.' However, in the same paper Sissons did present a classification based partly on environment of formation and partly on form and he distinguished six types of meltwater channel and discussed at great length the inadequacy of the class of channels, widely discussed in the British literature, termed overflow channels. Since the main purpose of geomorphological studies is the explanation of discrete forms, and the terminology adopted expresses the origins of the forms, it is inevitable that classification of forms is of prime concern to the geomorphologist. It is necessary to examine the various classifications that have been presented in the past, to examine the criteria that can be used in classification and to propose a classification based on our present knowledge that, it is hoped, can be refined as and when our knowledge of these channels increases.

Kendall (1902) and Rich (1908) both produced classifications of channels based on the assumption that they were associated with ice-dammed lakes. Channels were classified according to the method of drainage from the ice-dammed lakes, e.g. direct overflows, severed spurs, marginal overflows.

Mannerfelt's classification (1945, 1949), was based on the assumption that it was possible to determine the gradient of the ice surface and therefore the position of the ice margin during deglaciation, from the meltwater channels themselves. Mannerfelt was therefore able to distinguish col gullies produced as the higher ground emerged above the ice surface; lateral and sublateral drainage channels associated with the down-wastage of the ice surface along valley sides and subglacial chutes that were developed at right angles to and beneath the ice margin.

Derbyshire (1962) suggested a classification of meltwater channels on the basis of their morphological characteristics, topographical frequency and disposition, and their associated deposits. This classification also assumed that the position and slope of the ice margin was known and the two major classes were described as proglacial and sub-glacial. The sub-classes were based on the relationship between channels and the slopes on which they occurred, and it was assumed that the ice margin more or less paralleled the contours of the slopes.

Ives and Kirby (1964), strongly criticized some of Derbyshire's interpretations and his classification. They pointed out that channel position relative to the melting ice mass is determined by climatic and glacial conditions and particularly the thickness, temperature and regime of the ice. They state (p. 917), 'The final morphology, however, is decided by these factors in combination with local topography, lithology and structure. . . . Also it seems unsatisfactory to frame a genetic classification around morphology without considering details of the pre-glacial landscape.' Ives and Kirby support their criticism of Derbyshire's work by reference to channels in the process of development and recently formed along the margin of the Barnes ice cap. The aerial

Fig. 33. The north-west corner of the Barnes Ice Cap, Baffin Island. (Part of R.C.A.F. photograph: A-16293-30. Aug. 27, 1958.)

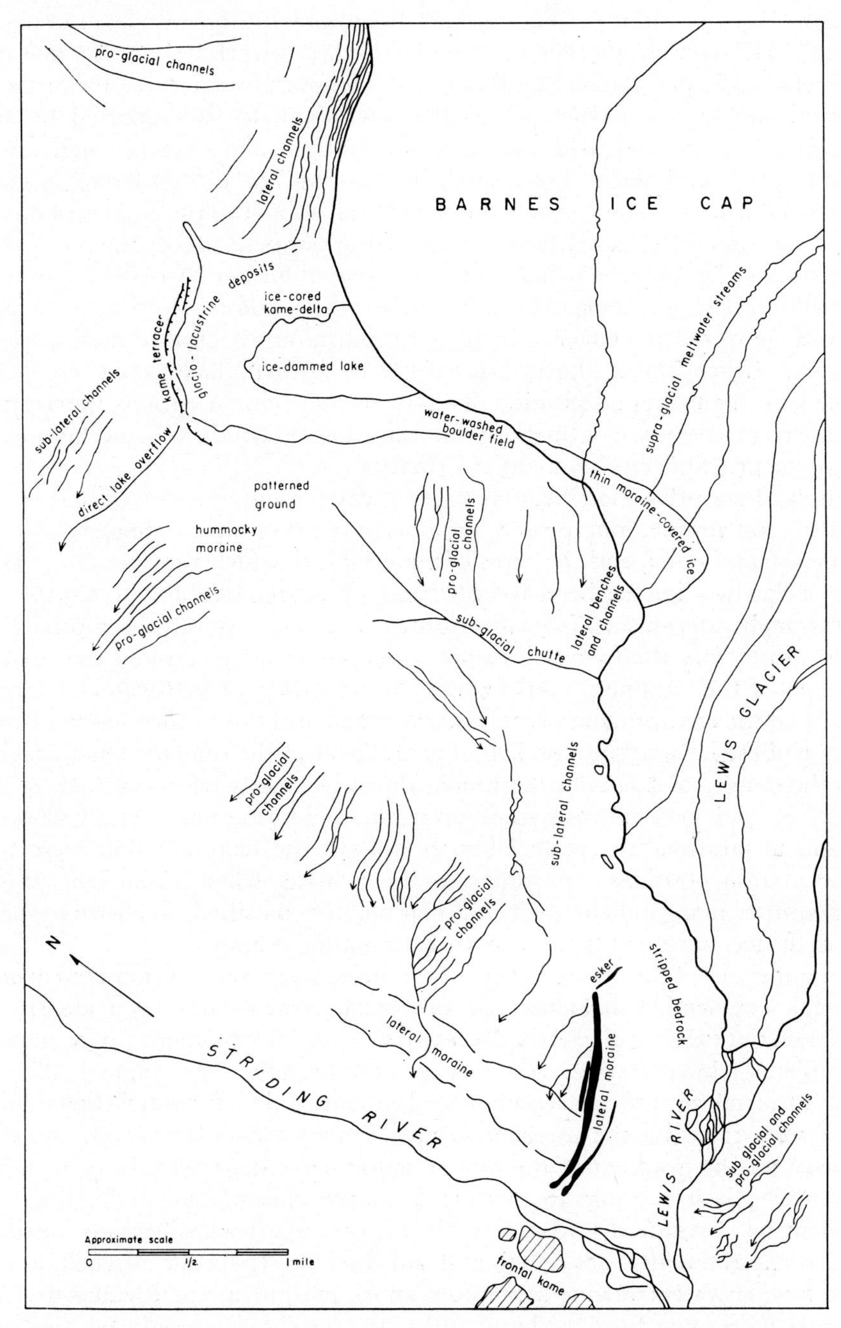

FIG. 34. Map based on R.C.A.F. photograph (Fig. 33) of a part of the Barnes Ice Cap, Baffin Island, and the Lewis Glacier. (After Ives and Kirby, 1964.)

photograph (Fig. 33) and map (Fig. 34) they use to present their evidence are extremely interesting. However, since they criticize Derbyshire's work because he does not give full consideration to pre-glacial morphology, the temperature and regime of the ice and the character of the ice margin, while their own evidence is supported by an aerial photograph of the ice and field information about ice temperature, their criticism is hardly fair. Ives and Kirby are correct in pointing out the problems of classifying channels when little is known about the pre-glacial topography, the position of the ice margin at the time of channel formation and the temperature of the ice. However, when meltwater channels are being studied in areas of former ice cover the only sources of information about glaciological conditions are the landforms and deposits produced by the wastage of the ice cover. In such circumstances a classification of meltwater channels can either ignore the interpretation of former glaciological conditions, be based solely on the form and position of channels and remain entirely descriptive, or it can include a consideration of the probable glaciological conditions and become genetic in terms of the probable environment of formation.

A purely descriptive classification of meltwater channels can be based on their dimensions, cross profile, long profile, position in relationship to slope, ridge crests and general pre-glacial relief, and the types of material into which they are cut. However, once such a classification has been accepted and it has been established that the channels were cut by meltwaters associated with a former ice cover, a purely descriptive classification is less satisfactory than a classification which attempts to consider the environment of formation of the channels. All genetic classifications of meltwater channels have been based on the environment of meltwater erosion and this in turn has been related to the position of the ice margins and extent of ice cover at the time the channels were cut. There is the danger of a circular argument that was clearly demonstrated by the wide acceptance of the lake-overflow hypothesis of channel formation. Many workers drew ice margins in positions to create lakes to explain the channels that were the basis of the ice margin positions. A similar problem arises when former ice margins are determined from marginal channels that can only be classified as true marginals if the position of the ice margin at the time of their formation is known.

The significance of the downwastage of ice surfaces and the development of marginal, supraglacial, englacial, subglacial and proglacial streams has been clearly demonstrated in areas of existing glaciers. Except in 'Alpine' areas where valley glaciers have steep gradients, downwasting ice surfaces usually have low angle surface slopes. Gradients of 1:100 to 1:200 are common. If it can be demonstrated that a channel or channel system required the presence of an ice mass for its formation, and it can be established that the gradient of the downwasting ice surface was likely to have been at a low angle, it is possible to develop a simple classification indicating probable environment of formation. Mannerfelt's (1945, 1949) classification distinguished two major environments: marginal and subglacial. In order to include channels produced by meltwaters moving away from an ice margin and meltwaters derived from ice-dammed lakes, two further classes must be established: proglacial and overflow. The following classification is suggested for channels associated with downwasting temperate ice.

Marginal/submarginal channels:
  lateral, frontal, in-out, col-channels.
Subglacial channels:
  marginal: chutes
  basal: normal, up-down profile, drainage of some ice-dammed lakes
  englacial: superimposed normal
    superimposed up-down
  col-channels
Open ice-walled channels

Proglacial channels:
  at right angles to ice margin
  parallel to ice margin

Overflow channels:
  in cols
  marginal
  proglacial

It must be pointed out that the above classification does not allow for the fact that some channels and channel systems develop in more than one environment. It is possible for a channel to be initiated by a subglacial stream, to be enlarged by the catastrophic subglacial drainage of an ice-dammed lake and to be subsequently utilized by a proglacial stream.

MARGINAL AND SUBMARGINAL CHANNELS

It has been stated above that it is possible to classify meltwater channels on the basis of their environment of formation defined in terms of the position of the ice margin at the

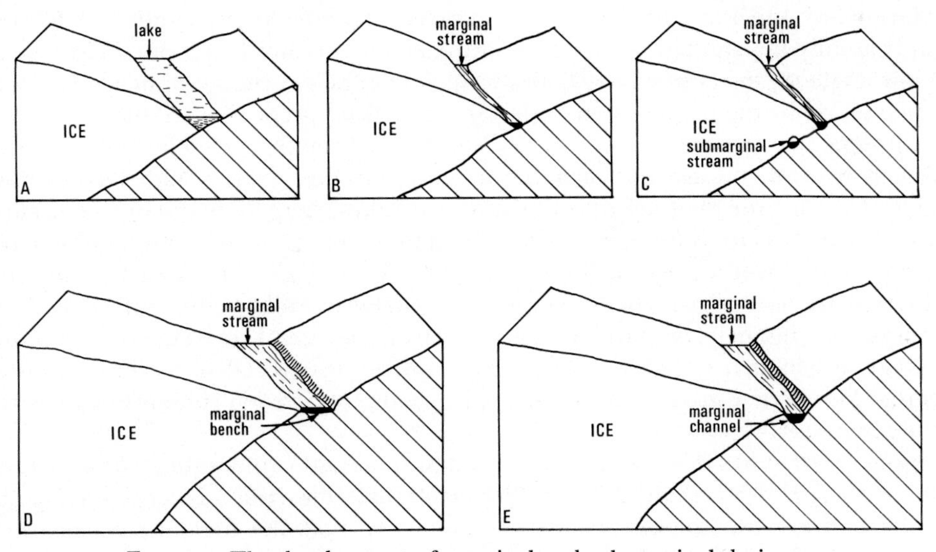

FIG. 35. The development of marginal and submarginal drainage.

time of their formation. As a result of ablation, meltwater near a lateral ice margin tends to accumulate in the trough formed by the sloping ice surface and the rock slope against which the ice rests. If there is no gradient parallel to the ice margin a lake may develop (Fig. 35*a*). If the ice margin slopes in a given direction then channelled flow will occur, with the ice forming one wall of the channel and the rock slope the other (Fig. 35*b*,*c*). This stream channel is liable to contain rock debris in transport and will therefore possess the tools of erosion. When the stream functions for a sufficiently long period and erosion of its bed occurs, two developments may take place. If the rock slope is steep the stream is more likely to cut into the ice and no evidence of meltwater erosion, apart from general washing, will be found on the slope after deglaciation. On the other hand, if the rock slope is gentle, a bench will be cut into the rock slope with the ice forming one channel wall (Fig. 35*d*). If this stream functions long enough the bench cut into the hillslope will develop into a channel with both banks formed in rock (Fig. 35*e*). Marginal meltwater streams are more likely to initiate benches and channels on slopes masked by drift deposits. However, it is known from observations near existing glaciers that such meltwater streams are capable of cutting channels in solid rock.

Schytt (1956) has described marginal meltwater streams associated with the Moltke Glacier in Greenland, which developed a series of parallel channels, controlled by differential ablation along outcrops of dirt-laden shear planes between active ice and stagnating ice along the glacier's margin (Fig. 36). The innermost channel was cut entirely in ice: the next was deeper and may have been in contact with bedrock and the outermost was a true marginal with one wall of ice and the other of rock. Schytt suggested that the terms to be applied to these channels should be: sublateral, lateral and extra lateral. However, it is very unlikely that evidence could be obtained to subdivide marginal channels in areas of former glaciation on this basis.

Marginal channels are usually short and rarely exceed 2 km in length. A marginal origin is usually interpreted on the basis of the channel either being parallel to, or at a very low angle to, the contours of slopes. Such channels certainly indicate the former presence of ice on their down-slope side but there is no proof that ice did not exist on their up-slope side (Sissons, 1961). They could have been produced by streams flowing in tunnels beneath the ice but near the ice margin (Fig. 37*a*). With further downwastage of the ice the channel that was initiated submarginally may later be occupied by a marginal stream (Fig. 37*b*). The fact that a channel has a low gradient is no guarantee that it was formed at the ice margin. Tarr (1909) and Von Engeln (1912) emphasized the importance of submarginal drainage parallel with, but beneath, the ice margin or obliquely beneath the ice. Such submarginal drainage often produces anastomosing channel systems leading down to truly subglacial or englacial drainage systems. Mannerfelt (1949) has stated that in parts of Sweden submarginal channels are the commonest type.

Sissons (1961a) has described a complex marginal and submarginal channel system on the south side of Strathallan, Perthshire, Scotland, (Fig. 38). Most of the channels in the system are less than 1·5 m deep. He states that the higher channels in the sequence have a gradient of 2% and may be truly marginal. Many of the lower

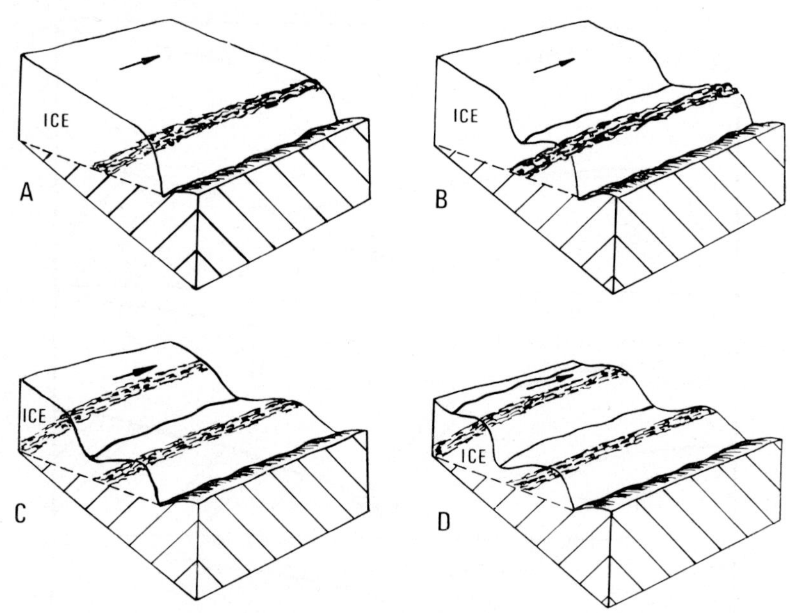

FIG. 36. 'Consecutive stages in the development of lateral drainage channels. *A*—During its retreat the glacier has melted down to a critical minimum thickness, and a lateral zone has stagnated, separated from the active glacier by a shear plane or a shear zone. *B*—The debris melted out in the shear zone largely protects the underlying ice from ablation and a lateral ice ridge develops between the "normal" lateral drainage channel and a new depression caused by melting and stream erosion on the glacier-side of the ridge. *C*—After some time the diminishing ice thickness will cause a new shear zone to develop, and the inner stream will go on to cut itself further down in the stagnant ice. *D*—Continued retreat will give rise to one more "lateral, superglacial" meltwater stream along the inside of the second lateral ice ridge. By now the first superglacial stream will be approaching the bottom of the ice, or it may even have reached this stage and started on the formation of the next lateral drainage channel to be preserved after the ice has melted away.' (After Schytt, 1956.)

channels have much steeper gradients and were cut by meltwaters flowing obliquely downslope beneath the ice. Sissons believes that the submarginal environment is not only indicated by the channel gradients but by sudden changes in direction of many channels so that meltwaters must have flowed obliquely down-slope, then at right angles to the slope and then returned to an oblique course. He also suggests that the changes in direction observed in the meltwater channels may be related to structures in the ice.

One of the four types of channel recognized by Kendall (1902) was described by him as an *in–out channel*. These are crescentic channels excavated in a hill slope. Many resemble meanders, and Common (1957) suggested that some may have been formed by meandering supraglacial or englacial streams that cut down through the ice on to the

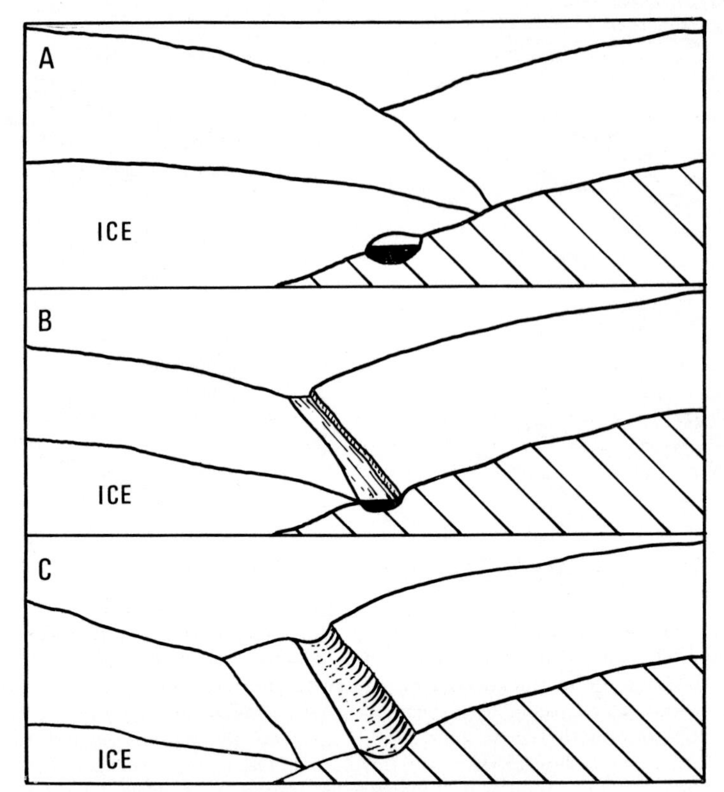

Fig. 37. The relationship between a downwasting ice margin and sub-marginal and marginal meltwater channels.

slope beneath. Similar origins were suggested by Sharp (1947) and Hoppe (1950). Such superimposition is most likely to occur near an ice margin where the ice is thin. Schytt (1956) implied superimposition by the sublateral streams of the Moltke Glacier. It must be admitted, however, that it is impossible to distinguish between channels superimposed by supraglacial streams near the lateral ice margin and the channels resulting from the superimposition of englacial streams at depth and at considerable distances from the ice margin.

The above discussion has been primarily concerned with marginal and submarginal channels developed along lateral ice margins. Similar forms can be produced along frontal ice margins when the ice front is retreating across a reverse slope (Fig. 39). Englacial or subglacial streams emerging at an ice front often flow in channels parallel to the ice front and can re-enter the ice via subglacial tunnels. All of the marginal and submarginal forms associated with lateral marginal drainage can occur in the frontal marginal and submarginal zone if there is no escape for meltwaters in a direction away from the ice front.

Mannerfelt (1945) described one type of meltwater channel that he called a *col gully*, which resulted from the changing relationship between the ice margin and the

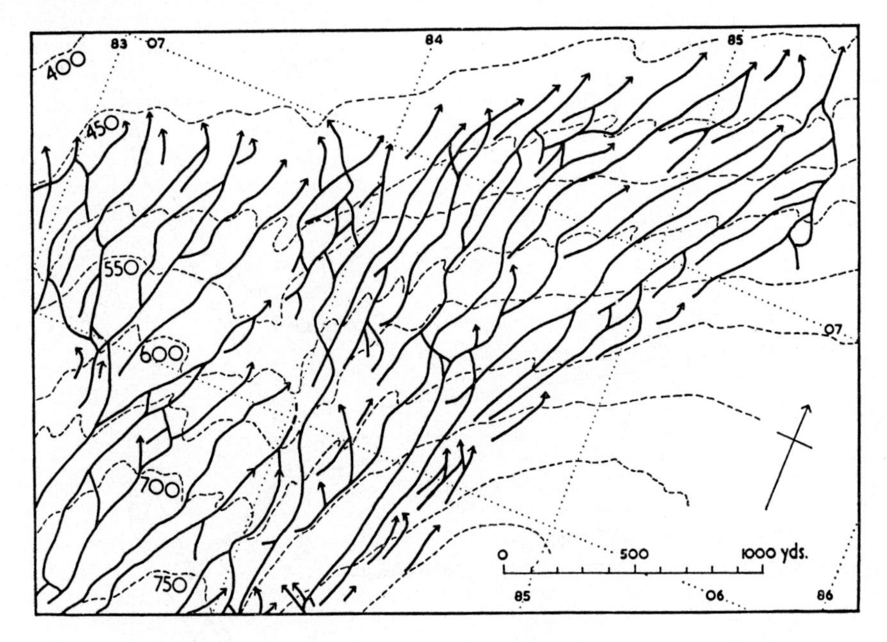

FIG. 38. Meltwater channels on the southern side of Strathallan, Perthshire, Scotland. (After Sissons, 1961a.)

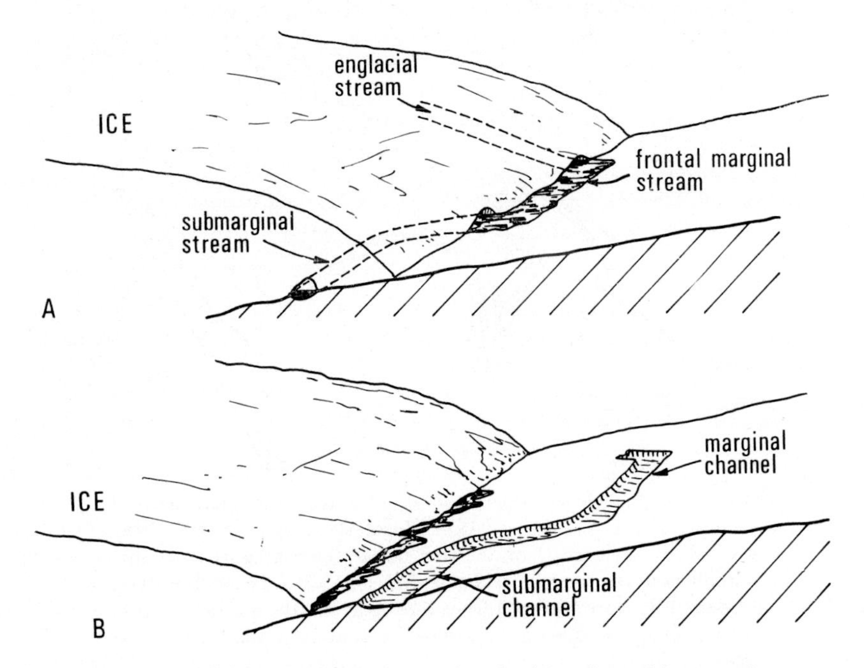

FIG. 39. Meltwater channels associated with a frontal ice margin.

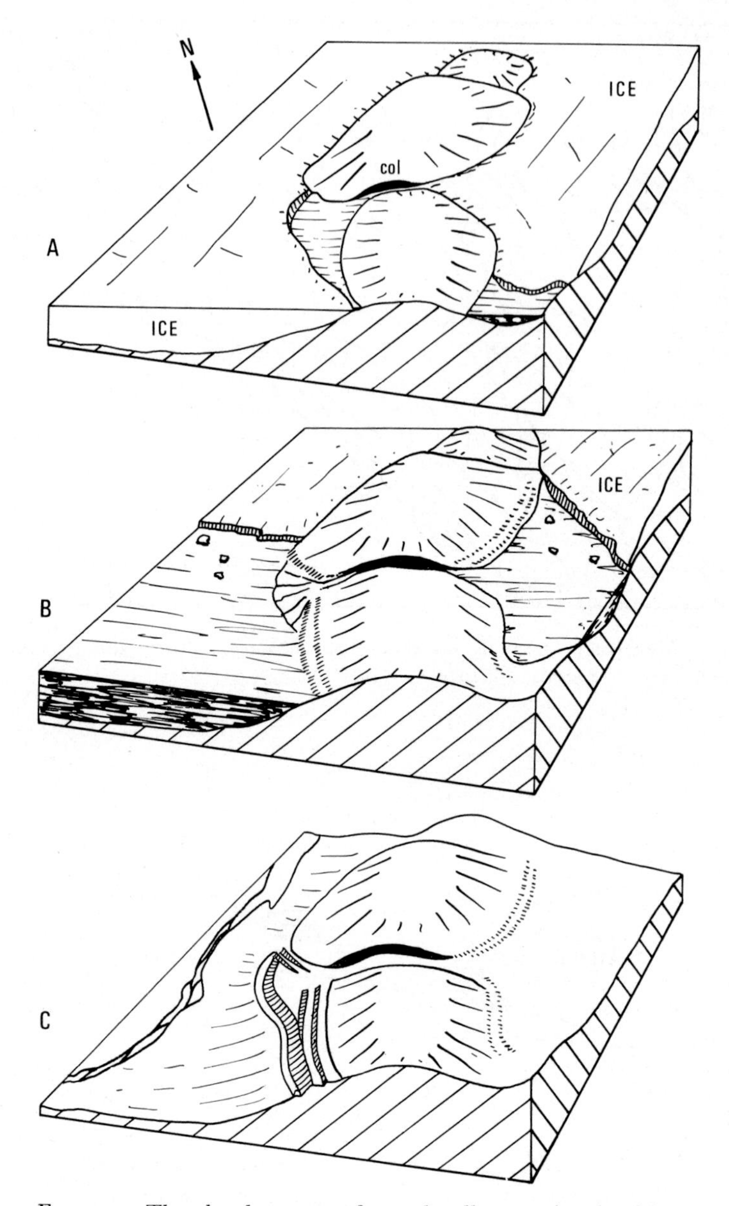

Fig. 40. The development of a col-gully associated with a downwasting ice surface. (After Mannerfelt, 1945.) *A*—Water flows off the ice surface on the east side of the emerging ridge into a small lake on the west side of the ridge; *B*—The col-gully is deepened by water flowing from a proglacial lake on the east side of the ridge; *C*—Deltas and shorelines are revealed when the lakes are drained.

underlying relief, as the surface of the ice sheet downwasted (Fig. 40). As the emergent relief begins to affect the supraglacial and englacial drainage of the sloping ice sheet some concentration of meltwater occurs in cols through the emerging divides. It is possible that the drainage through these cols may be initiated by englacial or subglacial streams (see section on subglacial channels) but it can also be initiated by marginal streams running along the slopes of the emerging hills, taking advantage of the lower cols. Mannerfelt describes col channels 10-20 m deep cut in solid rock. He states (1945, p. 224), '. . . they are not as a rule outlet drainage channels for ice-dammed lakes, but have been formed by the running water which flowed along through the mountain passes at the bipartition of the shrinking ice mass.'

SUBGLACIAL CHANNELS

Any meltwater that penetrates along the ice-rock interface beneath a glacier or ice sheet is subglacial and this adjective is also used to describe the channels in which that meltwater moves. In some instances meltwater flow occurs in tunnels, the walls and roofs of which are of ice, and the floors of rock or drift deposits. Subglacial channels that are strongly influenced by the proximity of the ice margin were discussed in the previous section. Both marginal and submarginal streams become subglacial when they continue at a high angle to the ice margin and descend beneath the ice. Supraglacial streams descend moulins and by means of englacial tunnels can eventually become subglacial. Ice-dammed lakes are sometimes drained by subglacial tunnels (DeBoer, 1949; Thorarinsson, 1939b; Howarth, 1968b; Price, 1971). Extraglacial streams directed towards an ice margin often become subglacial. The warmer waters of streams running across ice-free surfaces are capable of melting tunnels beneath the ice. As early as 1912 Von Engeln measured temperatures of 6·7°C in extraglacial streams in Alaska. Liestøl (1955) measured temperatures of 3-12°C in similar streams in Norway.

The depth of penetration of meltwaters into ice masses was discussed earlier in this chapter. In ice masses at the pressure melting point meltwater probably penetrates to depths of between 100 m and 150 m (Sissons, 1963; Embleton, 1964; Mathews, 1964; Clapperton, 1968). The depth of penetration will be a function of ice temperature, velocity and structure, the temperature of the meltwater and the presence or absence of an englacial water-table.

Meltwater streams have been observed disappearing down tunnels at lateral ice margins and emerging from tunnels at frontal ice margins. Channels produced by meltwaters plunging down slopes beneath the ice have been called subglacial chutes by Mannerfelt (1945). They may occur as single channels at right angles to the contours or they may be joined by right-angle bends to a series of marginal or submarginal channels (Fig. 41-2). Series of marginal or submarginal channels are often terminated by subglacial chutes in embayments in which extraglacial warm waters would be channelled towards the ice margin. Von Engeln (1912, p. 124) described such a situation on the margin of the Hidden Glacier, Alaska: 'On the lower side of a rock spur there was commonly an emergence of a marginal stream from an ice cave at the shelving ice edge, and this marginal stream then continued along the ice and rock contact until

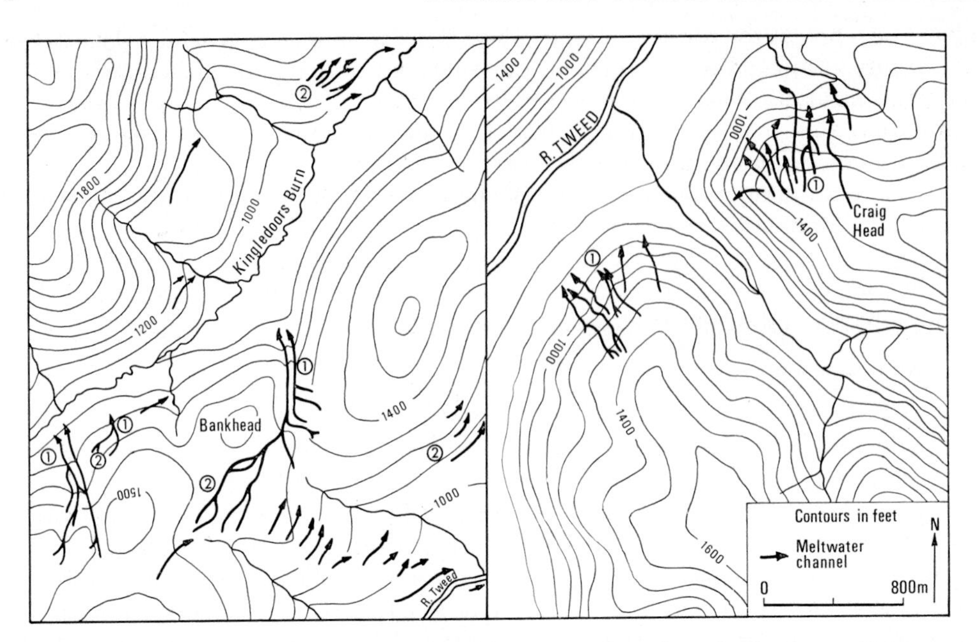

FIG. 41. Meltwater channels in the Upper Tweed Valley, Scotland. 1. Subglacial chutes. 2. Marginal or submarginal channels.

it became confluent with the major stream of the valley embayment (i.e., side valley) drainage. At the point of confluence both the valley drainage and the marginal drainage disappeared under the ice through a low ice cave.'

Some subglacial chutes occur on spur ends or on the lee side of spurs in terms of the direction of ice movement (Fig. 41-1). A small gap between the ice and the ground beneath allows supraglacial or englacial streams to develop subglacial courses.

Gillberg (1956) has suggested that some subglacial chutes are produced by the drainage of ice-dammed lakes. He presented evidence indicating the development of numerous small, narrow, marginal lakes on the flanks of the emerging nunataks of the Scandinavian ice sheet. Sometimes the outlets of these lakes were subaerial or submarginal, but sometimes they were subglacial and subglacial chutes were produced to carry the meltwaters from the marginal lakes to join the englacial drainage system. Sissons (1958b) pointed out that the lower ends of subglacial chutes in the Eddleston Valley, Scotland, all terminated at more or less the same altitude. He suggested that the lower level of meltwater erosion corresponded with the upper level of meltwater deposition and that they together reflected an englacial water-table.

There is little doubt that large volumes of meltwater can move in tunnels beneath the ice. The channels cut by such meltwaters can attain large dimensions even when cut in solid rock and they often occur in anomalous locations in terms of the post-glacial drainage pattern. The subglacial origin of these channels can be proved if they contain ablation moraine, but if they are subsequently occupied by proglacial streams any such deposits may be removed. It is their anomalous positions and lack of integration with the normal drainage pattern that usually indicate their subglacial origin. Such channels often have their beginnings part way along a uniform slope. The in-take may be in the form of steep, semicircular walls surrounding a plunge pool, indicating that the channel was fed by a stream descending through the ice and encountering the subglacial rock with considerable force (Fig. 42*a*). Other channels occur perched on hill-sides with no evidence of the englacial or supraglacial streams that fed them.

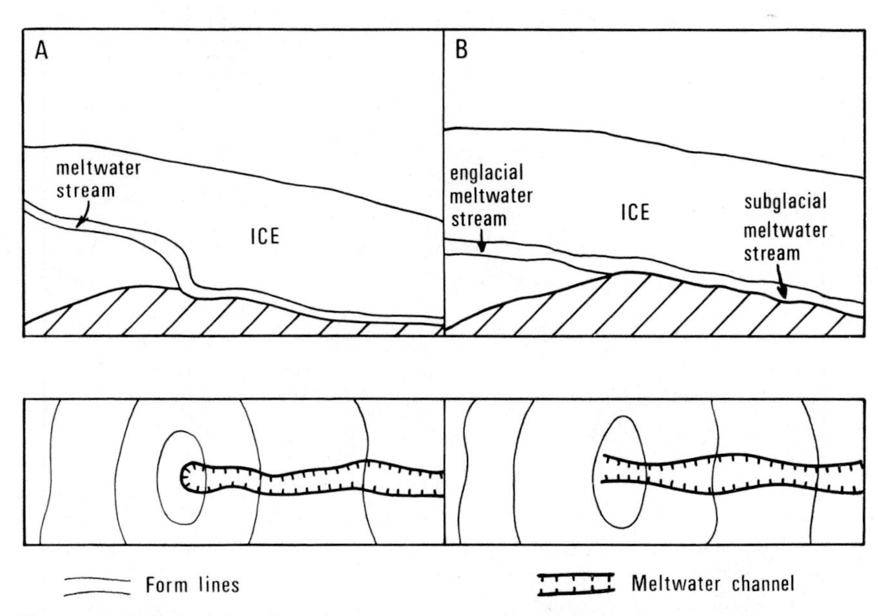

FIG. 42. Subglacial meltwater channels. *A*—An englacial stream descending to the base of the ice and cutting a plunge pool and subglacial channel; *B*—An englacial stream becoming subglacial when it encounters a ridge crest and cuts a channel through the ridge crest.

In some cases subglacial channels have unusual long profiles, in that the profiles of their floors are up and down in terms of the general direction of meltwater movement. Various theories have been put forward to explain these up-down long profiles (Sissons 1960b, 1961a) but if the stream that cut the profile was initially subglacial there is no reason why the up-hill section could not have been cut by meltwater flowing under hydrostatic pressure.

When an englacial stream encounters a spur or slope beneath the ice, it becomes subglacial. It is therefore possible for an englacial stream system to be superimposed on to the underlying land surface when the zone of meltwater penetration descends downward as the ice surface downwastes (Fig. 42*b*). In such a situation it is possible for part of the meltwater drainage system to be englacial, but when it is in contact with the land surface beneath the ice it is subglacial.

The author has studied a sequence of meltwater channels near West Linton in Peeblesshire, Scotland (Fig. 43), where the evidence suggests that an englacial drainage

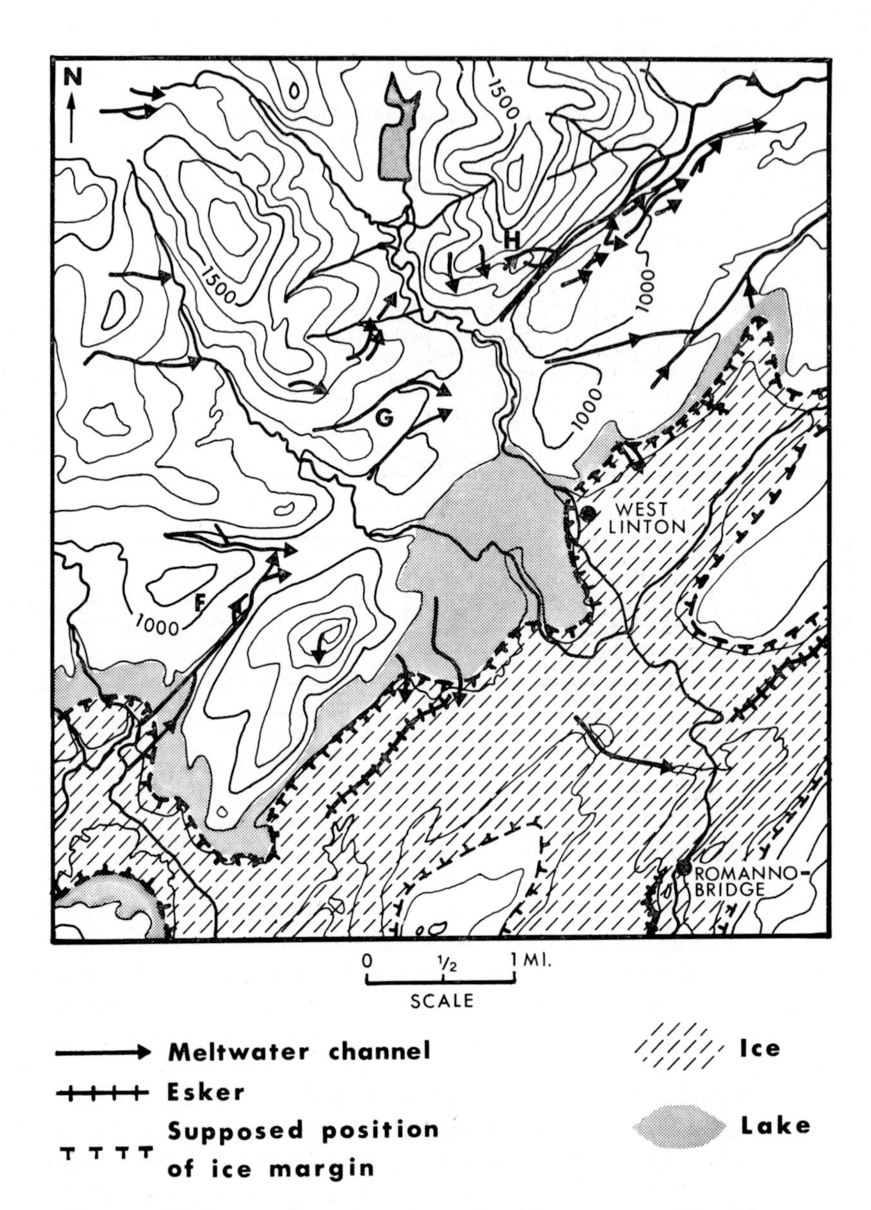

FIG. 43. Meltwater channels and associated features near West Linton, Peeblesshire, Scotland.

system carrying meltwaters from south-west to north-east became progressively superimposed from the north-east towards the south-west (Price, 1961). A system of meltwater channels mainly cut in solid rock and generally 5-30 m deep was mapped on the south-east side of the Pentland Hills. Several channels have up-down long profiles. The most complex part of the system is known as the Carlops channel system (*H* in Fig. 43) and is dominated by a channel with an up-down long profile and which attains a maximum depth of over 30 m. Towards the north-east end of the system an anastomosing pattern has developed with major channels separated by rock-islands. Sissons produced a detailed map of this system (1963, Fig. 2).

A series of channels also cuts through spurs *F* and *G* to the south-west of the Carlops system. The main channel on spur *F*, the Garvald channel, has a maximum depth of 30 m and cuts through the watershed between the Medwin and West Water valleys. The slope of the Garvald channel is towards the south-west but all the channels to the north and north-east of it indicate a flow of meltwaters towards the north-east. It seems most unlikely that the Garvald channel could have been cut by waters overflowing

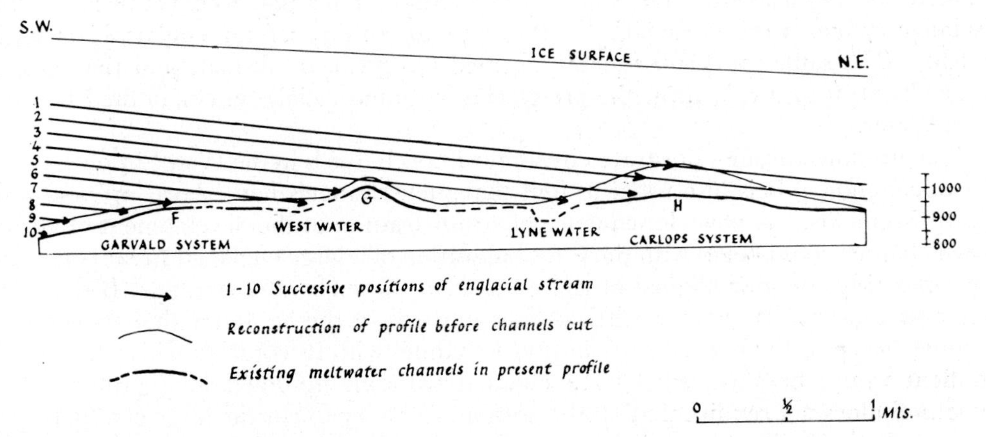

FIG. 44. The superimposition of an englacial stream to produce a channel sequence similar to that depicted in Fig. 43.

from a lake in the West Water valley, during the late stages of deglaciation, and it is much more probable that the Garvald system was related to the primary movement of meltwaters towards the north-east within the ice sheet that covered the area.

The altitudinal relationships of the channels on the three spurs is shown on Fig. 44. It can be seen that the Carlops system is generally above the channels forming the Garvald system and if altitudinal relationships alone are considered then a south-westward movement of meltwaters would seem more likely. However, the greater part of the Carlops system, the channels on spur *B* and the nature of the junction between the two tributary channels at the south-west end of the Garvald system all indicate a north-eastward movement. This general direction of meltwater movement is substantiated by evidence from the areas to the south-west and north-east of the West Linton area (Price, 1963a; Sissons, 1963).

The detailed chronological relationships between the channels on spurs *F*, *G* and *H* cannot be determined. However, it seems unlikely that they were all functioning at the same time. It is suggested that a major meltwater stream flowing in a tunnel, or tunnels, having a gradient which sloped down from south-west to north-east, was superimposed successively on spurs *H*, *G* and *F*. It seems unlikely that the channels on spur *G* carried the meltwaters that cut the Carlops system, because the south-eastward bend at the lower end of the second channel up spur *G*, and the general alignment of the lowest channel on that spur, suggest that the meltwaters turned south-eastwards down the Lyne valley. The final stage in the superimposition of the englacial stream system was the cutting of the Garvald system in a subglacial environment. The hydrostatic pressure produced by this movement of meltwaters from the south-west must have been sufficient to permit erosion along the line of the Pentland fault by a subglacial stream flowing up-hill. On reaching the West Water valley the meltwaters turned eastward and did not cut through spur *G*. It is therefore concluded that the channels on spurs *F*, *G* and *H* are the remnants of a meltwater drainage system that drained north-eastwards. The spurs of the Pentland Hills that were transverse to the drainage system were breached by the superimposition of an englacial drainage system. The same spurs subsequently formed too great an obstacle and they caused the meltwaters to use in turn, the pre-glacial (or interglacial) valleys, of the Lyne and West Water.

The up-down profiles illustrated by some of the channels in the West Linton area are not uncommon. It is also usual to find that such channels have short 'up' and long 'down' segments. A possible sequence of events leading to the development of an up-down channel could begin with the superimposition of a large englacial stream on to the crest of a ridge or spur aligned at right angles to the meltwater stream. If the meltwater stream cuts its channel vertically downwards in the ice faster than in the rock forming the spur, the normal superimposed channel with its constant width and gentle gradient would be superseded by a channel with an up-down long profile. This hypothesis does not require that all the erosion necessary to cut the large channels with up-down long profiles was carried out by meltwaters flowing under hydrostatic pressure. The major part of the channel was cut as the superimposition of the englacial stream proceeded and continued until the meltwaters flowing in the ice tunnel were able to cut down faster than the meltwaters flowing in the channel cut through the spur. The meltwater stream then had to flow up-hill under hydrostatic pressure if it was to continue along its course through the spur.

Clapperton (1968) has extended the concept of superimposition to account for the concentration of several small channels or one large meltwater channel in cols. He states, (p. 218), '. . . that while small channels may show no relationship to topography (i.e. they are superimposed across pre-existing ridges or spurs), it is remarkable how the largest channels and complex channel systems are consistently located in pre-existing cols and valley heads.' Clapperton explains this concentration of meltwater flow in cols by lateral migration, during superimposition, of meltwater streams down the steeper slopes of the cols, resulting in a concentration of meltwater flow along the floors of cols.

OPEN ICE-WALLED CHANNELS

Russell (1893, p. 239), described open ice-walled channels of the Malaspina Glacier; 'Each of these [rivers] issues from a tunnel and flows for some distance between walls of ice . . . one issues from the mouth of a tunnel and flows for half a mile in an open cut between precipitous walls of dirty ice 80-100 feet [24-30 m] high.' Sharp (1947) also described a stream that flowed between walls of ice of the Wolf Creek Glacier for a distance of 3 km.

Open ice-walled channels result from the collapse of the roofs of subglacial or englacial tunnels or from the downcutting of supraglacial streams. Sissons (1960a) explained certain channels in the Syracuse area, New York State, as a result of thinning ice and tunnel roof collapse. He also suggested that after the initiation of the channel system near West Linton by the superimposition of an englacial drainage system on to the underlying land surface that (Sissons, 1963, p. 110), 'As the ice surface continued to waste down, the roofs of the subglacial tunnels became thinner, until they finally collapsed, a process probably hastened as meltwater flow became concentrated in the very large channels.'

Meltwater channels cut by streams flowing between icewalls are not distinguishable from subglacial channels, channels produced by the superimposition of englacial streams or marginal or submarginal channels. They simply represent a late stage in the control of ice on the position of meltwater drainage.

PROGLACIAL CHANNELS

Whenever an ice margin occurs on a normal slope the meltwater from the ice will flow in channels across that slope and away from the ice margin. Meltwater drainage is often concentrated along pre-glacial valley systems. In the United States the Mississippi, Missouri and Ohio Rivers acted as proglacial meltwater channels for the great Wisconsin ice sheet. The problem of determining the amount of modification achieved by meltwaters during their movement through pre-glacial valley systems is difficult to solve. However, it is known that the catchment areas of some streams have been greatly increased by glaciation.

The writer has studied the evolution of meltwater channel systems in the proglacial areas of two glaciers (Price, 1964, 1965; Petrie and Price, 1966; Price, 1969; Price and Howarth, 1971). Both the Casement Glacier in Alaska and Breidamerkurjökull in Iceland have retreated considerable distances since 1900 and the ice wastage during the period 1900-1960 has produced large meltwater drainage channels cut in drift deposits.

The Casement Glacier (Fig. 45) became land ending in 1911. Meltwater channels produced during the retreat of the ice front over a distance of 5 km range from forms 30 m deep, 400 m wide and 4 km long, to minor but significant channels only 2 m deep, 10 m wide and 30 m long. Most of the channels are cut through till and gravel but channel 1 (Fig. 45) which still carried meltwater from the ice margin in 1963 had a rock gorge 10 m deep at the bottom of drift banks 30 m high.

It is significant that the three major meltwater channels numbered 1, 7 and 9 in

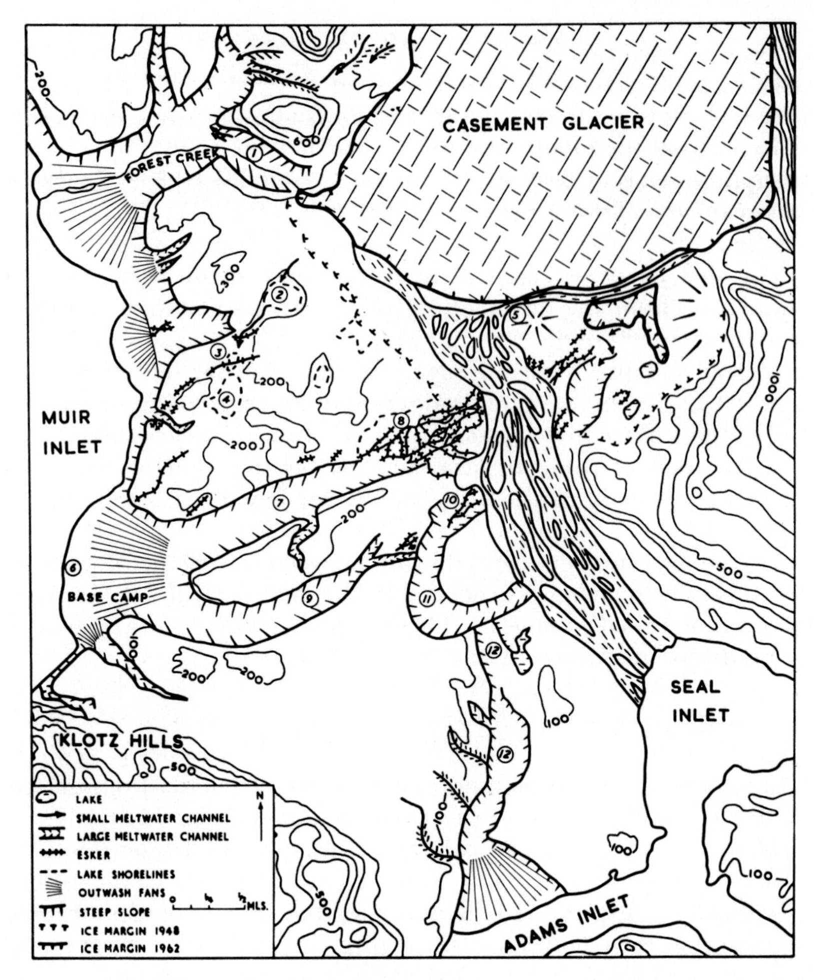

FIG. 45. The proglacial area of the Casement Glacier, south-east Alaska, in 1962. (After Price, 1965.)

Fig. 45 are all deepest where they cut through a north–south ridge in the proglacial area. This suggests that the courses of the meltwater streams were established before the ice front of the Casement Glacier had receded to the east of the ridge. An oblique aerial photograph taken in 1929 shows a subglacial stream emerging at the ice front near channel 7. As the ice continued to retreat, this routeway was sufficiently well established to be maintained and the channel became incised in the north–south ridge. It is possible that all these major meltwater channels functioned subglacially during the early stages of ice wastage. Their development continued during the retreat of the ice front from the 1935 position. It is believed that channels 7 and 9 functioned simultaneously and that they represent two major subglacial streams continuing their parallel courses in the proglacial area.

The next stage of fluvioglacial erosion resulted in the abandonment of both channels 7 and 9 in favour of a southward route to Adams Inlet. Three major channels were cut during this stage. Channel 11 is a steep-sided trough, 18 m deep, 400 m wide, cut through till and gravels. It swings in an arc across the head of channel 9. Meltwaters that had previously moved down channels 7 and 9 began to escape southward and eventually cut the floor of channel 11 below the level of channel 9. Water flowing down channel 11 at first continued southward down channel 12 but later abandoned that route in favour of a more direct route to sea level. The south bank of channel 11 truncates the head of channel 12 so that the floor of the former is approximately 12 m below the floor of channel 12. The most recent phase (since 1948) has seen the complete abandonment of channel 11 and the establishment of the present drainage system. During the period since 1948 the meltwaters moving off the southern part of the Casement Glacier have established a marginal stream parallel to the ice front which has destroyed eskers and other fluvioglacial deposits.

The evolution of the proglacial drainage of Breidamerkurjökull, south-east Iceland, during the retreat of that glacier from its position in 1890 is well documented (Price and Howarth, 1971). Maps of the system were made in 1904, 1945, 1951 and 1965 and aerial photographs of the system were taken in 1945, 1960, 1961, 1964 and 1965. Data obtained from these sources combined with information provided by local residents and by fieldwork has allowed the various changes in the system during the retreat of the ice front to be described. The drainage system has developed almost entirely on drift

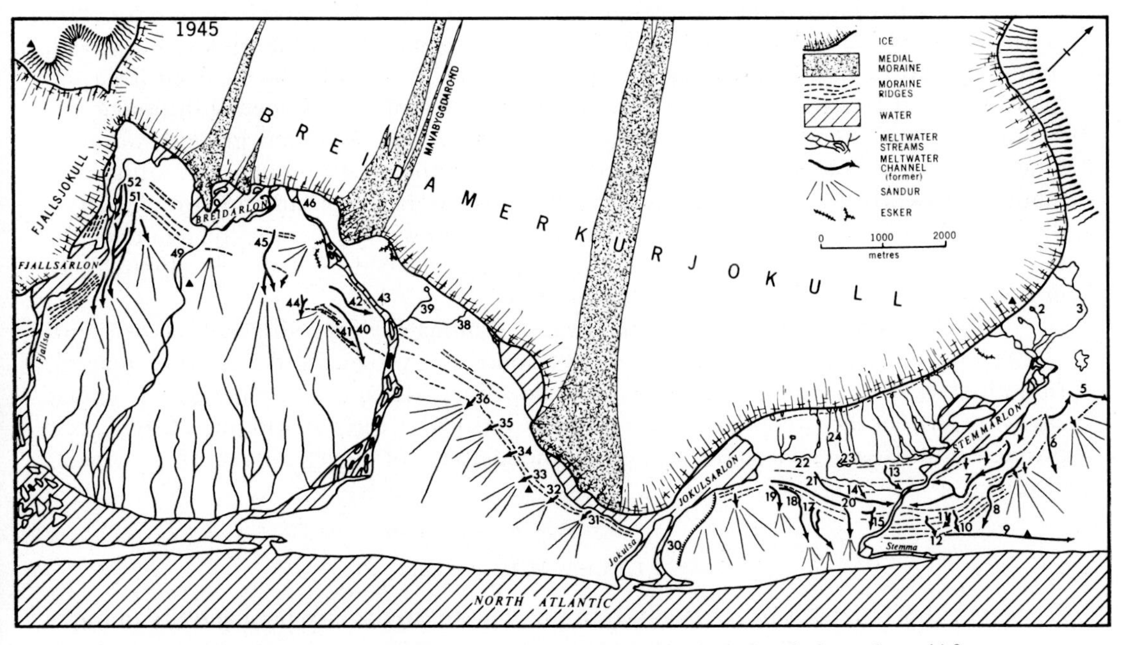

FIG. 46. The proglacial area of Breidamerkurjökull, Iceland, in 1965. (After Price and Howarth, 1969.)

deposits. In that part of the area inside the outermost moraine ridge (Fig. 46) most of the fluvioglacial erosion was confined to distinct channels ranging between 10 m and 200 m in width and 1 and 3 m in depth. Only where the channels have been cut through moraine ridges do they exceed 10 m in depth. There are also sandar inside the moraine system on which large numbers of anastomosing channels up to 1 m in depth occur. Beyond the outermost moraine, formed about 1890, fluvioglacial erosion was associated with the development of the drainage across the sandar surface. The channels on these sandar surfaces are complex and represent the changing pattern of the drainage system at the time the ice stood at the position of the 1890 moraine. The drainage system established beyond the moraine system was subsequently modified by those streams that cut through the moraine systems after 1890 and continued to function as the ice front retreated northwards.

In 1904 a great many streams issued from the ice front and followed braided courses to the coast. By the 1930s (Fig. 47) the number of channels carrying meltwaters beyond the oldest moraine had been reduced to seven and the effect of moraine ridges diverting meltwater flow, parallel to the ice front, can be clearly seen. By 1945 (Fig. 48) the large proglacial lakes, Stemmarlon, Jokulsarlon, Briedarlon and Fjallsarlon had begun to develop and apart from drainage from these lakes only one major meltwater stream flowed directly from the ice margin to the coast. Over the period 1945 to 1965 (Fig. 46) the continued development of proglacial lakes concentrated all the proglacial drainage into three systems, each associated with a major river: Stemma, Jokulsa and Fjallsa.

The sequence of the drainage evolution described above has largely been a function of the nature of the surface across which Briedamerkurjökull and Fjallsjökull have been retreating and the formation of the moraine systems. So long as the ice front has risen in altitude as it crossed successively higher ground, the southward flowing drainage was maintained. In some cases this drainage was controlled by fluted ground moraine. Once the ice front started retreating down a reverse slope, or if a sequence of substantial moraines blocked the seaward route of meltwaters, drainage parallel or subparallel to the ice front became dominant. Throughout the period 1890 to 1965 the tendency for the drainage pattern to become less complicated continued, and the development of proglacial lakes certainly facilitated that process.

The two examples of proglacial meltwater drainage systems discussed above tend to give an oversimplified picture of the development of proglacial drainage in general. In both cases the drainage systems were developed entirely on drift deposits. In many cases the form of the pre-glacial surface and in particular the outcropping of solid rock through a drift cover strongly influences the pattern of development of proglacial drainage.

OVERFLOW CHANNELS

A great deal of emphasis has been given in this chapter to the erosional activity of meltwater streams associated with meltwater drainage systems on, in, beneath and at the lateral and frontal margins of glaciers. Although it has been established that such

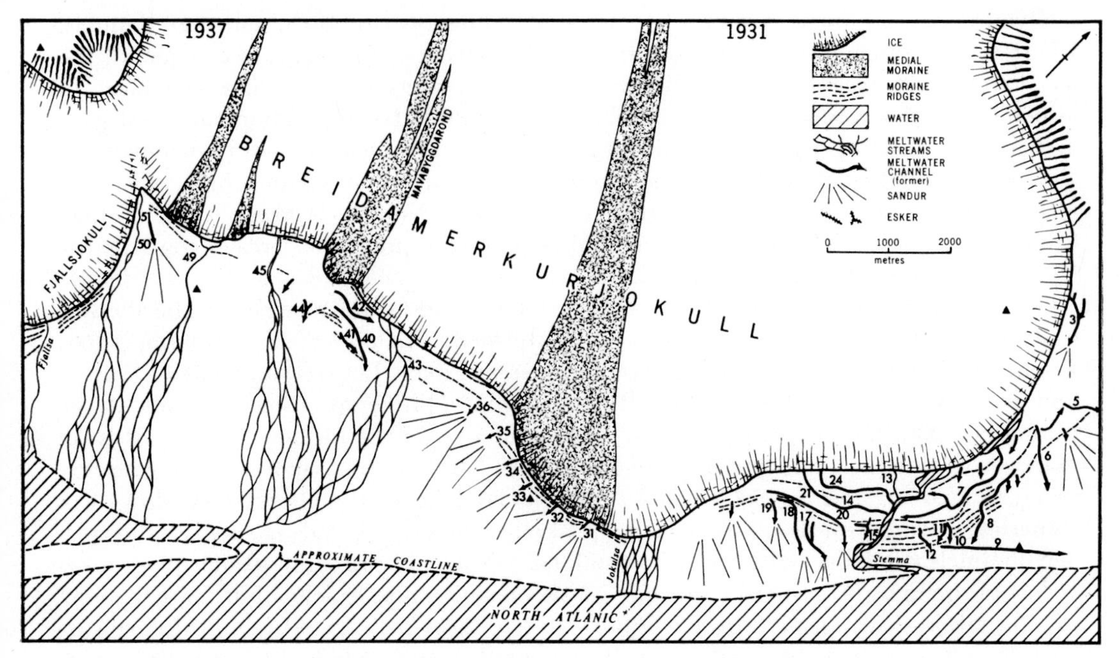

FIG. 47. The proglacial area of Breidamerkurjökull, Iceland, in 1931/1937. (After Price and Howarth, 1969.)

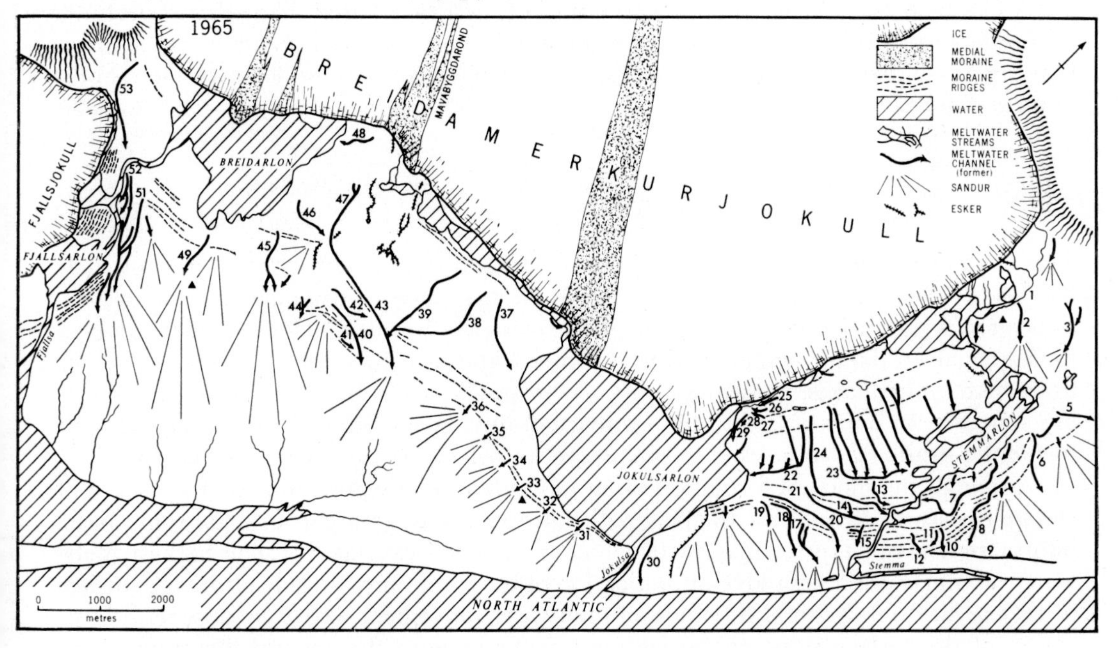

FIG. 48. The proglacial area of Breidamerkurjökull, Iceland, in 1945. (After Price and Howarth, 1969.)

meltwater streams are very important as erosional agents capable of cutting channels, where they impinge on the solid rock or drift deposits at the margins of or beneath the ice, it must not be forgotten that the overflow of waters from ice-dammed lakes in association with existing glaciers in Alaska (Stone, 1963); Iceland (Thorarinsson, 1939) and Norway (Liestøl, 1955) is well documented. Some of these lakes drain between the ice and the valley side, others drain across low cols, while others drain subglacially. In all cases erosion takes place along the route of drainage. Numerous lakes also occur along frontal ice margins whenever the regional slope of the proglacial area is towards the ice margin (Howarth and Price, 1969).

The existence of ice-dammed lakes along lateral and frontal margins of the Pleistocene ice sheets and glaciers is also well established. Lake Agassiz (Upham, 1896; Elson, 1957), the Pleistocene extensions of the Great Lakes (Leverett and Taylor, 1915), and the ice-dammed lakes of Glen Roy, Scotland (Jamieson, 1863), are all classic examples. Shotton (1953) provides detailed evidence for the former existence of Lake Harrison during the Saale glaciation of the English Midlands. Synge (1970) has clearly demonstrated the former existence of a proglacial lake in the Glen of Aherlow (Co. Limerick, Ireland) on the basis of successive delta levels at the ends of subglacial drainage channels carrying meltwater into the lake. Not only did large ice-dammed lakes occur on the margins of the Pleistocene ice sheets and valley glaciers but a great many smaller lakes occurred, perhaps for only relatively short periods of time, as the ice surface downwasted and the land surface reappeared from beneath the ice. So long as evidence for the existence of the ice-dammed lakes does not simply consist of the meltwater channels themselves, the interpretation of such meltwater channels as overflows cannot be disputed. The term 'overflow channel' should be restricted to those features that are associated with one or more of the following: lake-bottom deposits, lake shorelines, and deltas.

THE SIGNIFICANCE OF MELTWATER CHANNELS

This chapter has been primarily concerned with the origins of meltwater channels. It has been stressed that these channels owe their existence to drainage systems closely associated with an ice cover. Much of the evidence in areas of former glaciation of fluvioglacial erosion, whether directly by meltwater streams or by overflow of water from ice-dammed lakes, constitutes only a very small part of the total meltwater drainage system. In some situations the evidence of fluvioglacial deposition associated with meltwater drainage can be used to extend our knowledge of the system (see Chap. 6).

Although the mode of formation of an individual meltwater channel or a system of channels may be debatable, their overall pattern in any area can provide some basic information about the ice mass with which they were associated. The highest meltwater channels in an area give an indication of the minimum altitude of the ice surface in that area. The general direction of meltwater movement also gives a general indication of the surface slope of the ice mass and therefore the general direction of ice movement. The striking manner in which the meltwater channels in Peeblesshire,

Scotland (Fig. 49) reflect the direction of slope of the ice surface rather than the alignment of the pre-glacial (or interglacial) river network is a strong indication of the independent nature, that is in terms of the subglacial land surface, of the meltwater drainage system. Similar meltwater drainage systems have been described in New York State

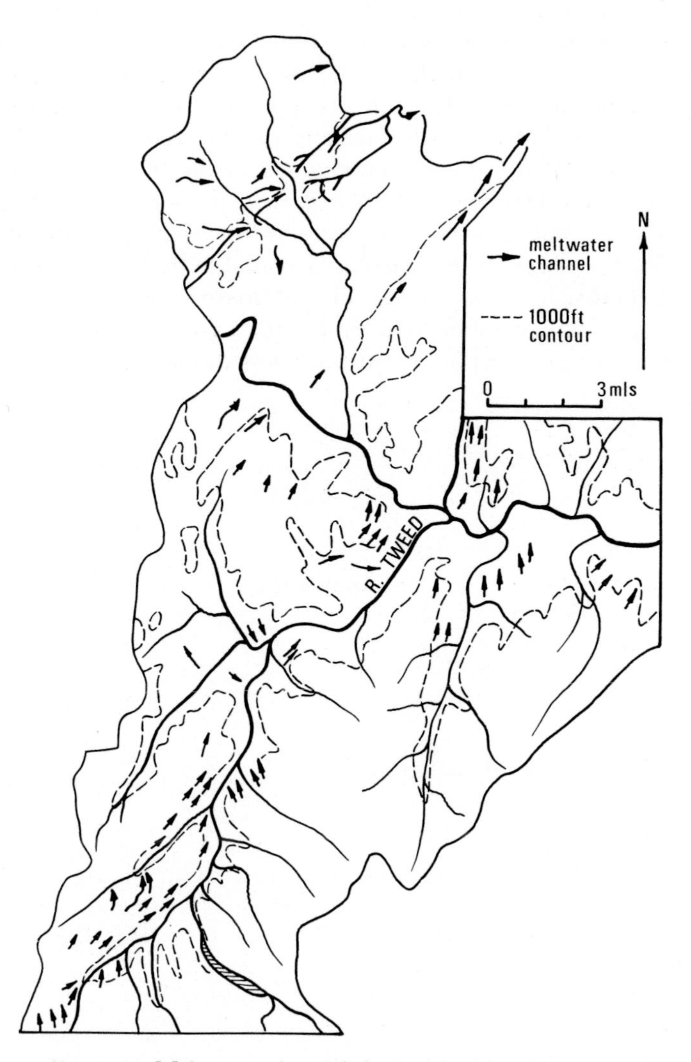

Fig. 49. Meltwater channels in Peeblesshire, Scotland.

(Sissons, 1960), and central Quebec-Labrador (Ives, 1958, 1960; Derbyshire, 1962). In some areas there is an uneven distribution of meltwater channels, with large numbers occurring in some situations and few or none at all occurring in intervening locations. Clapperton (1971, p. 376) suggests that 'Extensive systems of meltwater channels tend not to be found where meltwaters were directed to flow along steep-sided pre-existing valleys. Complex systems of meltwater channels are numerous where englacial stream networks (located chiefly in submarginal zones of the downwasting ice mass) were aligned discordantly across spurs and valleys.'

The occurrence of subglacial channels, or of channels resulting from the superimposition of englacial streams, also indicates that the ice mass was at the pressure melting point and therefore allowed the penetration of meltwater for considerable depths below the ice surface. In some areas it has been observed that meltwater channels do not occur below a certain altitude. In such situations it is likely that either an englacial water-table occurred or that the basal layers of the ice were sufficiently active to close any englacial tunnels. It has often been claimed in the literature that extensive submarginal and subglacial drainage will only occur if the ice mass is stagnant, and that marginal channels will only develop if the ice mass is active. Neither statement can be substantiated from the observations currently available from areas of existing glaciers. The relationship between the glaciological and hydrological aspects of glaciers and ice sheets still remains poorly understood, and needs a great deal of attention by research workers if many of the geomorphological problems discussed in this chapter are to be solved.

VI FLUVIOGLACIAL DEPOSITION

INTRODUCTION

Because the character of detrital sediments varies according to the properties of the environments in which they are transported and deposited, those sediments which accumulate as the result of transport and deposition by meltwaters can be expected to be distinctive. The actual mechanisms of transport and deposition are similar to those that occur in other fluviatile, lacustrine and marine environments, but the relative importance of the various mechanisms in the fluvioglacial environment is likely to be different. The fluvioglacial environment is characterized by fluctuations in discharge over both long and short terms. Depending on the character of the local climate, discharge may either cease entirely, or be very low, during the winter, whereas summer discharges that are the product of both meltwater and local precipitation can be very high. Short-term changes in discharge are related to diurnal variations in ablation and to the rapid drainage of both marginal ice-dammed lakes and englacial chambers. In an environment characterized by such variable discharges associated with large amounts of readily erodable material, it is not surprising that fluvioglacial sediments vary rapidly in particle size and sedimentary structures both laterally and vertically.

The relationship between the form and the internal characteristics of fluvioglacial deposits has been a subject of much interest to geomorphologists for many years. In terms of interpreting the sedimentary structures for the purpose of determining the environment of deposition the geomorphologist is very dependent upon the work of the sedimentologist. It is relatively easy to ascertain whether or not a deposit is fluvial in origin but it is often much more difficult to establish under what flow regime or in what depth of water the deposit accumulated (Allen, 1970).

Fluvioglacial deposits, being the result of meltwater transportation and deposition, have two characteristics that separate them from deposits resulting from transportation and deposition by the ice itself: they tend to be stratified and to consist of rounded or sub-rounded particles. The extent and type of stratification is largely a function of the environment of deposition, while the shape of the particles is a function of the source of debris, the type of debris and its mode and distance of transport.

The sedimentary structures exhibited by fluvioglacial deposits are very varied. They range from bed forms associated with channelled flow at differing velocities and involving bed-load of varying calibre, to the rhythmites that develop on the beds of lakes under very quiet water conditions. The relationship between the energy and particle sizes is of fundamental importance. In terms of bed forms, Allen (1970, p. 79) states that as stream power increases from 100 ergs/cm^2/sec, fine sands will first develop ripples, then at 1000-2000 ergs/cm^2/sec., dunes or cross bedding will develop to be followed by plane bedding at over 2000 ergs/cm^2/sec. In very general terms, channel deposits with their associated bed forms, and frequent changes in particle size both laterally and vertically are typical of fluvioglacial deposits that occur in front of, along the margins of, and beneath ice masses. The landforms produced by the accumulation of sediment by channelled flow range from eskers through to great outwash spreads known as *sandar* (Icelandic term meaning sand and gravel plain: singular = *sandur*).

When meltwater streams enter lakes or the sea they sometimes build deltas. The association of top-set, fore-set and bottom-set beds is characteristic of this environment. The glacial system is conducive to the production of large debris loads and high discharges and sedimentary basins tend to receive large amounts of sediment, particularly during periods of deglaciation. The development of deltaic accumulations is therefore a common occurrence.

The erosive power of glaciers and ice sheets results in large amounts of fine material occurring in glacierized areas. The subsequent reworking of this fine material by meltwater streams means that it is often transported considerable distances. Whenever lake-basins cause a decrease in the transporting energy, the fine sediments accumulate as lake floor deposits to produce laminated silts and clays. The higher energy environments of the lake shores are characterized by sand and gravel accumulations of either beaches or deltas. Each of these sedimentary environments produces distinctive landforms when the glacial lakes are subsequently drained. The same range of sediments and landforms occur in marine environments but they only become part of the land surface if changes in sea-level occur after their formation.

A great many techniques can be applied in the study of fluvioglacial sediments to reveal information about the environments in which they occur. From the geomorphologists point of view it is often necessary to distinguish between glacial and fluvioglacial sediments and to determine the direction of meltwater flow in the case of fluvioglacial deposits. King and Buckley (1968) claim that the lengths and shapes of pebbles enabled them to distinguish a variety of different depositional environments on Baffin Island. They showed that mean grain size alone provides a criterion by which deltas, eskers and ice-contact deposits can be differentiated. Roundness studies also indicated that there were significant differences in the roundness values of particles in kames, eskers and moraines.

Table 1. (After King & Buckley, 1968.)

Feature	Mean size (mm)	Roundness ($2R/a \times 1000$)
Moraines	386	138
Kames	—	238
Eskers	237	332
Deltas	121	334
Beaches	79	398

A similar study was carried out by the writer on sediments forming moraines and eskers at Breidamerkurjökull, Iceland (Table 2). Dolerite pebbles were measured and it proved impossible to recognize significant differences in roundness values between moraines and eskers. It is suggested that in this case the moraines were formed by the ice reworking fluvioglacial deposits and little change in particle roundness had taken place.

Table 2

Feature	Mean Roundness
Moraines	196
Eskers	205

There are numerous difficulties involved in using such measures as mean size and roundness to differentiate fluvioglacial deposits. Variations in the rock types constituting pebbles, the history of a pebble prior to glaciation and the range of environments through which a pebble has passed will produce changes in pebble size and shape that do not necessarily reflect the environment of deposition. It is very doubtful if mean size and roundness values can be meaningfully compared when samples are obtained from locations separated by considerable distances.

The location and form of fluvioglacial deposits are often very helpful in determining the general direction of flow of the water responsible for the deposits. Their relationship to other glacial and fluvioglacial landforms and deposits can also indicate likely environments of deposition.

FLUVIOGLACIAL DEPOSITION NEAR EXISTING GLACIERS—THE EARLY LITERATURE

Much of the early work on fluvioglacial deposits in areas of former glaciation was based on analogy with conditions observed along the margins of existing glaciers. The expeditions led by I. C. Russell to the glaciers around Yakutat Bay, Alaska, in 1890 and 1891, provided glacial geomorphologists with a great deal of information about fluvioglacial deposition. The observations recorded by Russell (1892, 1893), were the basis of several papers, published before the end of the century, that interpreted fluvioglacial deposits in areas of former glaciation in North America. Russell described marginal, supraglacial, englacial and subglacial streams and realized the significance of such drainage in the deposition of sands and gravels. He wrote (1892, p. 181), 'The formation of osars [eskers] seems fully explained by the subglacial drainage of the Malaspina ice sheet. On the north side of the glacier there are many streams which flow into tunnels and carry with them large quantities of gravel, sand and mud; while on its southern and eastern margin many streams emerge from tunnels and bring out large quantities of waterworn debris.' Russell also described the formation of 'alluvial cones' (sandar) at the exits of the subglacial channels and envisaged the deposition of materials on the tunnel floors. The increasing thickness of the deposits would bring a subglacial stream in contact with the ice forming the roof of its tunnel and further melting would take place. He concluded (p. 181), 'In this way deep narrow deposits of cross-stratified gravel might be formed within a glacier, which when the ice melted would assume an anticlinal structure, owing to the displacement of the material along its sides.'

Towards the end of the nineteenth century, British geomorphologists were also provided with information about subglacial fluvioglacial deposition as a result of expeditions to Spitzbergen by Sir Martin Conway and E. J. Garwood. Garwood (1899) described supraglacial streams containing rounded material, and of one stream, that drained a lake formed on the lee side of a nunatak, he wrote (p. 685), '. . . the water of the lake escaped with great impetuosity down . . . a tunnel 3·6-4·5 m in diameter excavated in the solid ice. The debris . . . had evidently been carried with great force along the tunnel during the bursting of the lake in the previous spring and . . . it still strewed the mouth and floor of the tunnel. As the climate ameliorated this waterworn glacial material would be deposited as a gravelly ridge, running roughly at right angles to the long axis of the valley and forming a kame [esker].'

Further publications by Tarr (1909) and Tarr and Butler (1909), on the Alaskan glaciers included descriptions of fluvioglacial deposition. The formation of eskers in subglacial tunnels in stagnant ice and the development of kames, as a result of ice melting irregularly from beneath stratified gravels, were discussed. Kame terraces, high above the existing ice surface, were described and the fact that such gravels can survive overriding by ice during a readvance was established. Studies of the margins of the Alaskan glaciers indicated that conditions changed rapidly, and that the nature of the material available and the environment in which it was deposited were very variable.

FLUVIOGLACIAL DEPOSITION IN AREAS OF FORMER GLACIATION

After the publication of Russell's papers on the glaciers of Yakutat Bay, Alaska, (1892, 1893) papers were published by Stone (1893) and Hershey (1897), applying the principles of marginal, and subglacial fluvioglacial deposition, outlined by Russell, to similar deposits found in Maine and Illinois. Stone realized the significance of relatively warm waters flowing from the land surface above the ice into tunnels which they were capable of cutting and enlarging beneath the ice. Hershey (1897), pointed out that, 'If instead of mapping the deposits by their present areal extent, lines are drawn through the crests of the ridges, the analogy with an ordinary drainage system will be very close.' Hershey also suggested that meltwater streams flowing in subglacial tunnels were capable, because of hydrostatic pressure, of depositing gravel 100 feet above the point of its erosion.

During the thirty years, 1902-1932, a large volume of literature relating to fluvioglacial deposition, was published. Crosby (1902), Gregory (1912), Trowbridge (1914), Giles (1918), Chadwick (1928), Flint (1928, 1930), and Brown (1931) are a few of those who published descriptions and explanations of kames and eskers. The debate about the subglacial, supraglacial, or pro-glacial origins of eskers and kames was concluded in 1928, when Flint stated that both eskers and kames are polygenetic. The outcome of all the detailed work on these fluvioglacial landforms in areas of former glaciation produced new concepts of deglaciation and in turn started another major controversy amongst geomorphologists. Flint (1929, p. 256) outlined two methods of ice dissipation: 'These are, *1*) retreat of ice due to excess of melting over alimentation, with the preservation of a well-defined glacier front, and *2*) dissipation of the ice as a 'dead' or stagnant mass resulting from the total loss of forward motion while at its maximum southward extent.' The concept of downwastage implied the appearance of nunataks and the washing of debris by meltwaters on to the ice occupying the lowlands and therefore protecting it from melting. The shrinkage of the ice was inward from the valley sides, resulting in the formation of marginal lakes and the development of ice-contact kame terraces. It was possible that paired terraces could develop on either side of a valley, due to hydrostatic connection through crevasses in the ice linking two marginal lakes. Kettles occur in the almost horizontal terraces, some of which would be related to spillways that drained the marginal lakes in which deposits accumulated.

The concept of the downwastage of an ice mass with the resultant development of characteristic fluvioglacial forms, sometimes related to an englacial water-table, was applied by Flint (1930) to the deglaciation of central Ireland. In 1931 Andersen published similar opinions based on work in Denmark and he enlarged the concept by suggesting that when the ice surface became so thin that the water-table was high enough to cause the ice to lift, that there was a rushing, and irregularly streaming, meltwater drainage that would produce kames and pitted plains.

In 1942, Flint and Demorest published a more detailed account of the evidence that glacier thinning occurred during deglaciation and the forms that developed as a result of such thinning. They pointed out that evidence from existing glaciers indicated that thinning does take place but that it is not necessarily accompanied by stagnation. The authors concluded (p. 132) ' . . . the available evidence indicates that separation or stagnation of vanished ice can rarely be proved and that at best it can be inferred to have only affected small areas at any one time.' Rich (1943) also favoured local stagnation without regional stagnation. He envisaged burial of ice, under aggrading outwash at the margin of a continental glacier, that became detached and stagnant as a result of its burial.

Much has been written about fluvioglacial deposition in Britain. The memoirs of the Geological Survey and papers by Charlesworth (1926a, 1926b), Trotter (1929), Hollingworth (1931), and others, all utilized the conclusions mainly developed in North America and discussed above. However, the interpretations of the fluvioglacial deposits in Britain have almost always been influenced by the concept of normal horizontal retreat of ice fronts rather than by the idea of downwastage associated with frontal retreat. The subglacial origin of eskers has long been accepted by British geomorphologists but the other forms of fluvioglacial deposition have always been associated with the proglacial environment and frequently with ice-dammed lakes.

Mannerfelt (1945, 1949) emphasized the importance of subglacial deposition in areas of stagnant ice. He pointed out that meltwaters made their way down beneath the ice, taking with them sand and gravel which was deposited in tunnels under the ice. The ridges of sand and gravel covered by ablation moraine Mannerfelt called *subglacially engorged eskers*. He also described marginal terraces (i.e. kame terraces), formed as the deposits of marginal streams.

The work done in Scandinavia by Mannerfelt was the stimulus for a new appraisal of fluvioglacial deposition in Britain. Sissons' work in East Lothian (1958a) and the Eddleston Valley (1958b) was based on the work of Mannerfelt and other Scandinavians. In East Lothian the fluvioglacial deposits form kame terraces, kames and eskers. Sissons concluded that some kame terraces, consisting of coarse deposits, were formed by a depositional phase of a marginal stream, while others were formed in long narrow lakes, between ice and rock. Kames with no definite orientation were explained by the melting out of ice from beneath fluvioglacial deposits.

In his paper on the Eddleston Valley (1958b), Sissons described fluvioglacial deposits that were laid down at the margin of, on top of, and beneath a mass of stagnant, downwasting ice. He pointed out that some kames and eskers were the result of deposition by meltwater streams that had cut channels elsewhere. The meltwaters flowed from the sides of the Eddleston Valley into a mass of stagnant ice occupying the bottom of the valley. Within the mass of stagnant ice a water table developed that was controlled by outlet channels to the north. Sissons observed two distinct levels of fluvioglacial deposition, each controlled by a water-table which was in turn controlled by an outlet channel.

Since 1957, the importance of subglacial deposition has been stressed by Scandinavian geomorphologists. Holdar (1957) described a terrace of fluvioglacial deposits at the base of a subglacial chute, and stated that the terrace was deposited about 4 m under the ice. He stated (p. 500) that 'This shows that distinct terrace forms can be submarginal which in turn plays a great part for the determination of the lateral terraces as chronological indicators.' Holdar also attributed mounds of fluvioglacial material to deposition in subglacial holes and eskers to deposition in subglacial tunnels. The emphasis on fluvioglacial deposition in a subglacial environment was epitomized by J. Gjessing (1960) in his detailed and extensive work in east-central-southern Norway. He described terrace flats that were found undisturbed and concluded (p. 454), 'Thus the deposits under the flats must have rested on their present under-layer, i.e. the ground moraine or the bedrock, from the first moment of their accumulation. Consequently the accumulation must have begun at the bottom of the ice . . . From these deposits the ridges continue often with regular forms and clear limitations. Obviously these ridges must have been accumulated from below, from the bottom of the ice upwards.' Gjessing also discusses how the subglacial drainage system formed fans, terraces and esker-like ridges beneath the ice. In the areas he was considering, the direction of subglacial drainage was opposite to the normal drainage and Gjessing envisaged a situation, whereby, at a critical thickness of the ice the pressure of the subglacial water system was sufficient to lift the ice above it and a sheet-like drainage developed. The importance of Gjessing's conclusions lies in the stress upon the subglacial environment for large scale fluvioglacial deposition.

There is an extensive literature on fluvioglacial deposition based both on work in areas of existing glaciers and on landforms and deposits occurring in areas of former glaciation. The relatively recent realization of the significance of extensive subglacial deposition combined with further developments in the field of sedimentology mean that a simple correlation between forms and origin is no longer acceptable. The complex origins of many fluvioglacial deposits have tended to undermine the genetic terminology used in the past. Owing to the reliance on a genetic classification of fluvioglacial depositional landforms the development in our understanding of depositional processes and environments has made classification and terminology difficult to standardize.

CLASSIFICATION OF FLUVIOGLACIAL DEPOSITS

Fluvioglacial deposits can be classified on the basis of their form and, or, environment of deposition. A simple classification based on form and dividing deposits into groups according to whether they occur as spreads, mounds or ridges is a useful first approximation. However, such a classification does not permit the separation of forms that develop in very different environments, e.g. an outwash plain produced by fluvial processes and a lacustrine plain consisting of lake bottom deposits would occur in the same group. A classification based solely on sedimentary characteristics implies that such characteristics can always be used to determine the environment of deposition. Lack of exposures on the one hand, and the frequent lack of conclusive sedimentary structures on the other, often make it difficult to determine the environment of deposition, except in very general terms.

Numerous attempts have been made to establish a definite link between form and origin (or sedimentary environment), and much of the so-called genetic terminology that occurs in the literature relating to fluvioglacial depositional forms is based on this assumption. However, it is now generally accepted that similar forms can be produced by different mechanisms. As will be demonstrated in the section on eskers, there are several different mechanisms that result in the formation of the sinuous ridges of sand and gravel which are usually called eskers. The term esker can no longer be used to convey one mode of origin.

Reference has already been made to the use of sedimentary characteristics as a means of determining sedimentary environments and as a basis for the classification of deposits. Meltwater can transport and deposit material in stream channels, in deltas and into both marine and lacustrine basins. Both the internal characteristics and the surface form of these deposits can be greatly affected by their relationship with the ice mass from which the meltwater is being derived. This relationship has been used as the basis of a general subdivision of fluvioglacial deposits into proglacial and ice-contact deposits. Proglacial deposits result from meltwater transport and subsequent deposition in locations some distance from, and undisturbed by, the actual presence of glacier ice. Ice-contact deposits result from deposition by meltwater of sediments against, on, under and in an ice mass. If the relationship between the deposit and the ice mass can be determined then adjectives such as 'frontal-marginal', 'lateral-marginal,' 'subglacial', and 'englacial' can be applied to the deposit.

Work that has been done during the last 100 years along the margins of existing glaciers and ice sheets has revealed just how complicated fluvioglacial environments can be. Simple analogy between forms and deposits studied in areas of former glaciation and those observed in the process of development along existing ice margins has become increasingly difficult. Any classification system and terminology used in connection with forms and deposits in areas formerly glaciated must be amended as our understanding of fluvioglacial environments and deposits associated with existing glaciers is increased. Bearing this in mind, the classification outlined below is based on the assumption that fluvioglacial sedimentary environments are closely related to glaciological conditions. It is therefore of fundamental importance to establish not only the position of the ice margin at the time any set of deposits is accumulated but also the glaciological characteristics of the ice, and the nature of the debris and proglacial and subglacial surfaces with which they are associated. Some of this data can be derived from the characteristics of the sediments themselves, their surface form and the relationship between the forms of fluvioglacial deposits themselves and with other glacial and fluvioglacial landforms in the area. However, there is a danger of circular argument involving the determination of glaciological and sedimentological environments from the deposits and the subsequent classification of the deposits on the basis of their environments of accumulation.

Table 3

Dominant Sediment	Environment	General Form	Relationship to Ice	Genetic Term
		ICE-CONTACT DEPOSITS		
Sand and gravel	Fluvial	Ridge	Marginal, subglacial Englacial, supraglacial	Esker
Sand and gravel	Fluvial	Mound	Marginal, subglacial Englacial, supraglacial	Kame Kame complex
Sand and gravel	Fluvial	Spread with depressions	Marginal	Kettled sandur
		PROGLACIAL DEPOSITS		
Sand and gravel	Fluvial	Spread	Proglacial	Sandur
Silt and clay	Lacustrine	Spread	Proglacial/marginal	Lake plain
Sand and gravel	Lacustrine	Terraces, ridges	Proglacial/marginal	Beach
Clay, sand and gravel	Lacustrine	Terrace	Proglacial/marginal	Delta
Silt and clay	Marine	Spread	Proglacial/marginal	raised mud flat
Sand and gravel	Marine	Terrace, ridge	Proglacial/marginal	raised beach
Clay, sand and gravel	Marine	Terrace	Proglacial/marginal	raised delta

From Table 3 it can be seen that there are primarily two types of depositional environments in which fluvioglacial sediments accumulate: the channels of meltwater streams or in bodies of standing water. A further subdivision of the second category is possible on the basis of the standing water being either marine or lacustrine. However, the forms and the sedimentary characteristics are likely to be similar, except perhaps in terms of overall size, whether the sediments accumulate in marine or lacustrine conditions.

Another subdivision of fluvioglacial deposits is possible in terms of their relationship to the ice. Deposits which accumulate some distance away from the ice margin retain undisturbed sedimentary characteristics and forms that develop primarily as a result of the extent and position of the deposits. Other deposits accumulate in close association with the ice mass and are laid down against it, beneath it, within it, or on top of the ice surface. Such deposits are referred to as ice-contact deposits and often exhibit disturbed sedimentary structures. The form of the deposits is often produced either by the ice mass determining the position and extent of deposition or by subsequent slumping resulting from loss of support of ice walls or blocks of buried ice. In some situations ice-contact deposits merge with proglacial deposits (see the section on pitted sandar).

ICE-CONTACT FLUVIOGLACIAL DEPOSITS

Meltwater streams flowing on, in, under, and along the margins of masses of ice, transport and deposit sediment. The landforms that result from the accumulation of this sediment on, in, against or beneath ice are known as ice-contact landforms. The majority of these landforms result from deposition by meltwater streams and are characterized by channel deposits. Some of these landforms result from accumulations of sediment in lateral marginal, frontal marginal, subglacial, englacial or supraglacial lakes and may exhibit lacustrine rather than channel flow characteristics.

The terminology that has been applied to ice-contact fluvioglacial deposits is very confusing. The confusion has arisen because in several instances one term has been used to refer to several different forms, each with different mechanisms of formation. The greatest confusion has occurred with reference to the terms 'kame' and 'esker'. Some authors have used the term kame in connection with any accumulation of fluvioglacial deposits, whether it has ridge or mound form and regardless of the processes involved in the formation of the ridges or mounds. Others have not distinguished between primary ridge and mound forms resulting from the accumulation of sediment over restricted areas and secondary ridge and mound forms resulting from the destruction of a uniform surface by the melting out of buried ice. In this book the term 'kame' is applied to primary accumulations of sediment in water either beneath, within or at the margins of an ice mass, that have a mound form resulting from the removal of supporting ice. When accumulation results in a terrace or delta form the necessary adjective is used to produce the terms kame-terrace and delta-kame. The term 'esker' is used for ridges of fluvioglacial deposits laid down by meltwater streams as channel deposits.

The above separation of eskers and kames is difficult to apply to an area of channel deposits accumulated on top of a sheet or blocks of stagnant ice, which develops secondary ridges and mounds as a result of the subsequent melting of the buried ice. When mounds and ridges of this type occur it is preferable to emphasize the initial form of the deposit and refer to it as kettled-outwash or kettled-sandur.

Eskers. Narrow, sinuous ridges of silt, sand, gravel and cobbles, have been of interest to geomorphologists for many years. At various times their origins have been related to a wide range of phenomena, e.g. the Biblical flood, medial and push moraines and in Denmark they have even been associated with a goblin with a leaky sand bag. It is now generally accepted that ridges of sand and gravel are related to the former presence of an ice cover and they represent the deposits laid down in stream channels associated with the wastage of glaciers and ice sheets.

The term esker originates either from the Irish word *eiscir* or Welsh word *escair* meaning crooked or winding. In Scandinavia the term *ose* is used for similar features.

Eskers can range from a few metres to 400 km (with breaks) in length. The height and width of the base of an esker ridge appear to be functions of the overall length of the ridge in so far as the longer ridges are usually higher and wider than the shorter ridges. In general terms it appears that eskers between 50 and 200 m in length range between 3 and 10 m in height. Those eskers over one kilometre in length range between 10 and 100 m in height.

Eskers can occur as single continuous ridges of uniform dimensions, ridges with varying height and width or ridges consisting of a series of mounds linked by low ridges. Eskers can also occur in the form of complex systems with a series of tributaries and distributaries. Eskers appear to be best developed on low angle slopes but they are also formed on divides and descending quite steep valley sides.

The material of which esker ridges are formed is usually rounded gravel and cobbles with some silt and sand. Some eskers consist mainly of sand. The usual features associated with fluvial deposits have been observed in sections in eskers. Stratification of the deposits, cross bedding and the usual range of bed structures are quite common. In sections at right angles to the ridge crests of eskers anticlinal structures have been observed and are usually interpreted as being the result of slumping consequent to the removal of supporting ice walls. From a limited number of investigations (Hellaakoski, 1931; Trefethan and Trefethan, 1945; Lee, 1965) it can be stated that the gravel of an esker is usually derived locally and rarely travels more than 10 km. Lee (1965), working on the Munro esker in Canada, that can be traced for 400 km and is up to 88 m high and 1 to 6 km wide, noted the sheet bedding and cross bedding and concluded that the deposit was laid down by a braided stream. A study of the lithology of the esker indicated that the maximum abundance of any mineral did not occur at the point of outcrop but at some distance down-stream depending on the particle size and density of the material. This distance ranged between 3 and 15 km.

The term *esker* has been applied to a group of deposits varying in form, composition and size, the variations appearing to indicate that this group is polygenetic. The deposits that form eskers can be accumulated in four environments: subglacial, englacial, supraglacial and proglacial.

Eskers can be produced by the accumulation of sediment in the channels of subglacial streams. Meltwaters descend from the ice surface down crevasses and moulins, and tunnels develop at the base of the ice to allow movement of meltwater towards the ice margin. Normal conditions of sediment transport and deposition can occur in these subglacial tunnels, and channel floor deposits can accumulate between walls of ice. The actual cross-sectional shape of the tunnels will have an important part to play in determining the final cross-sectional shape of the esker and its internal structural characteristics (Fig. 50). In most cases, the removal of the tunnel walls as a result of melting determines the cross-sectional shape of the esker, and the ratio of tunnel width to tunnel height therefore determines whether or not the eventual deposit has a flat top. If the subglacial tunnel is wider than it is high, a flat-topped esker with a considerable proportion of undisturbed sedimentary structures will be produced (Fig. 50*a*). If, on the other hand the tunnel is high and narrow, a sharp-crested ridge consisting of slumped

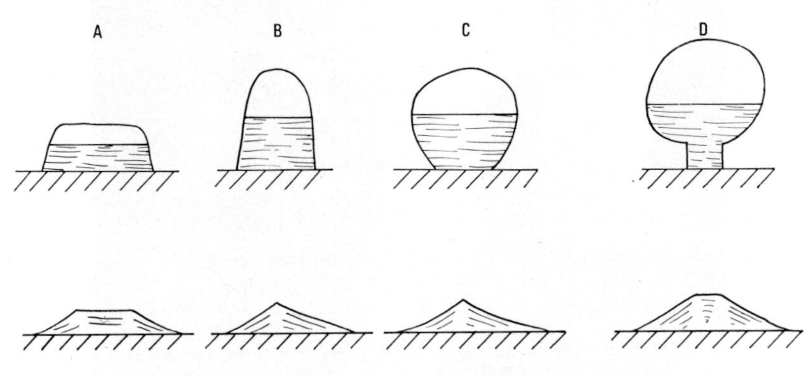

FIG. 50. Cross-sections of subglacial tunnels.

sediment will result (Fig. 50*b*). An almost circular tunnel (Fig. 50*c*) will also tend to produce a sharp-crested ridge of slumped sediment. A keyhole-shaped tunnel (Fig. 50*d* and Fig. 51) can produce a flat-topped ridge. Whether these tunnels occur at the base of an ice mass or beneath the lateral margin to produce what Mannerfelt (1945) called 'subglacially engorged eskers,' the principle remains the same.

The meltwater flowing through the tunnels at the base of a temperate ice-mass can develop a considerable hydrostatic head simply because the meltwater largely originates on the ice surface and descends to the base. Pressure can also be built up by constrictions developing in the tunnel as a result of ice movement. Water flowing under hydrostatic pressure can flow up-hill and this can result in deposits occurring on up-grades. Such deposition, under hydrostatic pressure, has been used by various authors to explain the way in which some eskers run across country, largely ignoring the occurrence of local reverse gradients.

It is fairly easy to envisage a tunnel system developing in the marginal zone of a decaying temperate ice mass and the sediment accumulating in these tunnels to produce eskers up to 5-10 km long. However, it is hard to imagine the existence of a tunnel system hundreds of kilometres in length, and it has been suggested that the very long eskers that developed in association with the great Pleistocene ice sheets must be the products of the inward migration of the stagnant decaying marginal zone and its associated tunnel system with the retreat of the ice margin. Another problem associated

FIG. 51. A subglacial tunnel at Breidamerkurjökull, Iceland. (Photo: R. J. Rice.)

with the subglacial hypothesis is why considerable thicknesses of sediment are not subjected to erosion by the same meltwater system that is responsible for their accumulation. The fact that there is little mention in the literature of evidence of breaks in the stratigraphy or erosional episodes in the sedimentary sequence in esker deposits may simply reflect the limited opportunities to study undisturbed sedimentary sequences in eskers.

R. F. Flint (1930) proposed another type of subglacial deposition beneath a mass of stagnant ice, which is (p. 628) '... seamed with crevasses and tunnels, the former filled with standing water, and some of the latter furnishing conduits through which flowing water connected the crevasses.' This situation produces a large area in which fluvioglacial deposition beneath the ice is controlled by a water table, and in which both narrow ridges and large flat-topped masses of sand and gravel could accumulate.

It can be seen, therefore, that the subglacial hypothesis is really made up of two separate theories, the first proposing deposition in subglacial tunnels at the base of a thick, either stagnant or slowly moving ice mass, and the second envisaging a 'honeycomb' arrangement of tunnels and cavities in a thin stagnant ice mass in which the fluvioglacial deposition is controlled by a water-table.

The englacial and supraglacial environments of esker formation can be treated together. As early as 1902 Crosby suggested that the deposits of sand and gravel formed in supraglacial glacial and englacial channels may be let down on to the subglacial surface without obliteration of the distinctive form of an esker. He argued that the meltwater responsible for these supraglacial and englacial deposits would also seep through the deposits and melt the floor of the channel, resulting in the eventual emplacement of the deposits on to the subglacial surface. Such an hypothesis had the advantage of explaining the lack of conformity between the trend of eskers and the underlying topography observed by Stone (1893) and other workers. Crosby (1902) stated: 'Eskers and esker systems . . . exhibit a tendency to conform in trend with the movement of the ice. . . . Eskers are to a good degree independent of the topography and often do not hesitate to forsake or to cross at all angles, large and well-accentuated valleys, in order to adhere to their normal courses . . . they freely climb slopes and cross ridges from 100 feet to 200 feet, and more rarely to 300 feet and 400 feet . . .' This hypothesis of superimposition of ridges of fluvioglacial deposits, initially deposited on or in the ice, on to the subglacial topography also had the advantage of not requiring deposition under hydrostatic pressure to explain the 'up-hill' sections of some eskers believed to have accumulated subglacially.

The superimposition hypothesis of Crosby contained two major difficulties. Some workers found it difficult to believe that ridges of sand and gravel would survive the process of being let down on to the subglacial surface. Other workers pointed out that for long esker ridges to develop on the ice surface, or in tunnels in the ice, it was necessary for large areas of the ice mass to be without crevasses, otherwise the supraglacial or englacial streams required to produce the esker deposits would become subglacial. Obviously, it was necessary to turn to areas of existing glaciers, to ascertain if eskers did occur on the ice surface and whether or not they could survive the process of being let down on to the subglacial surface.

Three papers were published between 1949 and 1958 describing eskers in the process of formation. Lewis (1949), described an esker in Norway that had been deposited in ponded water and a part of which had an ice core. In 1954 Meir described a series of eskers near glaciers in the Wind River Mountains in Wyoming which he believed to have originated in subglacial tunnels, and Stokes (1958) described an esker emerging from a subglacial tunnel, but again there was evidence of an ice core. It was with this background that my own work on the eskers near the Casement Glacier, Glacier Bay, Alaska, began in 1962. (Price, 1964, 1965, 1966; Petrie and Price, 1966.) Further opportunities were presented to study eskers in the process of formation at Breidamerkurjökull in south-east Iceland during the summers of 1965, 1966 and 1968 (Price, 1969).

The Casement Glacier is located on the east side of Muir Inlet, south-east Alaska, in an area that has experienced rapid deglacierization during the last 150 years

(Fig. 52). Since 1911 Casement Glacier has been land-ending and has retreated 5 km. During three months of fieldwork in 1962 all the eskers in the area, across which the Casement Glacier had retreated, were mapped and examined. When new aerial photography, taken under the direction of Austin Post, became available in 1963 and the

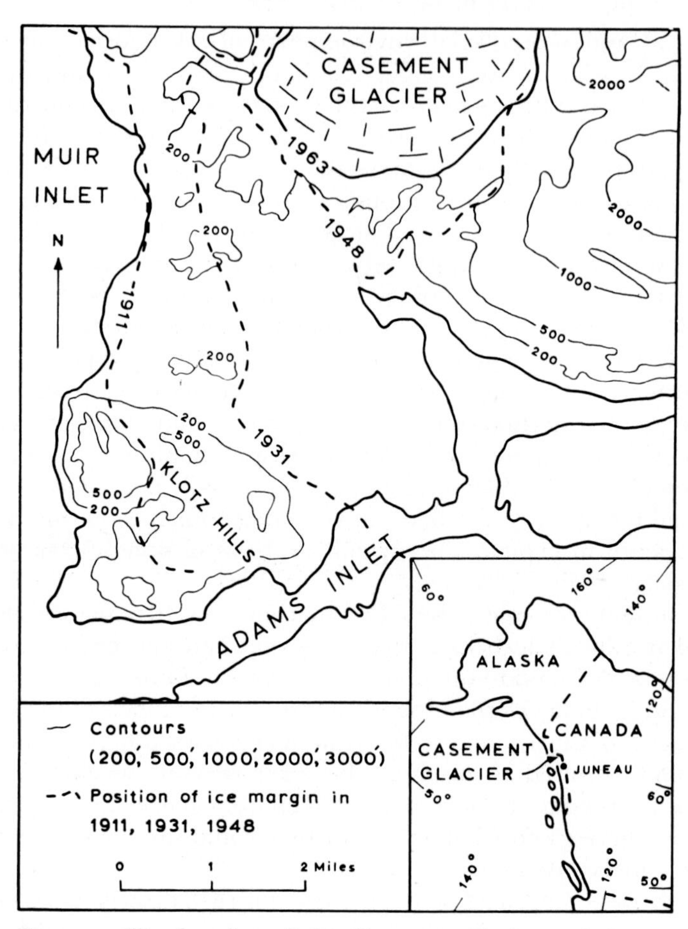

FIG. 52. The location of the Casement Glacier and former positions of its terminus. (After Price, 1966.)

author had access to photogrammetic plotting machines at the University of Glasgow, it was decided to produce large scale photogrammetric maps from the 1948 and 1963 aerial photography in order to study the changes in the glacier, its drainage system and the, eskers. The methods used in the photogrammetric work, the accuracy of the maps the results of the measurements of the ice wastage and a description of the general morphological changes between 1948 and 1963 are given in Petrie and Price (1966).

In front of the Casement Glacier there is a total of 17 km of esker-like features, consisting of sharp-crested ridges 3-18 m high composed of coarse gravel and cobbles

(Fig. 53). Both single ridges and complex systems occur. The longest single continuous ridge is 1·6 km long. There were no cross sections in these ridges which revealed sedimentary structures, and the constituents were so coarse that attempts to dig out sections resulted in the collapse of any structures that may have existed. The

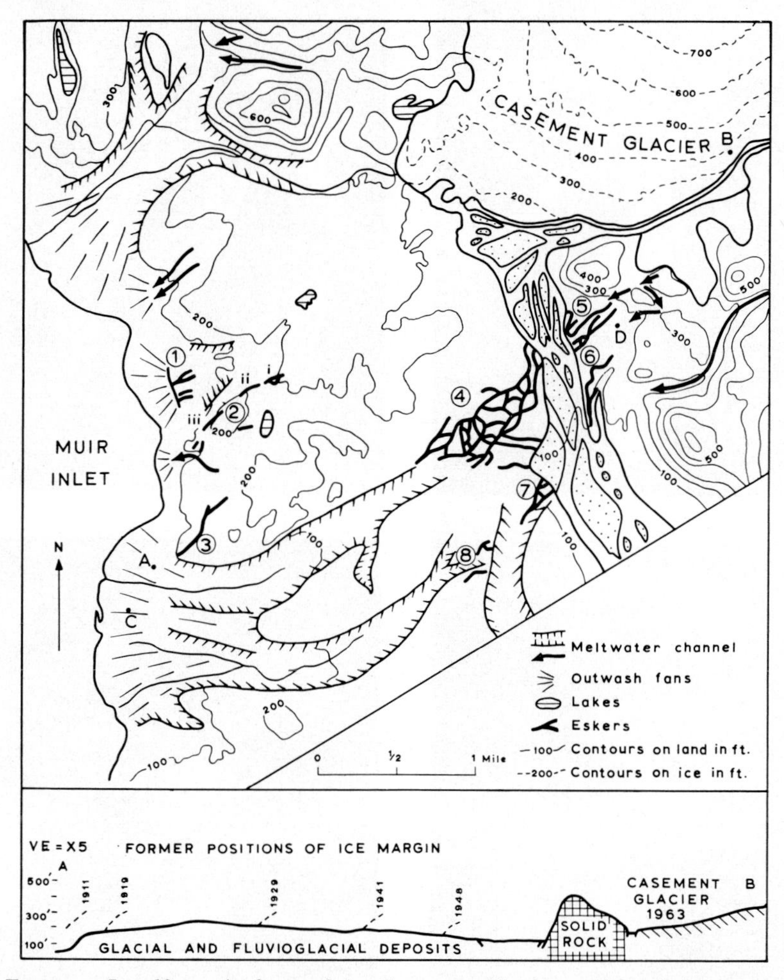

FIG. 53. Landforms in front of the Casement Glacier. (After Price, 1966.)

angles of the slopes constituting the sides of the ridges were between 25° and 30°. All ridges have a general north-east to south-west trend which is parallel to the direction of former movement of the ice across the area.

Apart from the fact that esker 2 (Fig. 53) at its eastern end descends a minor west-facing slope, crosses a dried-out lake floor and ascends an east-facing slope, there is nothing very remarkable about the single esker ridges in the area. It is the complex systems (Figs 53-8), of eskers that provide the most interesting data. It can be seen, by examination of the 1948 aerial photography (Fig. 54) that systems 4 and 5 are actually one single system, part of which was visible on the glacier surface in 1948 and part of which must have been englacial at that time. Between 1948 and 1963 a major melt-

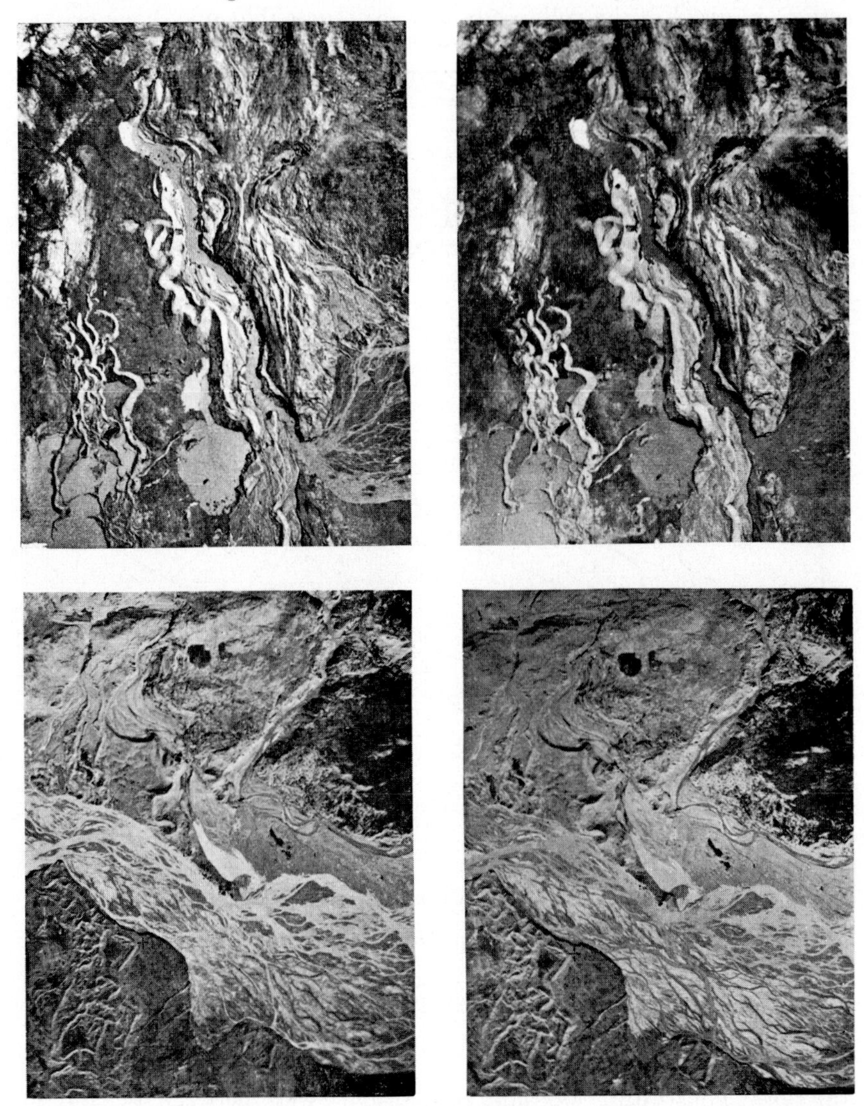

FIG. 54. Stereoscopic pairs of vertical aerial photographs of the same area taken in 1948 (upper) and 1963 (lower). (See Fig. 55.) (After Price, 1966.)

water stream (Fig. 55*a*, *b*), flowing from the Casement Glacier south to Adams Inlet, cut through the esker system so producing systems 4 and 5 (Price, 1965).

Esker system 4 (Fig. 55) consists of a series of anastomosing ridges formed of gravels and cobbles and situated at the eastern end of a major meltwater channel. The system is spread over an area 1·6 km from north-east to south-west and 0·8 km from north to south. Most of the ridges are between 6 and 12 m high with a few ridges attaining 12 to 15 m. The system includes sharp-crested linear and circular ridges and flat-topped ridges. Associated with this ridge system is evidence of the former existence of a lake that surrounded some ridges and submerged others. Shorelines of this lake were clearly visible on the sides of some of the ridges in the south-western part of the system. Aerial photographs taken in 1948 show the existence of the lake (Fig. 54) and at that time all ridges below 53 m were beneath the surface of that lake. The lake was dammed by the margin of the Casement Glacier.

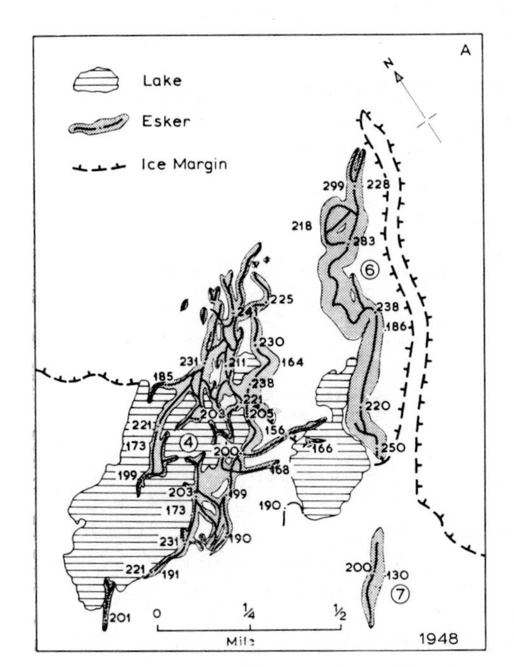

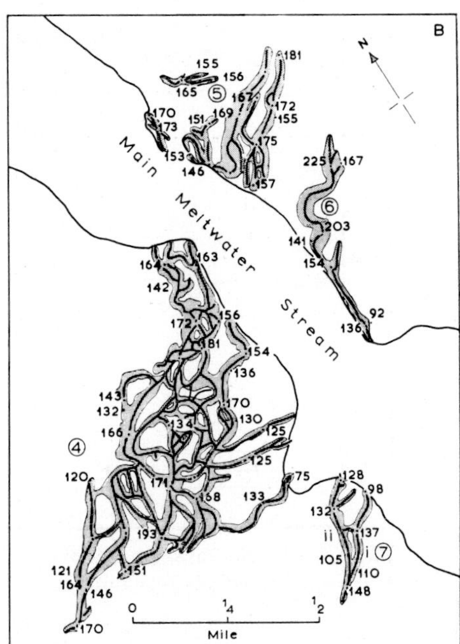

FIG. 55. An esker system at the Casement Glacier, in 1948 (*A*) and 1963 (*B*). Both maps were plotted on a Wild B8 Aviograph Plotter in the Dept. of Geography, University of Glasgow. (After Price, 1966.)

In 1962, 4 m of buried ice was seen exposed in one of the gravel ridges at the extreme north-eastern end of system 4. Also, at three localities at the truncated ends of the ridges forming system 5, ice could be seen forming the cores of the esker ridges. However, it was in system 6 that the greatest thickness of buried ice was observed. Most of the south-western part of this esker system had been undercut by the present meltwater stream and an ice cliff some 6-12 m high and covered by fluvioglacial deposits could be seen.

The occurrence of ice cores in some of the eskers and the availability of aerial photography taken in 1948 and 1963 prompted the author to try and establish, by means of photogrammetric measurements, the changes that had taken place in the esker system over that fifteen-year period. (Price, 1966; Petrie and Price, 1966.) It proved possible to undertake spot-heighting of eskers with an accuracy of approximately ± 1 m.

There were no detectable changes in the position, altitude and extent of the eskers in systems 1, 2 and 3. Esker system 4 is clearly visible on the 1948 photography (Figs 54 and 55) and when this system was plotted in the photogrammetric machine it was seen to have a very different character compared with its 1963 condition. In 1948 (Fig. 55*a*), the margin of the Casement Glacier was crossed by part of this esker system and several of the eskers can be observed standing above the ice surface.

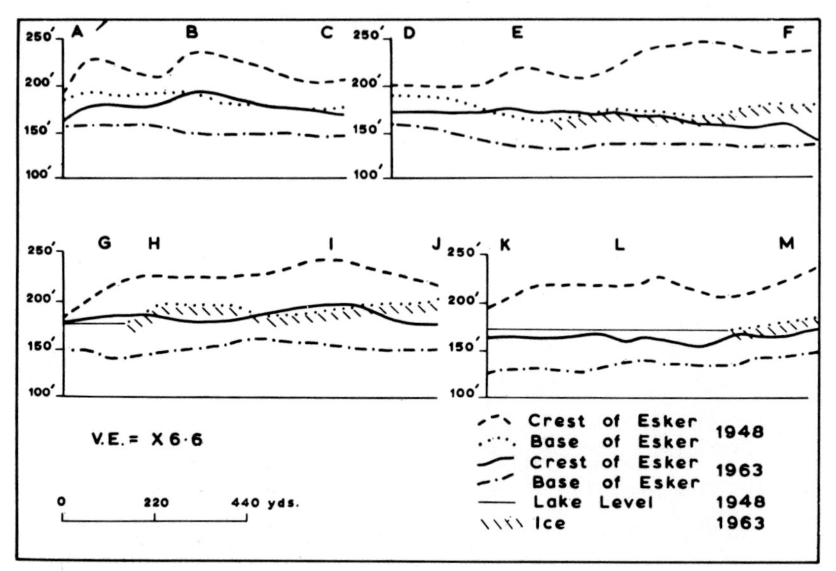

FIG. 56. Profiles of segments of esker system 4, (Fig. 55) constructed from spot heights obtained photogrammetrically. (After Price, 1966.)

Profiles have been plotted (Fig. 56) along the crest and base of segments of the esker system that appear on both the 1948 and 1963 aerial photographs. It can be seen that each of these esker segments was lowered between 1948 and 1963 and the extent of this lowering increased towards the north-eastern end of the system. In general, the crests of the eskers had been reduced in altitude by between 6 and 24 m, whereas the bases of the same eskers had been lowered by between 6 and 18 m. These altitude reductions are consistent throughout the system and are far in excess of the probable error in the photogrammetric spot-heighting method. It is suggested that these changes in altitude are due to the wastage of buried ice from beneath the eskers over the period 1948–63.

Esker system 6 (Fig. 55) is one of the most noticeable features on the 1948 photographs. It is just over 1·6 km long and is on the glacier surface throughout its length. The ridges forming the system are generally 15 to 24 m high, have a sinuous pattern and an irregular crest line. Two-thirds of this esker system was destroyed between 1948 and 1963 by the southward flowing meltwater stream but the remaining third survived a considerable lowering as a result of ice wastage (Fig. 57) from beneath the esker.

It can be seen from the above evidence that the change in altitude of some of the eskers near the Casement Glacier clearly indicates that eskers can be let down from the surface of a glacier on to the subglacial surface without being destroyed. Only small amounts of ice remained in 1962 in the extreme north-eastern end of system 4 and it can be clearly demonstrated that parts of this esker system were underlain by ice that ranged in thickness between 6 and 18 m.

There is little evidence from the Casement Glacier to prove whether the gravels forming the eskers were laid down in supraglacial channels or in englacial tunnels. The fact that a few eskers were not visible on the 1948 photographs but occurred on the ground in 1962 would suggest that at least some of the deposits accumulated englacially. However, whether the deposits accumulate supraglacially or englacially, if they survive the early stages of ice wastage they will prevent the ice beneath them from melting at the same rate as the gravel-free ice on either side (Fig. 58). This will result in an ice-cored gravel ridge. Such a feature will have little of its original stratigraphic character, as slumping will take place down the sides of the ice core. If this process continues long enough a series of either very small ridges or mounds would be the end product, but if the deposits arrive at the subglacial surface at an earlier stage, distinct ridge forms will be preserved. Unless the stream deposits are very wide it is also unlikely that flat-topped features will be preserved. The dominance of sharp-crested ridges in the esker systems near the Casement Glacier support this hypothesis. The irregular crest lines

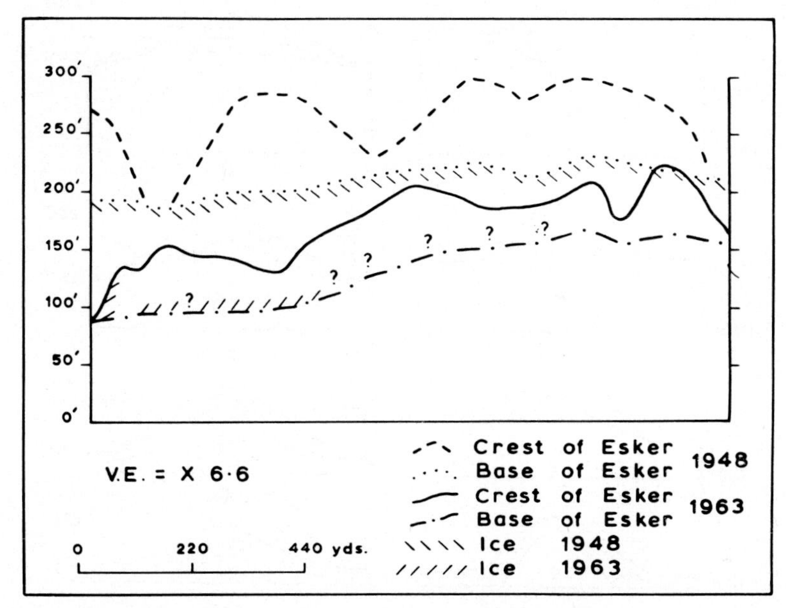

FIG. 57. Profiles of esker 6 (Fig. 55) based on spot heights obtained photogrammetrically. The extent of the ice under the esker in 1963 is unknown. (After Price, 1966).

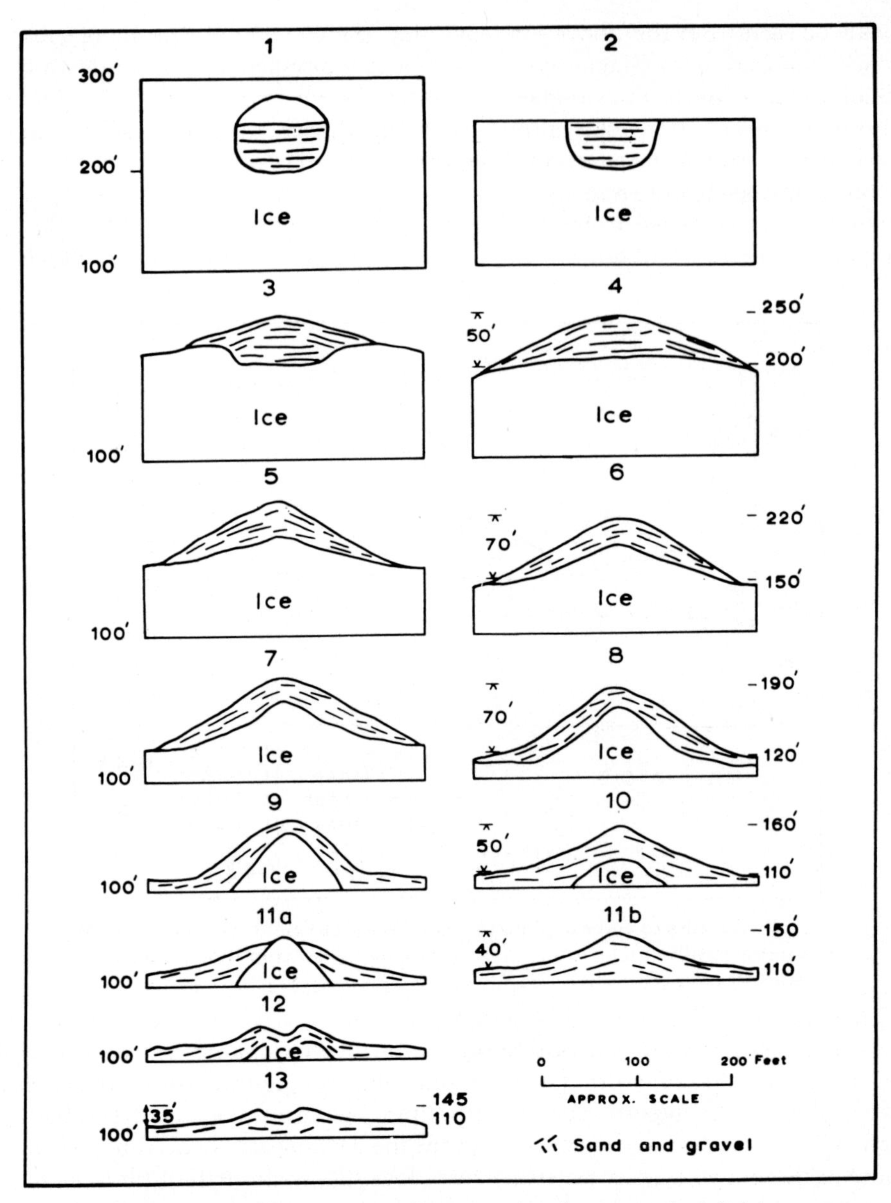

FIG. 58. The probable sequence of events whereby an englacial (1) or supraglacial (2) stream deposit 50 feet (15 m) thick, can be lowered on to the subglacial floor. The fluvioglacial deposits protect the ice beneath them from wasting as rapidly as the clean ice on either side (3-7). If the ice core remains long enough for sufficient slumping to take place down each side of it (8, 9, 11*a*) two esker ridges will be produced (12, 13). Alternatively, if the ice core is wasting away rapidly, only one ridge, about 40 feet (12 m) high will be produced (10, 11*b*). (After Price, 1966.)

of the eskers in these systems could also be the result of the varying thickness of the ice forming the ice cores.

In all of the complex esker systems in front of the Casement Glacier there are examples of tributary, distributary and meander patterns. It appears that the meltwater stream system that was responsible for these deposits must have developed on stagnant ice and must have had access to a considerable amount of morainic debris either on or in the ice. It is significant that the complex esker systems occur on the lee side of a series of solid rock ridges that are between 60 and 90 m high. These rock ridges were just beneath the ice surface in 1948 but were completely ice free in 1963. The ice on the lee side of these ridges must have been stagnant for a considerable period of time and permitted the development of a complex drainage system that was not subsequently destroyed by ice movement. The meltwater streams in the system must have been of considerable size, as they permitted the accumulation of up to 15 m of sediment on their channel floors. The source of this material was the extensive cover of ablation moraine that occurred on the surface of the Casement Glacier above the eskers.

Eskers occurring on the surface of a glacier were also studied in south-east Iceland (Price, 1969). Once again a zone of debris-covered ice in the form of three medial moraines was the source of material that, as ablation proceeded, was washed into the englacial tunnels to produce esker ridges with ice cores. Breidamerkurjökull is a piedmont glacier some 20 km wide which has experienced rapid wastage since 1890, with the ice front retreating over 3 km between 1890 and 1965. The major esker system at Breidamerkurjökull is located in front of a group of three medial moraines (Figs 59 and 60). They also occur in an area in which the topography is arranged in a series of steps between the ice margin and the 1890 moraine. There are over 5 km of sharp-crested, steep-sided gravel ridges within this area. The ridges consist of sand, gravel and cobbles and in some localities the deposits can be seen to be stratified.

Esker 4 (Figs 60 and 61) is a very large esker with a sharp, undulating crest line. It has two distinct parts (*a* and *b*) which appear to be related and are in fact connected by low ridges 1-2 m high. Esker 4*a* is generally between 5 and 10 m high, sinuous in plan, and at its distal end the main ridge bifurcates to form a loop. A pit dug near the almost right-angle bend and some 3 m below the crest revealed good stratification in fine sand, gravel and cobbles. On the surface of parts of this esker there was a covering of brown, angular fragments very similar to the material composing one of the medial moraines on the glacier surface to the north. The base of esker 4*a* generally falls in altitude from the proximal towards the distal end but there are irregularities in the sub-esker surface that are reflected in the crest profile (Fig. 61). The proximal part of the base of esker 4*a* is some 10 m higher than the distal part of 4*b*.

Esker system 4*b* is a complex one consisting of a series of sub-parallel ridges 3-5 m high, often separated by deep circular or oval kettle holes. In the wall of one of these kettle holes buried ice was observed in the summer of 1966. The proximal end of the system merges into an area of meltwater channels which contains several kettle holes.

Esker 5 (Figs 59, 60, 62, 63, 64), is the biggest in terms of height (up to 20 m) and cross-sectional area, near Breidamerkurjökull. It is a sinuous, sharp-crested ridge with steep sides (30°). In several places along this esker, recent slumps revealed good stratification in the fluvioglacial material. The esker continues on to the glacier surface and is definitely underlain by ice at its proximal end. Detailed maps made of this esker in 1965 and 1966 revealed that the crest of the proximal part of this esker had been lowered by between 3 and 7 m over a period of 12 months (Price, 1969, p. 37).

Eskers 6, 7, 8 and 9 (Figs 60, 65, 66) can be regarded as forming one system which is sub-parallel to that made up of eskers 4*a*, 4*b* and 5. In plan, eskers 6, 7, 8 and 9 clearly resemble an anastomosing stream pattern. The ridges are larger (10-15 m high), and more simple in plan at the proximal end of the system, and smaller and more

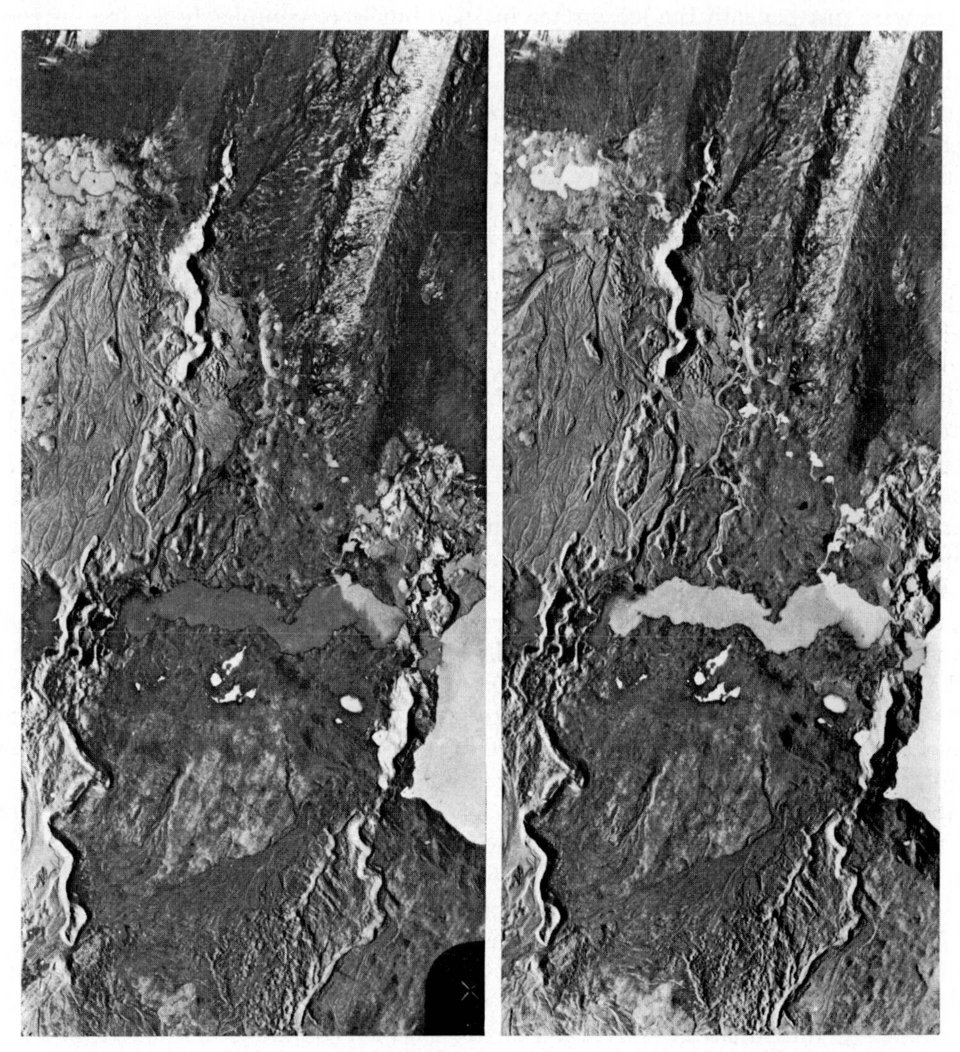

FIG. 59. Stereoscopic pair of air photographs of a part of the proglacial area of Breidamerkurjökull, Iceland (see Fig. 60). (Photography taken in 1965 by Landmaelingar Islands for University of Glasgow.)

complicated in plan at the distal end of the system, where they merge into a sandur surface with numerous kettle holes. Within this ridge and kettle-hole topography (Fig. 60, *E*9) there are a few flat areas upon which remnants of the former channel systems can be seen. Many of the channel systems are now truncated by kettle holes up to 15 m deep. There can be little doubt that the meltwaters which deposited the

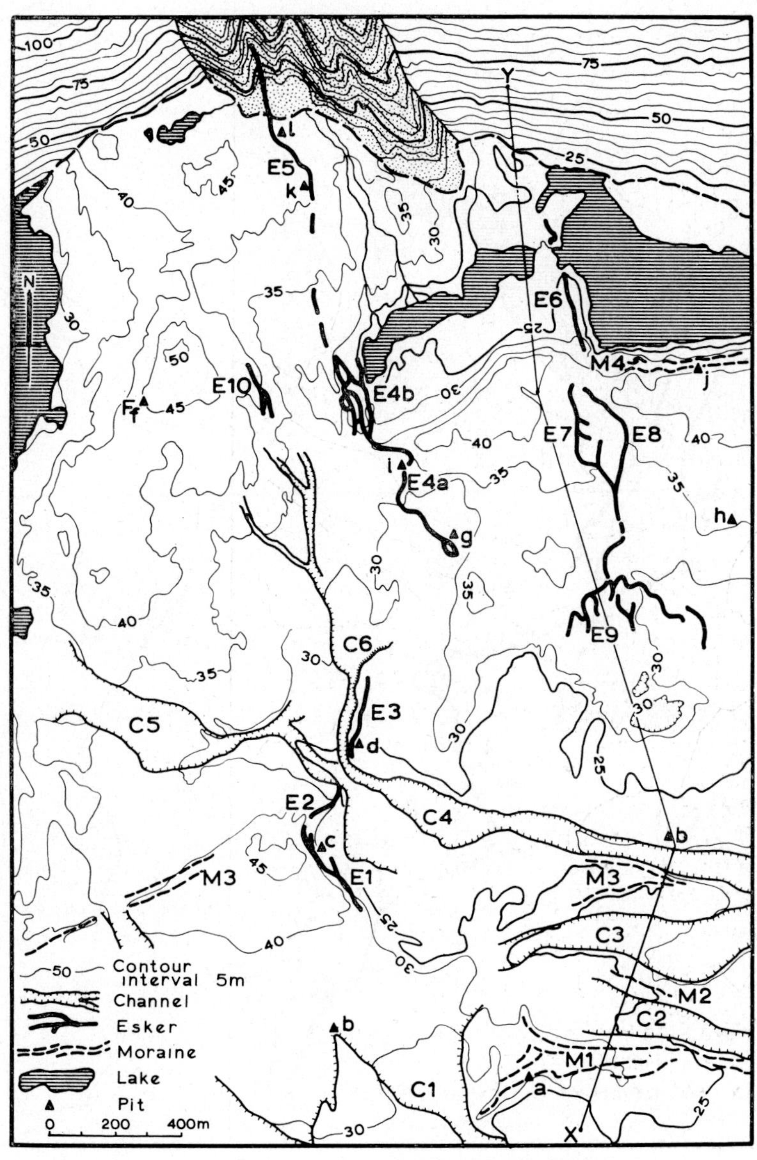

FIG. 60. Eskers, moraines and meltwater channels near Breidamerkurjökull, Iceland. (See Fig. 59.) (After Price, 1969.)

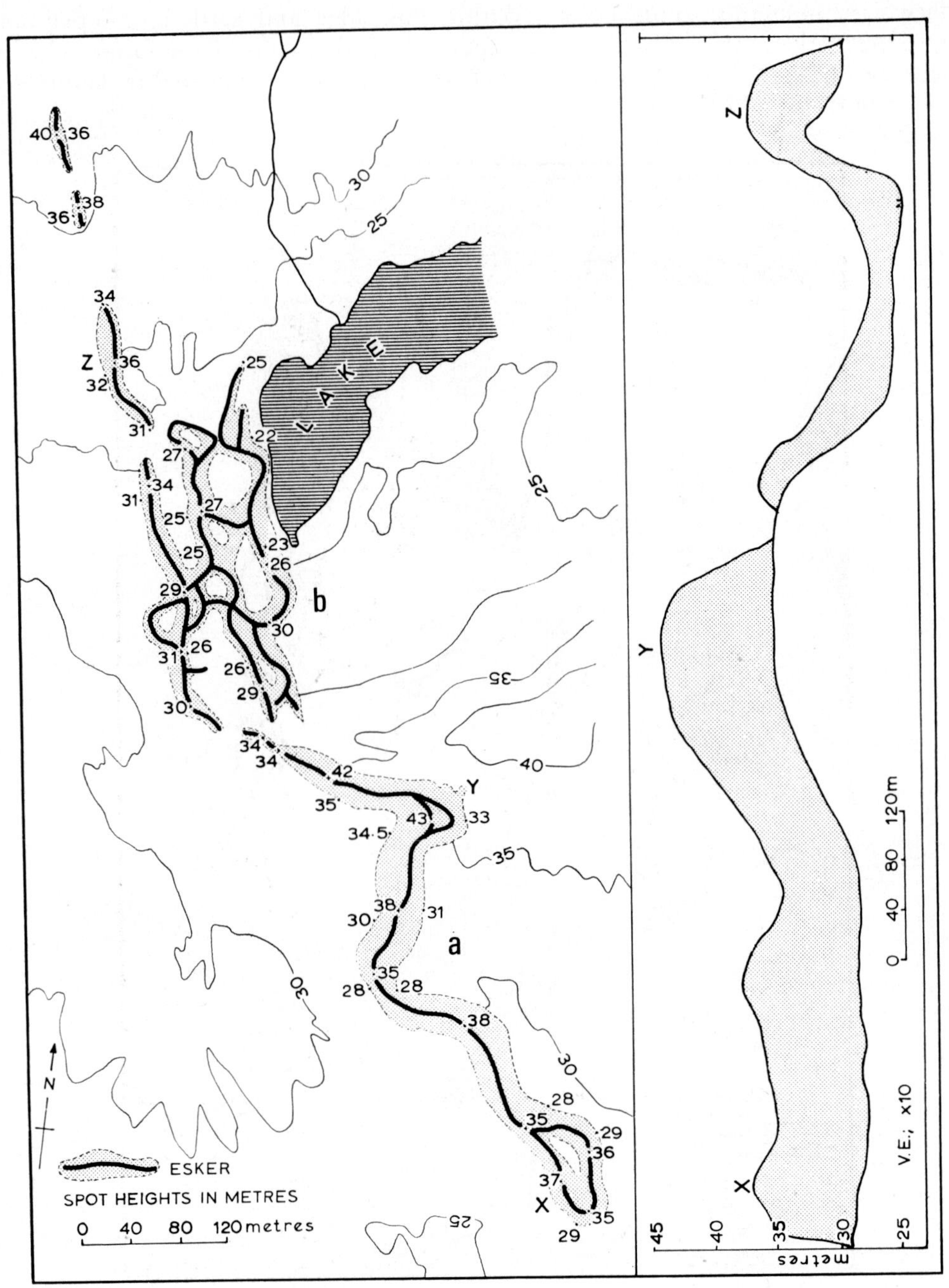

FIG. 61. Map and profile of esker system E4 (Fig. 60). (After Price, 1969.)

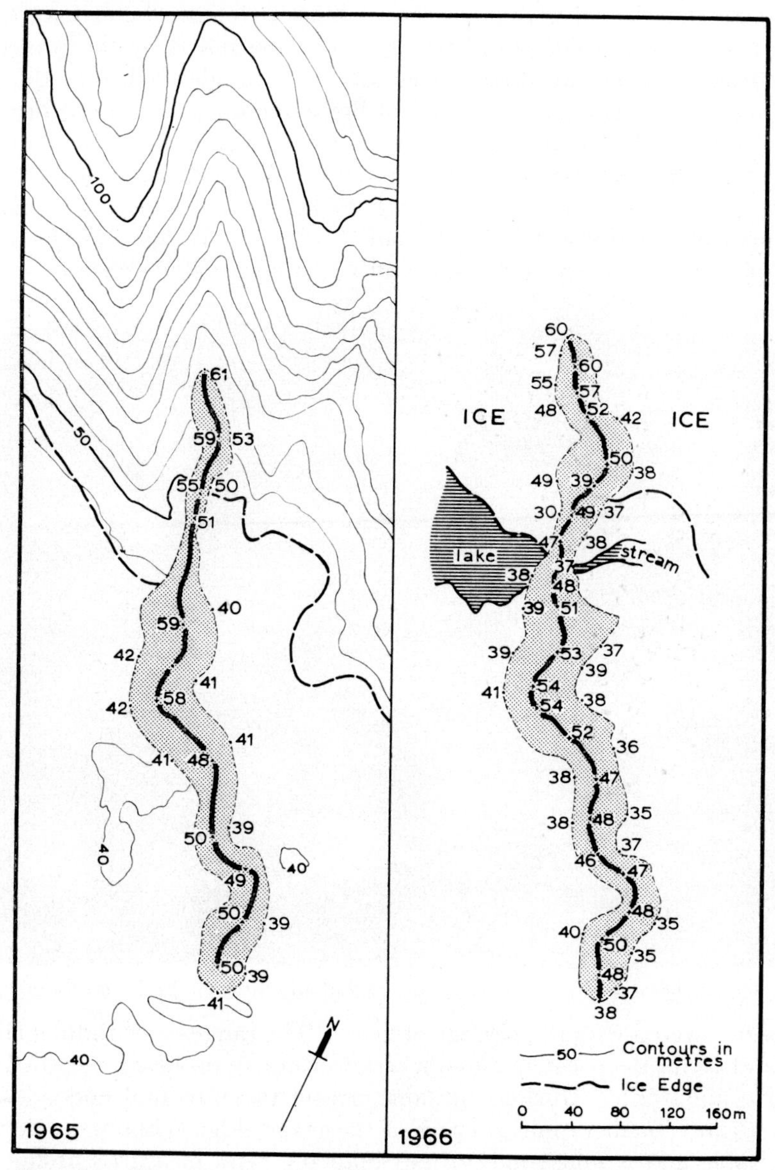

FIG. 62. Esker system E5 (Fig. 60). 1965-photogrammetric map. 1966-Tachyometric map. Spot heights in metres. (After Price, 1969.)

gravels forming eskers 6, 7 and 8 in definite channels or tunnels in the ice, were more widely dispersed in numerous interconnected channels at the time when esker system 9 was deposited. The sandur, to which esker system 9 is linked, was deposited in a proglacial environment but the sandur was underlain by buried ice that subsequently melted out to produce numerous kettles.

It is probable that the ridges of fluvioglacial deposits described above were laid down by meltwater streams, the courses of which were determined by the presence of ice. The material which makes up these ridges was probably derived from the very large medial moraines that occur on the surface of Breidamerkurjökull at the proximal ends of the esker systems. This material was carried by supraglacial, englacial or subglacial streams and deposited in these channels or tunnels. In 1945, only eskers 1, 2, 3 and 9 (Fig. 60) were exposed. By 1961, all the eskers were visible. Photogrammetric measurements with accuracies of ±2 m, made by Welch (Welch and Howarth, 1968) on photography taken in 1945, 1960, 1961 and 1965 indicate that eskers 4*b*, 5, 6, 7, 8 and

FIG. 63. Esker E4 (Fig. 60) from the medial moraine on Breidamerkurjökull.

9 have all been lowered with the passage of time. The range of altitudinal changes was between 2 and 13 m, the greatest amount of lowering, 13 m, being recorded at esker 5 between 1960 and 1961. During the same period the proximal end of esker 6 was lowered by 10 m. Such rapid lowering of the esker ridge without its destruction is quite remarkable and is much more rapid than the rates measured at the Casement Glacier (Price, 1966; Petrie and Price, 1966).

These measurements and the fact that ice was actually observed beneath eskers 4*b* and 5, and that esker 5 continued on to the glacier in 1966, strongly suggests that at least parts of the eskers in this area were deposited by supraglacial or englacial streams. The fact that irregularities in the sub-esker surface are often reflected in the profiles of the ridge crests supports the hypothesis that at least parts of these eskers were superimposed from an ice surface on to the subglacial topography. It may be that some parts of an esker ridge were deposited subglacially while other parts were subsequently let down by the melting out of ice which once formed the floors and sides of englacial tunnels or supraglacial channels.

In parts of eskers 4*b*, 7 and 9, the juxtaposition of kettle holes and gravel ridges raises the question of whether esker-like ridges can in fact be produced by the melting-out of buried ice from beneath a spread of fluvioglacial deposits. Particularly in the case of esker system 9, which merges into a sandur containing many kettle holes, it is difficult to decide whether a ridge separating a series of kettles is the product of the formation of

Fig. 64. Ice exposed in the side of esker E4 at Breidamerkurjökull.

the kettles or is a deposit of a meltwater stream confined between walls of ice. As will be demonstrated in the section on kettled-sandar, it is possible for esker-like ridges to be produced in the proglacial environment as a result of the melting of buried ice from beneath a spread of fluvioglacial deposits (Howarth, 1971). It therefore must be concluded that the form, disposition and internal characters of eskers, cannot reveal whether the initial deposition took place in a supraglacial, englacial, subglacial or proglacial environment.

Some authors have suggested that eskers can be produced in the proglacial environment but in close association with the ice margin. The simplest example is the marginal crevasse filling where a meltwater stream of the supraglacial, englacial or subglacial type discharges both meltwater and sediment into a re-entrant along the line of a crevasse at the ice margin (Fig. 67). The deposition takes place either within a channel

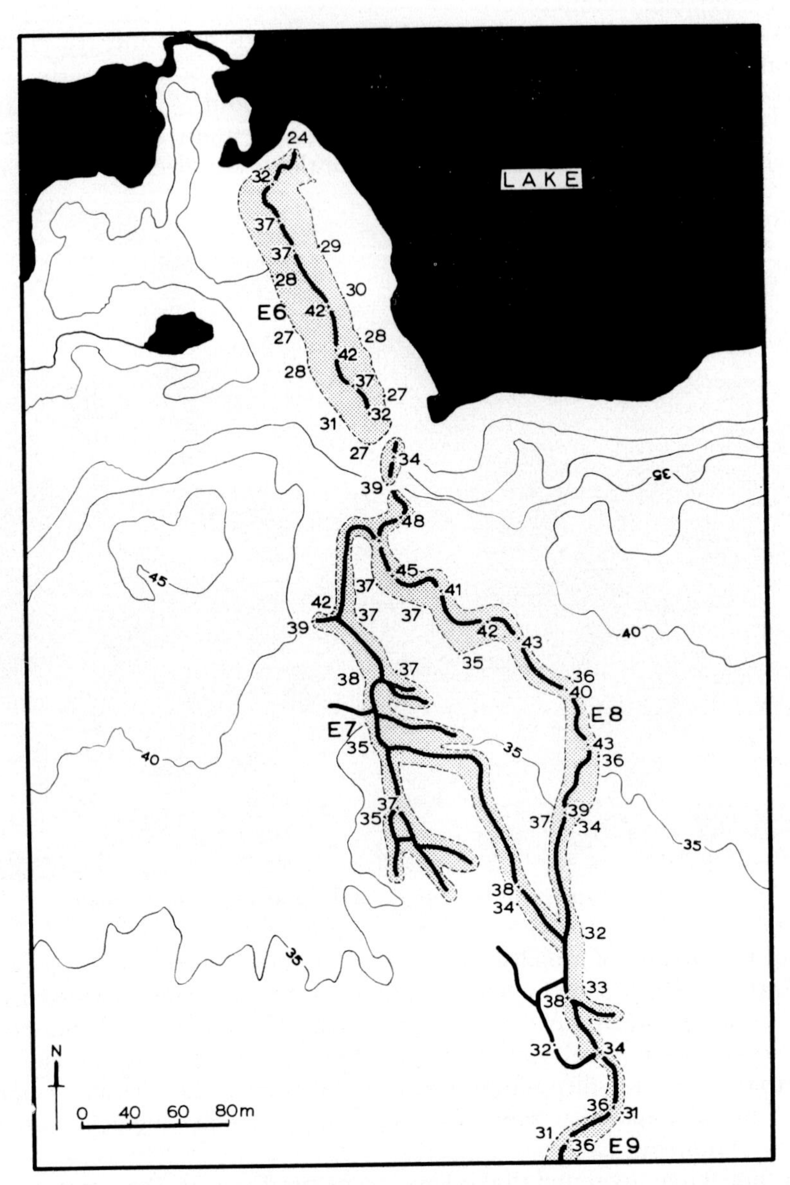

Fig. 65. Esker systems E6, E7, E8 and E9 (Fig. 60) at Breidamerkurjökull. Spot heights and contours in metres. (After Price, 1969.)

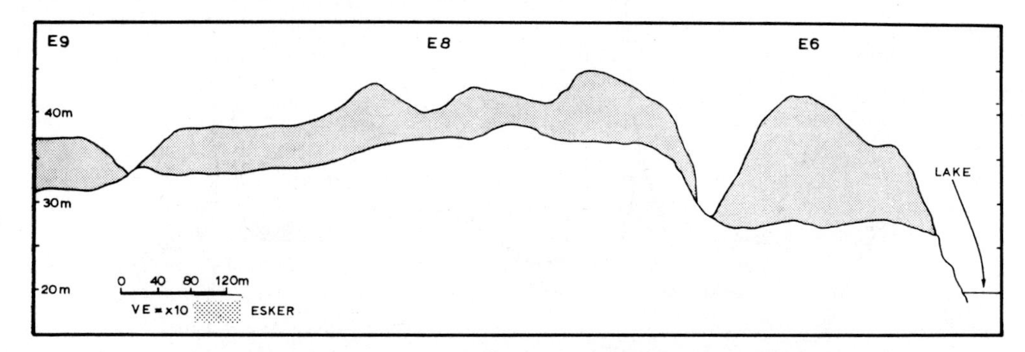

Fig. 66. Long-profile of eskers E6, E8 and E9 based on Fig. 65.

with ice forming the two sides or in a lacustrine environment if there is ponding of water in the proglacial area. In either case a ridge of fluvioglacial sediment will be produced when the ice walls melt away. The concept of esker formation in the proglacial/marginal environment was suggested by De Geer (1897), when he envisaged a series of fans being extended into a large ridge as the ice front retreated allowing growth of the deposit in line with the depositing stream. Trowbridge (1914, p. 217) stated this concept very simply: '. . . eskers are but kames drawn into long lines by slow retreat of the edge of the ice while kame deposition is in progress.' However, Giles (1918, p. 230), pointed out that, 'The constant association of eskers with terminal and recessional moraines indicates their formation at a time when the ice front was nearly or quite stationary building the moraine, not when it was in rapid retreat.' Such a view indicates a lack of appreciation of the time scale involved in the retreat of an ice margin and the rate of formation of both moraines and eskers. A far more serious problem associated with the proglacial/marginal hypothesis of esker formation is that of explaining why the meltwater streams that deposit the fans at the ice margin, either in a crevasse or in an ice marginal lake, do not destroy their own deposits by subsequent erosion.

In 1943, Hanson designed a laboratory experiment to simulate esker formation in a proglacial lake. He produced model eskers by discharging water and sediment from streams both above and below the water level in his proglacial lake. He therefore demonstrated the feasibility of the proglacial/marginal hypothesis of esker formation.

It appears from the above discussion that ridges of fluvioglacial deposits can be produced in a variety of ways:

1. Subglacial deposition in tunnels.

2. Subglacial deposition in caverns linked by tunnels and controlled by a water table.

3. Deposition in englacial tunnels and subsequent lowering on to the sub-ice surface.

4. Deposition in supraglacial channels and subsequent lowering on to the sub-ice surface.

5. Deposition in ice-walled re-entrants at the ice margin.

Fig. 67. Sediment transported by a supraglacial meltwater stream is deposited in a marginal crevasse that contains an extension of a proglacial lake. After deglaciation these sediments form a crevasse filling linked to a delta.

6. Deposition at the ice margin to produce a fan which becomes elongated as accumulation accompanies the retreat of the ice margin.

7. Deposition at the ice margin to produce a delta in standing water, the delta being elongated as the ice margin retreats.

8. Ridges of fluvioglacial deposits can be produced by the wastage of buried ice from beneath a spread of sand and gravel.

It is possible that any one esker ridge could be produced by the combination of several of the above mechanisms. It is almost impossible to determine which of the above mechanisms were responsible for any particular esker when the only evidence upon which such a determination can be based is the form, disposition and sedimentary characteristics of the esker.

Kames. The word *kame* (*cam* or *kaim*) is of Scottish origin and has two meanings: crooked and winding or steep-sided mound. In the literature of glacial geomorphology the word has been used either as a noun to refer to mounds or ridges of fluvioglacial deposits or as an adjective as in kame-terrace, kame-delta or kame-moraine. The confusion regarding the use of kame and esker to refer to either mounds or ridges of fluvioglacial deposits has already been discussed. In the previous section all ridges of fluvioglacial deposits were referred to as eskers. The most widely adopted definition of a kame is that of C. D. Holmes (1947, p. 248): 'A mound composed chiefly of gravel and sand, whose form has resulted from original deposition modified by a slumping incident to later melting of glacial ice against or upon which the deposit accumulated.' The actual process involved in the formation of such mounds of fluvioglacial deposits is not clearly understood. Cook (1946*a*,*b*), believed that some mounds were formed by deposition, in holes in stagnant ice, of masses of sand and gravel derived from the ice surface. The columns of sand and gravel would form a cone when the supporting ice walls melted (Fig. 68*a*). There is no reason why fluvioglacial deposits should not accumulate in englacial and supraglacial cavities (Fig. 68*b*) and be subsequently let down on to the sub-ice surface in a similar manner to eskers. Some of these deposits

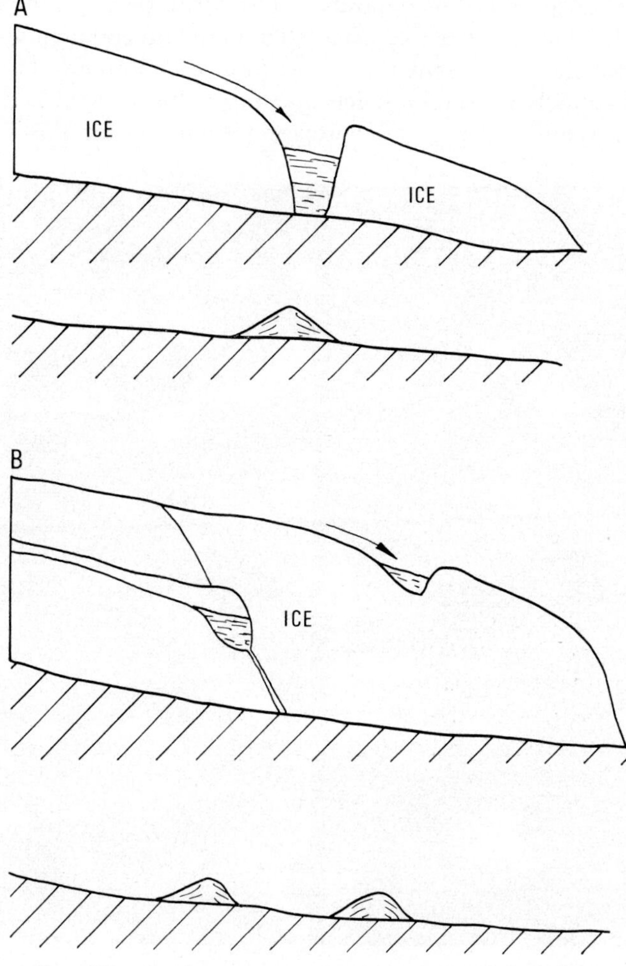

FIG. 68. *A*—A crevasse open to the ice-rock interface allows the development of a cavity in which sediments accumulate. *B*—Supraglacial and englacial cavities produce similar forms to the crevasse filling in *A*.

would have lacustrine characteristics because they were accumulated in pools of water in the cavities under, in or on the ice. Similar mounds could be produced by accumulation in cavities at the ice margin and they could be very similar to crevasse fillings or eskers, except that they are isolated mounds rather than ridges. Where a delta form rather than a more generalized mound form is produced the term delta-kame is applied.

Cook's papers (1946*a* and *b*) remain the most useful interpretation of the manner in which kames are produced. The penetration of stagnant ice by tubes and cavities filled with meltwater and sediment is carefully argued by Cook and he refers to descriptions of Alaskan glaciers with supraglacial lakes to support his hypothesis for the formation of what he describes as perforation deposits. The formation of individual mounds of sediment of considerable size (possibly up to 50 m high and 400 m in diameter) in association with the wastage of stagnant ice seems quite acceptable. The development of lakes, caves, and tunnels on stagnant ice masses has been described as 'glacial karst topography' by Clayton (1964). The process responsible for this topography is not

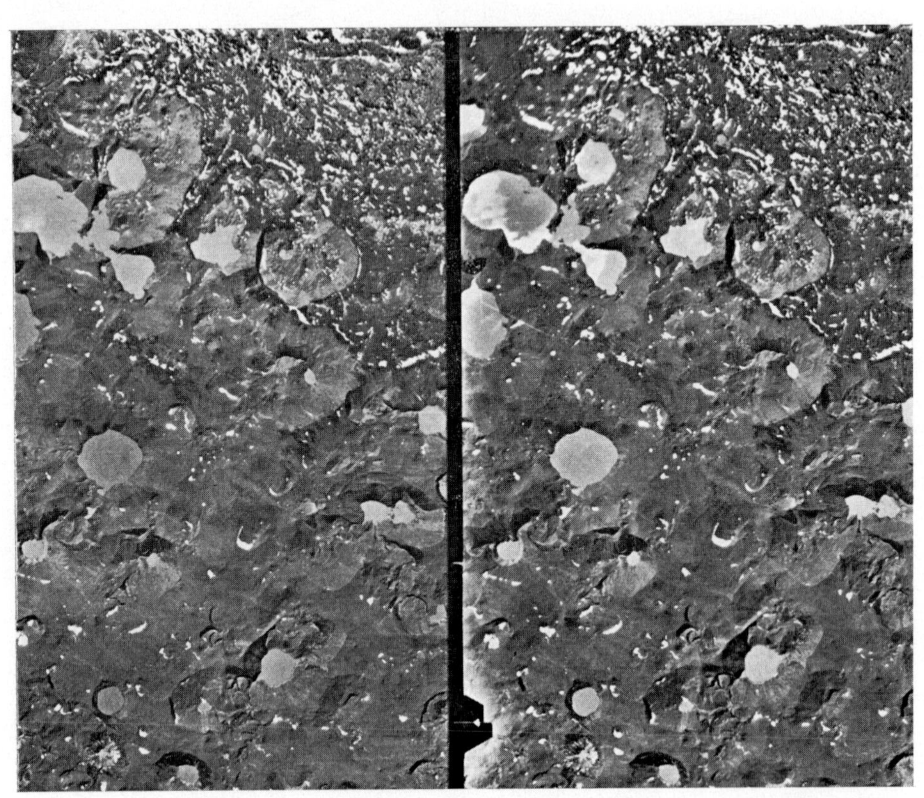

FIG. 69. Stereoscopic pair of air photographs of 'glacier karst' on stagnant part of Martin River Glacier, south-central Alaska. (Parts of photographs EEV-15-81 and 82, June 12, 1959, by U.S. Dept of Agriculture, Forest Service.)

solution as is the case in limestone areas, but melting. Clayton refers to the descriptions of glacial karst on the stagnant and drift-covered margin of the Malaspina Glacier by Tarr and Martin (1914, p. 205-28) and Russell (1901, p. 112-21), and then describes similar topography on the Martin River Glacier in south-central Alaska. The terminal, stagnant zone of this piedmont glacier is covered by a layer of ablation moraine that ranges in thickness from several centimetres near the centre to 3 m along the outer edge. Clayton states (p. 109), 'The most striking karst features on the Martin River Glacier are funnel-shaped sink-holes . . . most are 60 to 370 m across, and there are as many as thirty in a square kilometre. They are 15 to 90 m deep and many hold small lakes'. (Fig. 69.)

It appears, therefore, that not only single supraglacial depressions and shafts can develop in stagnant ice but also numerous cavities in which fluvioglacial sediments can accumulate. Cook (1946a) suggested that when such sediments are let down on to the sub-ice surface they form numerous mounds and that such a series of mounds should be called a kame-complex (Fig. 70). This is preferable to 'kame-moraine' which tends to imply a glacial rather than a fluvioglacial origin. Another term that has been used to refer to a number of mounds associated with depressions in fluvioglacial deposits is 'kame and kettle topography', the depressions or kettle holes representing the location

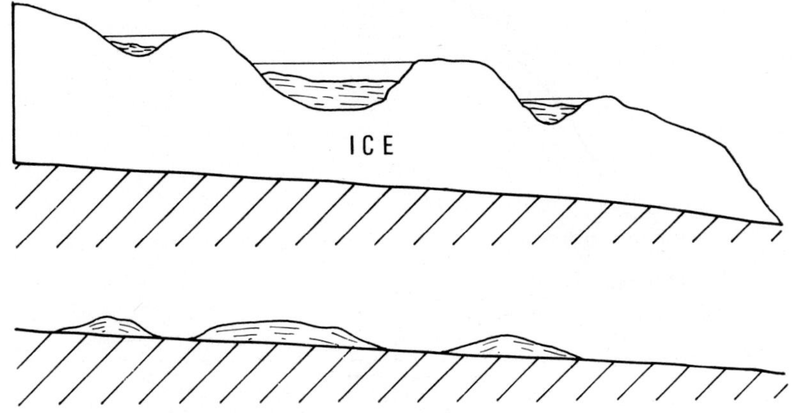

Fig. 70 The development of a kame complex after the deposition of fluvioglacial sediments in lakes on a downwasting ice surface.

of glacier ice upon or against which the sediments accumulated. When it can be established that the fluvioglacial sediments forming the mounds represent separate 'basins' of accumulation then the term kame-complex (or kame and kettle topography) should be applied. However, similar morphology can be produced by the accumulation of fluvioglacial sediments in the form of a sandur or delta on top of stagnant ice, the buried ice then melting out to produce a kettled-sandur or delta (see below).

Kame-terraces result from the accumulation of fluvioglacial sediments either in lateral or frontal marginal, lacustrine or fluvial environments (Fig. 71.). A flat-topped, steep-sided feature can result from deposition either in a marginal stream or narrow marginal lake. Since deposition occurs against the ice margin, when the ice wastes away, slumping occurs and a steep, crenulated slope develops to form the 'scarp face' of the terrace. Small ice blocks can be buried by the sediments forming the kame-terrace and when they melt out kettle holes develop on the terrace tread. McKenzie (1969) has studied an ice-cored kame-terrace in south-east Alaska and calculated that glacier ice beneath 4 m of gravel was being lowered by 24 cm per year. The survival of kame-terraces during and after deglaciation is largely a function of the angle of the slope on which the fluvioglacial sediments accumulated, the extent of buried ice, and whether or not extraglacial streams, or mass movement processes, destroy the terrace form.

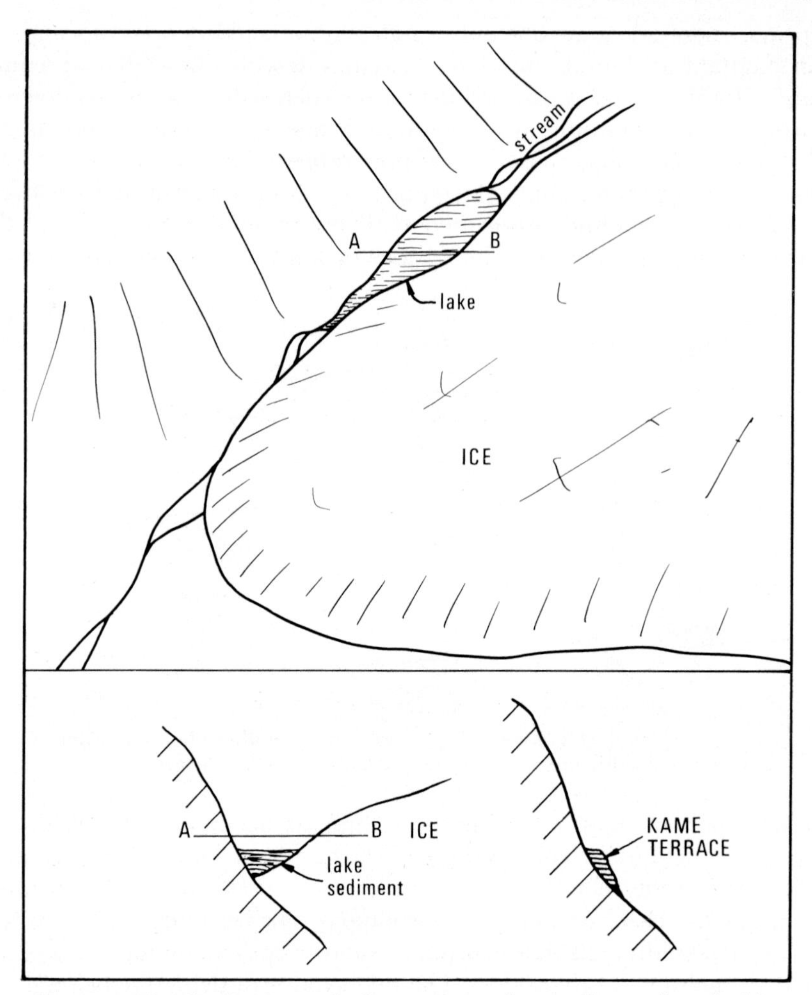

FIG. 71. The development of a kame terrace from sediment deposited in a marginal lake.

Kettled-sandar. These features are difficult to fit into any classification of fluvioglacial deposits. They are both proglacial and ice-contact deposits. They are proglacial in that they are the proximal parts of sandar that are, in the main, the products of proglacial anastomosing stream systems (see below). They are ice-contact deposits in that the fluvioglacial deposits are either laid down on top of ice or contain large blocks of ice. The eventual surface form of a kettled-sandur is the product of the modification of a sandur surface by the melting out of glacier ice leading to the development of kettle holes on the sandur surface.

The development of kettle holes as a result of the burial of stagnant ice by fluvioglacial deposits was discussed by Thwaites (1926) and Rich (1943). They used the term 'pitted-outwash' to describe the resultant form. The extensive development of

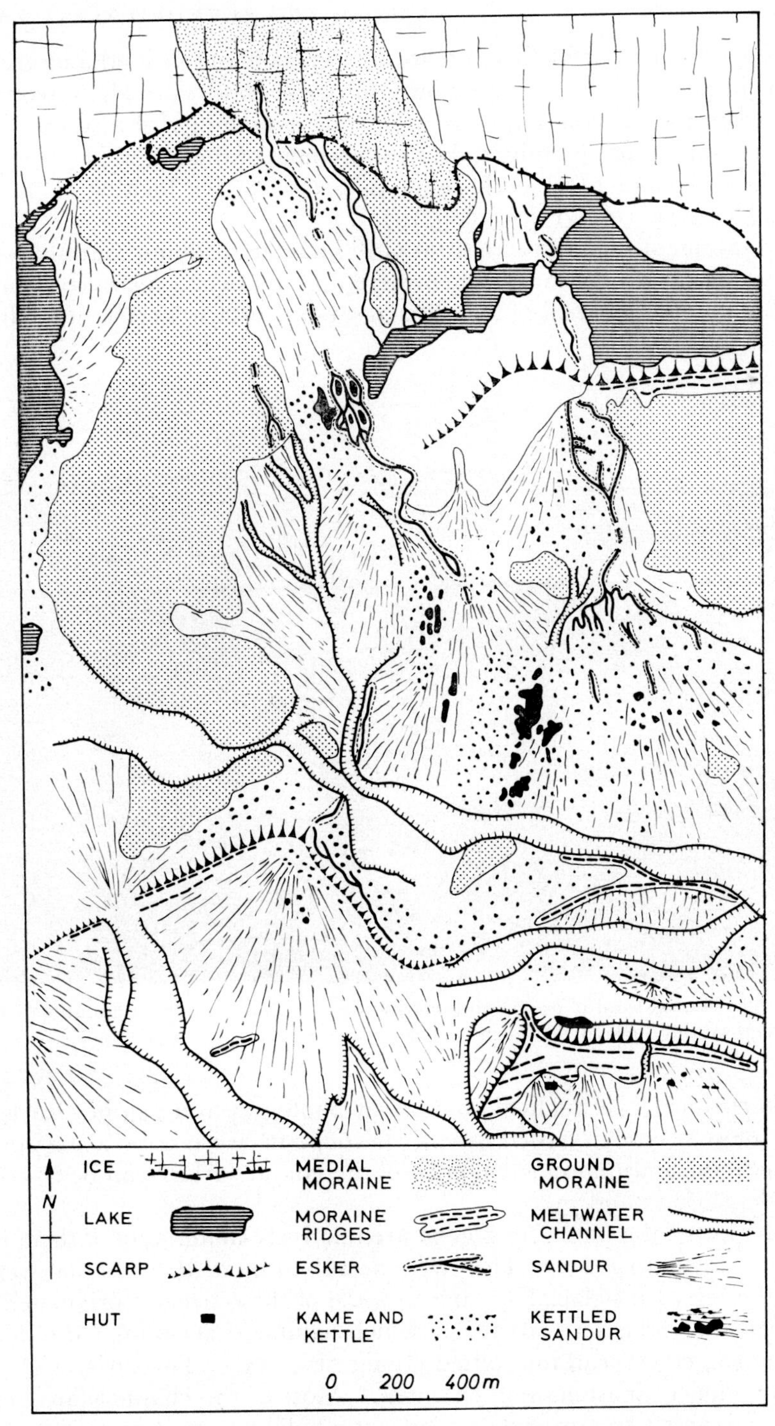

FIG. 72. Landforms in front of a part of Breidamerkurjökull, Iceland. (After Price, 1969.)

proglacial deposition in Iceland in the form of great plains of sand and gravel has led to the wide adoption of the Icelandic term for these features. The proximal parts of sandar are often pitted with kettle holes and the availability of a series of air photographs of some sandar has permitted the measurement of the lowering of the sandar surfaces and the development of kettle holes as a result of the melting of the buried ice (Welch and Howarth, 1968; Price, 1969, 1971).

Detailed studies of the sandar in front of Breidamerkurjökull, south-east Iceland by Price (1969), revealed that many of them were built up over a relatively short period of time and that quite large areas of the original sandar surfaces had been modified by the

FIG. 73. Pitted-sandur near Breidamerkurjökull. Note the flat remnants of the original sandur surface.

melting of buried ice. In the field, and on air photographs it can be clearly seen that even in the most severely 'kettled' areas, the deposits were once a part of a sandur surface because the anastomosing channels of the sandur surface can be traced between the kettle holes (Figs 72 and 73).

Although most of the sandar in Fig. 72 are relatively small, most of them have quite steep gradients (1:30 to 1:45). They were deposited by meltwaters that were moving away from the ice as proglacial streams. Some of these streams originated as supraglacial, englacial or subglacial streams in which sediments accumulated to form eskers. In some instances eskers lead into pitted sandar (Fig. 74). The surfaces of the sandar consist of abandoned anastomosing channels. Most of the channels are between 0·3 and 2·0 m deep, and 1 to 5 m wide. The kettle holes are most numerous, larger and deeper towards the proximal part of any sandur and although many are between 1 to 3 m deep and 3 to 10 m in diameter, some are up to 15 m deep and 100 m in diameter. The progressive destruction of sandar surfaces by the development of kettle holes can be seen by comparing aerial photographs of sandar. In 1945 sandur *A* (Fig. 75) contained very few kettle holes but by 1961 (Fig. 76) a large area of the proximal part of *A* had been destroyed by kettle development.

Detailed study of a sandur near the extreme western margin of Breidamerkurjökull (Price, 1971) also showed progressive destruction of a sandur surface between 1961 and

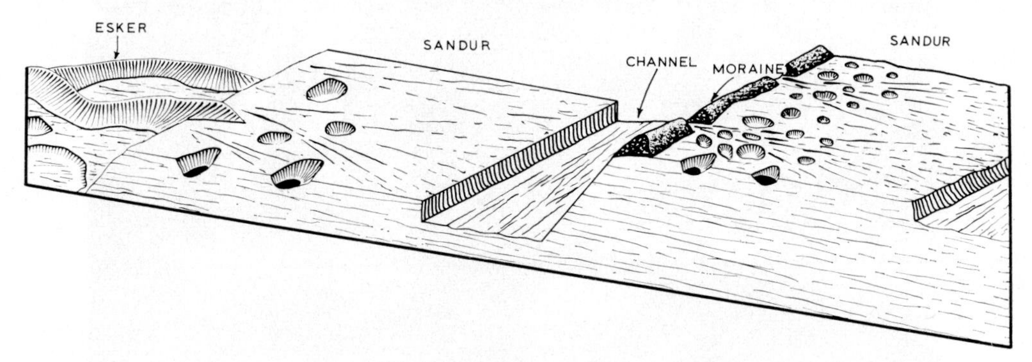

FIG. 74. An idealized block diagram of the relationship between sandur, pitted sandur, moraines and eskers. (After Price, 1969.)

1965 (Fig. 77). However, this particular sandur also received an addition of sediment to its surface over the period 1961-1965 associated with the drainage of an ice-dammed lake. Even so, it was demonstrated that kettle holes 2 to 7 m deep (Figs 78 and 79) developed between 1962 and 1965 and parts of the sandur surface were lowered by between 2 and 4 metres.

There can be no doubt that pitted-sandar surfaces can develop quite quickly after the deposition of fluvioglacial materials on top of ice. The main problem lies in explaining how quite extensive areas of ice became buried by the fluvioglacial deposits. Frodin (1954) has pointed out that individual streams on a sandur surface are not usually deep enough to carry large blocks of ice away from the ice margin, to allow their burial within the sandur sediments, except when sandur stream discharge is supplemented by catastrophic drainage of ice-dammed lakes. Such a catastrophic drainage was observed on Skeidarasandur in 1934 (Askelsson, 1936). On that occasion small ice blocks were carried over 20 km, but at 2 km from the ice edge ice blocks 54 m in circumference and 16 m high were observed on the sandur surface and sinking into it. Such catastrophic events can account for some kettled-sandar but many other sandar have developed on top of detached sheets or blocks of stagnant ice. If fluvioglacial deposits have ice beneath them, and if that ice was a part of the glacier at the time the deposits were laid down, then the stream that laid down the deposits must have originated in an englacial or supraglacial environment (Fig. 80). It is possible, however, for a subglacial stream to deposit material on top of detached stagnant ice if the outlet of

FIG. 75. Air photograph taken in 1945 of the area depicted in Fig. 72. (Photograph by U.S.A.F.)

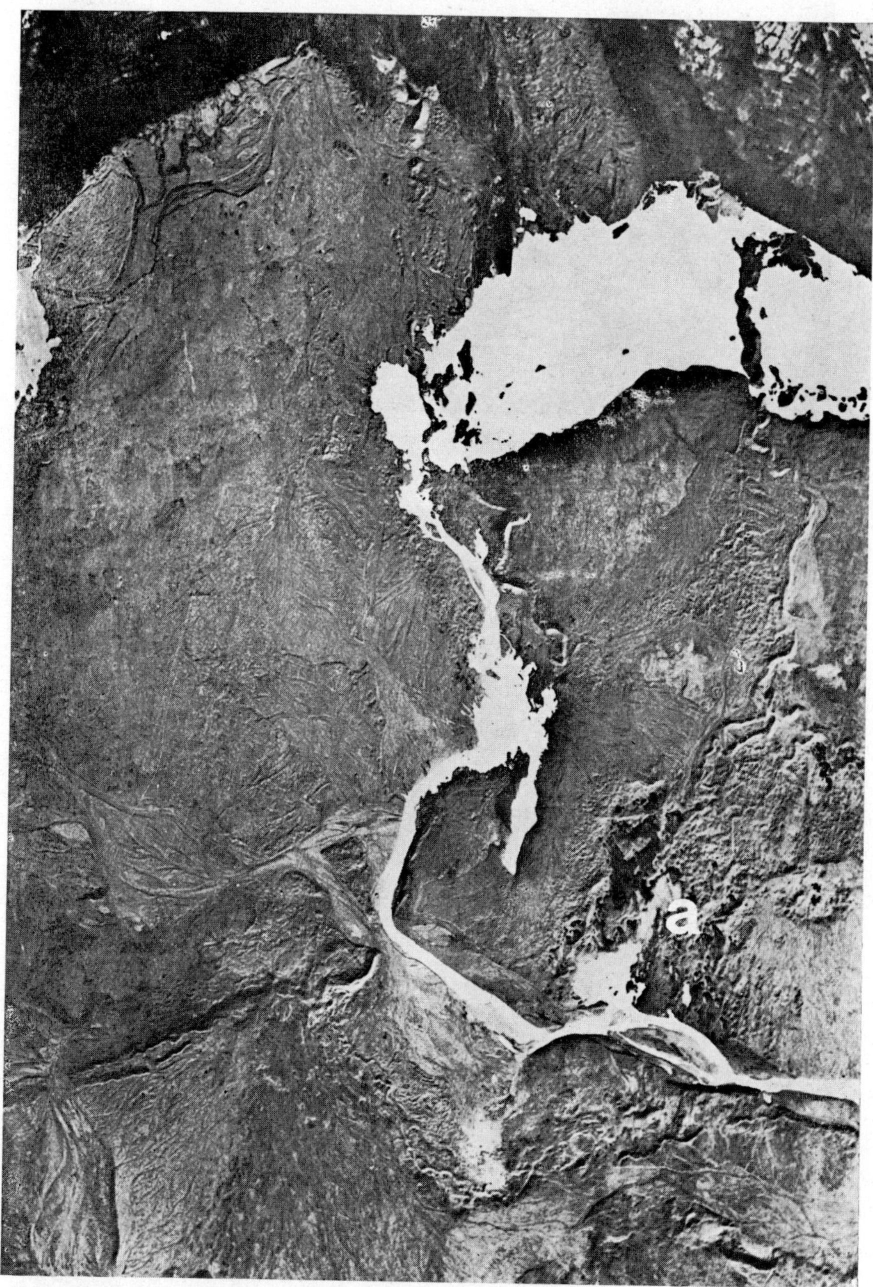

Fig. 76. Air photograph taken in 1961 of the area depicted in Figs 72 and 75. (Photograph by Landmaelingar Islands.)

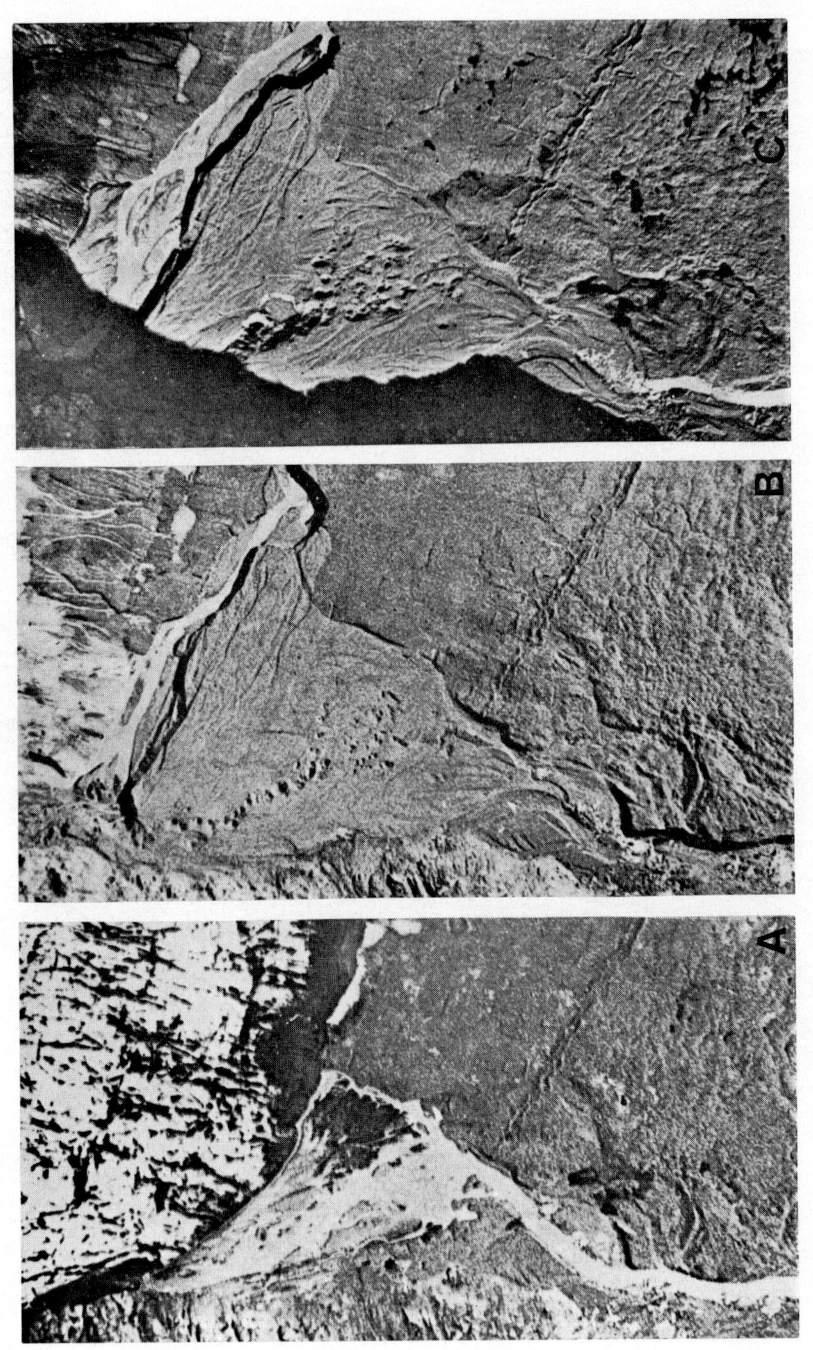

FIG. 77. Air photographs of a sandur at Breidamerkurjökull, taken in 1961 (*A*), 1964 (*B*), and 1965 (*C*). (After Price, 1971.)

FIG. 78. Glacier ice (black centre ground) beneath sandur gravels near Breidamerkurjökull, 1968. (After Price, 1971.)

the subglacial tunnel is at a higher altitude than the surface of the detached stagnant ice.

When remnants of the initial sandur surface survive the development of kettle holes, it is relatively easy to distinguish between a kettled-sandur and a kame complex. In some cases, however, the number of kettles is so great and the slumping of the steep sides of kettle holes so considerable that none of the original sandur surface survives and the end result is a series of hummocks and depressions in fluvioglacial material. In such cases it is impossible to distinguish between kettled-sandar and kame-complex.

PROGLACIAL FLUVIOGLACIAL DEPOSITS

The discharge of large volumes of meltwater from the margins of glaciers and ice sheets results in the transportation and deposition of sediment in various environments beyond the limit of the ice cover. The deposits accumulated within the channels, and by overbank flow associated with the channels, produce the fan-shaped accumulations known as sandar. When meltwater carried by proglacial streams discharges into lakes or into the sea, coarse sediments form beaches and deltas and fine sediments accumulate on the lake or sea floors.

Sandar. These extensive accumulations of gravel, sand and silt have been divided

FIG. 79. Kettle holes in a sandur at Breidamerkurjökull. Compare with Figs 77 and 78.

into two groups by Krigstrom (1962). Plain-sandar develop with very few constraints on their lateral extent and develop as a result of aggradation of a network of braided streams that produce fan-shaped features which may join and overlap to produce an extensive plain. Valley-sandar develop between valley walls and they usually only have one main channel that branches out to create a network of smaller channels. The network does not usually fill the whole valley floor at any one time but often winds back and forth across it. The development of a valley-sandur is closely related to the provision of meltwater and debris in a relatively constricted location. The constriction, usually in the form of valley walls, causes transportation and deposition to be concentrated in a relatively narrow zone. The slopes of these valley-sandar can be very steep or very gentle. In one section in front of the Emmons Glacier the White River sandur has a gradient of 1 in 8 (Fahnestock, 1963). Fahnestock's detailed study of this sandur revealed that its slope was related to particle size and discharge. There was a systematic decrease of 60 mm in median diameter of the sandur deposits over a distance of 1260 m. Discharge was essentially constant through the reach. Discharges of 200 to 500 cfs were capable of transporting all sizes of material present and therefore of being able to modify the form of the valley-sandur.

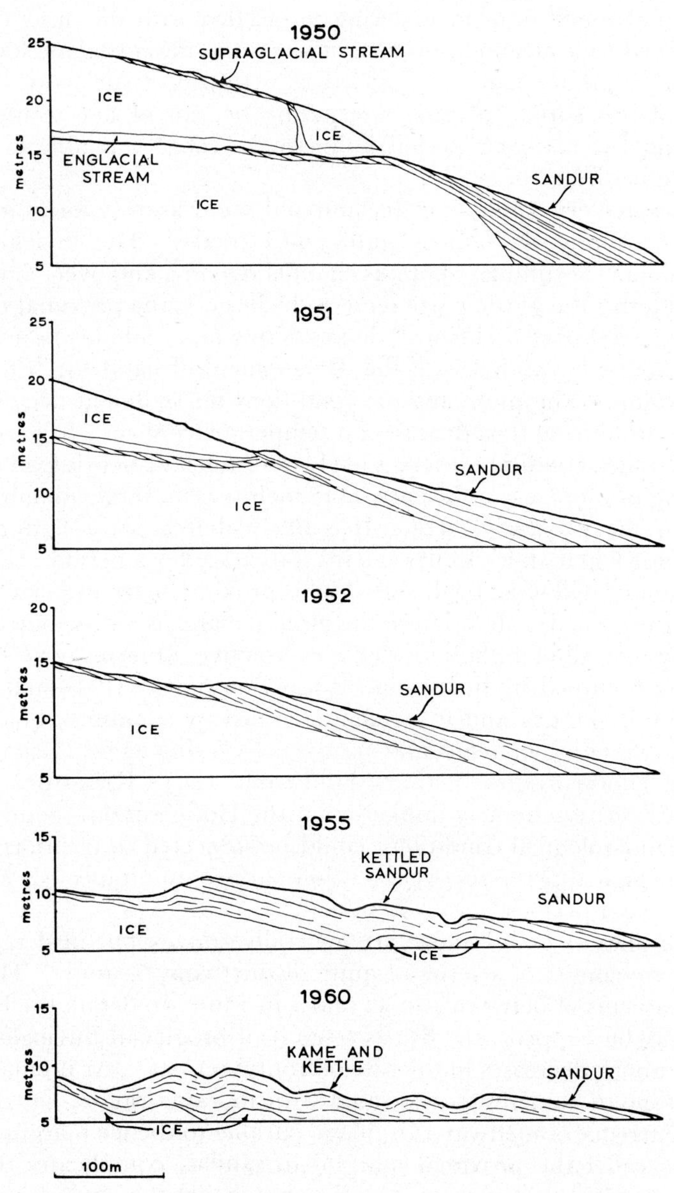

FIG. 80 The forms resulting from the deposits laid down by supraglacial and englacial streams on top of stagnant ice. Dates are hypothetical. (After Price, 1969.)

The meltwater stream associated with the White River sandur derived most of its load from morainic debris, mud-flow deposits and earlier valley-train deposits. The channel pattern changed from meandering to braided with the onset of high summer flows and returned to a meandering pattern when the flows became lower in autumn. The valley-sandur surface can be regarded as a transport surface with variations in discharge and debris supply causing alternating periods of net accumulation or net erosion. During one two-year period Fahnestock (1963) was able to measure a net gain to the sandur surface of 0·36 m.

Plain-sandar are very common in Iceland and occur as fossil forms in West Jutland, Poland, North Germany, the United States and Canada. The fluvioglacial sediments forming plain-sandar accumulate both as channel deposits and over-bank flow deposits. In very general terms the particle size tends to be large at the proximal part of a sandur and small in the distal part. However, lenses of fine material may be seen in the proximal part of a sandur associated with the development of bars within the channel network of the sandur. The most suitable conditions for sediment accumulation in the proglacial area are during the advance of a temperate ice sheet when erosion by the ice is prevalent. In such conditions there is likely to be a heavy debris load in the marginal zone and so long as there is a sufficiently long melt season, large amounts of debris will be washed into the proglacial area. It is likely that a plain-sandur will reach its maximum altitude when an ice front remains stationary for a period. During a retreat phase a plain-sandur will often be dissected in its proximal part and new sandar surfaces will develop as the eroded material from the proximal area is redistributed.

It is unlikely that all of a plain-sandur is ever active, in terms of all the channels on its surface being occupied by meltwater, at any one time. It is usual for quite large parts of a sandur to be dry and to be characterized by a maze of abandoned stream channels with intervening plateaus and terraces of varying ages. Using data provided by Ahlman and Thorarinsson (1939) and Hjulstrom (1954), Krigstrom (1962) deduced that (p. 342), '. . . it is extremely unlikely that the Hoffellsandur (south-east Iceland), under present climatological conditions, could be subjected to transformation through fluviatile morphological processes over its whole surface simultaneously even at the time of extreme high water periods.'

The cross profiles of plain-sandar are generally convex but in detail can be quite irregular and may consist of a series of quite distinct convex units. The long profiles often exhibit gradients of between 1 in 30 and 1 in 150. In detail, the long profile of a plain-sandur may be stepped (Fig. 81) as a result of proglacial fluvioglacial deposition being linked to minor decreases in the rate of frontal retreat. At Breidamerkurjökull a series of small sandar, sometimes associated with moraine ridges, has developed since 1890. The occurrence of meltwater drainage parallel to the ice margin and producing channels that parallel the proximal margins of sandar, complicates the relationship between the sandar and the ice margin. To what extent the north-facing slopes of the sandar (Fig. 81) are erosional and to what extent they are true ice-contact slopes is difficult to determine. If they are ice-contact slopes then it is likely that at some stage meltwater had to move up-hill from beneath the glacier to reach the upper surface of the sandur.

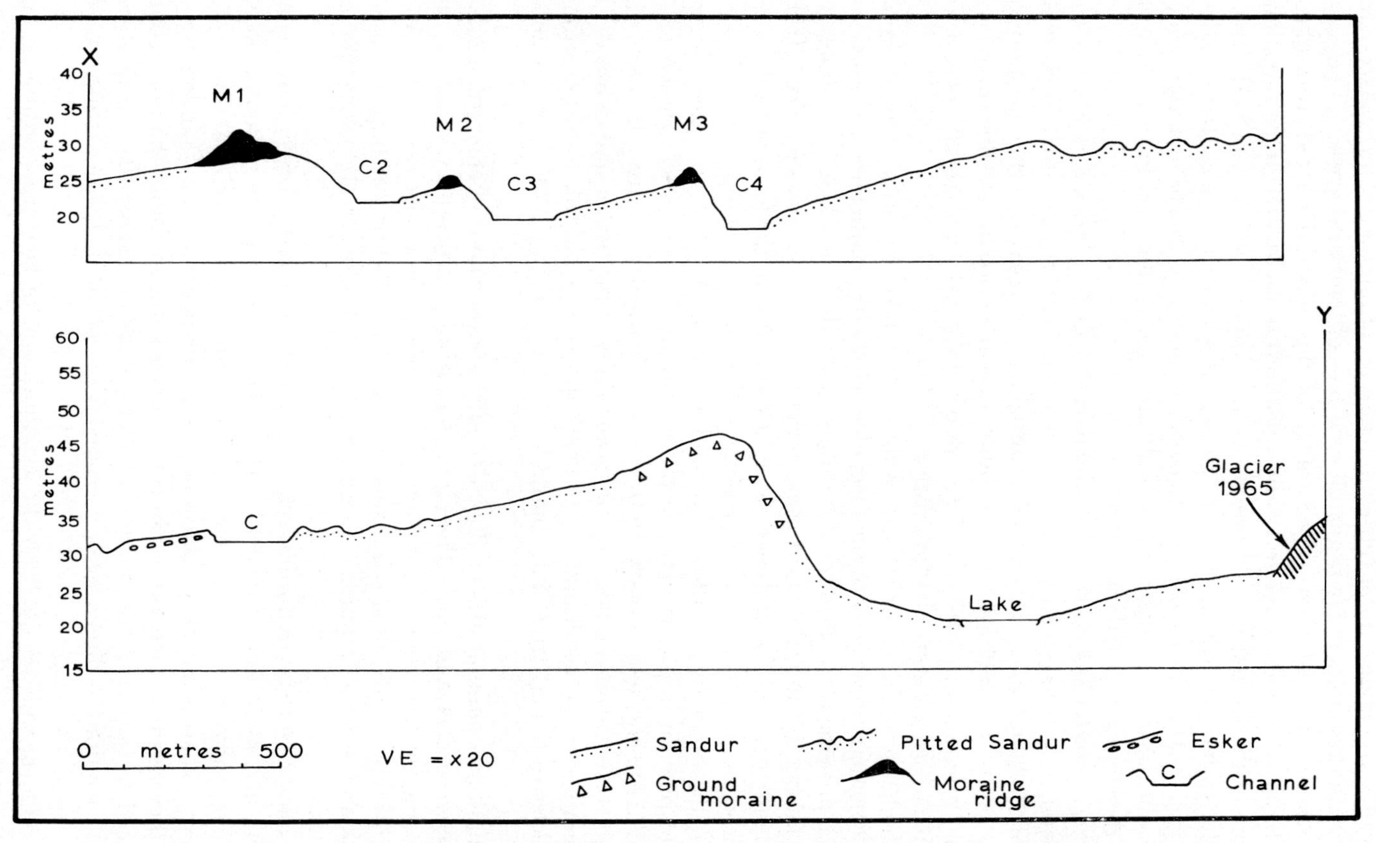

FIG. 81. A profile across the proglacial area at Breidamerkurjökull. Points x and y are indicated on Fig. 60. (After Price, 1969.)

Both valley and plain-sandar represent major accumulations of fluvioglacial deposits. In many ways they are similar to the alluvial fans of semi-arid areas. They represent the final resting place of a large percentage of the material eroded from glacierized drainage basins. The seasonal and daily variations in meltwater discharge, changes in the position of the ice margin, variations in the supply of debris and variations in the form of the surface on which they are deposited affect their size, shape and surface characteristics. It is possible for sediment accumulation to begin in a marine or lacustrine environment and for deltas to be built up to the water level in either case and for sandar to develop on top of these deltaic deposits.

Lake plains, beaches and deltas. The discharge of meltwaters into lakes is a common phenomenon. The lakes can be strictly marginal, with the ice itself forming at least one side of the lake basin, or they can be proglacial, occurring either in depressions in the drift deposits or in rock basins. Accumulation of fluvioglacial sediment in lacustrine environments tends to produce three groups of landforms, each group being associated with particular sedimentary characteristics.

The discharge of debris-laden meltwater into a lake usually results in the coarser fraction being deposited either in the form of a delta with characteristic top-set and fore-set bedding or being re-worked by wave action to produce beaches. The finer material is carried farther out into the lake basin to accumulate as bottom deposits. If the lake is subsequently drained the bottom deposits are revealed in the form of a lake plain.

The processes involved in the spreading of fluvioglacial sediment across lake basins have been discussed by Kindle (1930) and Kuenen (1951). Turbidity currents are believed to play an important part in the development of lake-floor deposits. Many factors influence the sedimentary character of lake-floor deposits. In particular the depth of the lake, variations both annually and daily of meltwater and debris discharge, the temperature of the water and the presence of icebergs capable of transporting large rock particles to the centre of lake basins will affect the character of the accumulated sediment.

Laminations, common in lake-floor deposits, are the result of changes in grain size, usually between clay and silt. They generally display differences in colour, the fine laminae being on the whole darker and thinner than the coarse laminae. Alternate layers of this type in lake-floor sediments are called 'rhythmites', a couplet consisting of one coarse and one fine lamina. In glacial lakes couplets are often between 0·5 and 5·0 cm thick.

The assumption that rhythmites reflect seasonal variations in amount and condition of sediment accumulation was used by De Geer (1912, 1940) in Scandinavia and by Antevs (1925, 1957) in the United States and Canada as a basis for establishing a 'varve' chronology. There is a considerable amount of evidence that many couplets are annual and that they can be called varves, but it is also possible for more than one couplet to be produced in one year and for no sediment at all to accumulate in another year because of variations in discharge or even diversions of the meltwater streams supplying the lakes.

The landform that results from the accumulation of lacustrine sediments and the subsequent draining of the lake or complete infilling of the lake basin by sediments is the lake plain. They may be of great extent and may be major landform elements in areas formerly glaciated. In North America the lake-floor deposits of Lake Agassiz and the Pleistocene extensions of the Great Lakes, and in Europe the floor deposits of the Ancylus Lake in Sweden are major landform units.

The lateral extent of lake-floor deposits reveals the size of the water-bodies in which they developed. The discharge of large meltwater rivers into these lakes produced deltas. Upham's (1896) study of Lake Agassiz contains descriptions of very large deltas that were built out into the lake. The Assiniboine delta covers an area of over 5000 km^2 and has a front face over 100 m high. The occurrence of both large and small deltas is a useful indicator of the former presence of glacial lakes.

The shore features of glacial lakes vary greatly depending upon the size of the lake, the length of time the lake existed, the nature of the material forming the sides of the lake and the supply of debris to the lake basin. Short-lived lakes may develop small depositional terraces that are destroyed within a very short period of time. Large lakes may develop major beaches with storm ridges, and both wave-cut platforms and cliff lines may develop, at least in drift deposits if not in solid rock. The mechanics of beach formation and general shoreline development around lakes are the same as in similar marine environments.

Glacio-marine and estuarine sediments. The discharge of meltwater streams into fiords, bays and estuaries results in the accumulation of fluvioglacial sediments in both marine and estuarine environments. These sediments only constitute landforms of interest to the geomorphologist if, after the sediment has accumulated to produce bottom deposits, deltas or beaches, there are changes in sea-level that result in these sediments appearing as parts of a land surface. Such sea-level changes are common in areas that have been glaciated because of the isostatic adjustments that take place during and after deglaciation. It is beyond the scope of this book to discuss the interpretation of raised beaches, raised deltas and mud flats (see Sissons, 1967; Andrews, 1970). The recognition of the marine environment of deposition for these features is usually dependent upon the occurrence of marine as opposed to fresh water organic materials. The forms of the raised mud flats, raised beaches and deltas are similar to their lacustrine equivalents.

The geomorphologist who is interested in understanding the genesis of landforms that consist of fluvioglacial deposits is heavily dependent on workers in allied fields. The sedimentary characteristics of the deposits and the glaciological environments with which the deposits were associated are of fundamental importance. The forms that these deposits exhibit certainly suggest their genesis in general terms, but without examination of their constituent materials and interpretation of the environment of accumulation, little progress can be made to a full understanding of their origins.

VII MODELS OF GLACIATION

INTRODUCTION

A MODEL can be defined as '. . . a simplified structuring of reality which presents supposedly significant features or relationship in a generalized form.' (Hagget and Chorley, 1967, p. 22.) Models are therefore selective approximations of reality and represent a synthesis of data. Descriptive models involve a stylistic description of reality whereas normative models are means of indicating what might be expected to occur under certain stated conditions. Much of what has been written in the previous six chapters of this book has involved descriptive models, in association with historical and spatial analogies. On numerous occasions it has been assumed that the key to the understanding of Pleistocene glacial landforms is to be found in areas of existing glaciers and it has also been assumed that geomorphological relationships established in one area can be applied to other areas, at least in general terms. The previous chapters have also involved the division of the 'glacial system' into its component parts so that the operation of each part could be examined and descriptive models developed. It now remains to reconstitute the individual descriptive models of the earlier chapters into a general systems approach. Chorley (1967, p. 77), defines a geomorphic 'system' as, ' . . . an integrated complex of landforms which operate together according to some discernible pattern (e.g. a drainage basin); energy and matter input into the system giving rise to a predictable system response in terms of internal organization and the resulting energy and matter output.' Comparable with the drainage basin cited by Chorley is a glacierized drainage basin; the input is in the form of snow that is subsequently transformed to ice, and the output is expressed in terms of the modification of the drainage basin by the ice and its associated meltwaters, either by erosion or deposition. An individual glacierized drainage basin may in fact be part of a 'super-system' involving a large area of mountain and valley glaciation or an ice sheet. Much of the material in earlier chapters has dealt with sub-systems, whereas this and the next chapter will take a much more comprehensive view and examine the glacial system as a whole, firstly during the development of glaciation and secondly during deglaciation.

Models have been used by geomorphologists for very many years. Chorley (1967) has pointed out that many of the important developments in geomorphology during the nineteenth and early twentieth centuries used a 'black box' approach to general systems model building. This early work was characterized by large intuitive decisions involving speculation about many of the processes involved in the system. Chorley (1967, p.

85) goes on to point out that '. . . the lack of such information has not been a disadvantage, its very deficiency allowing the theoretical model builders much greater scope than they would otherwise have had.' Such a situation enabled W. M. Davis (1900) to present a model for glacial erosion and to suggest the possibility of a cycle of glacial denudation comparable to the cycle of denudation associated with fluvial activity.

The early models of glaciation reflect an approach that many geomorphologists have adopted in terms of the significance of climatic controls in determining landform development. The concept that each climatic regime produces characteristic geomorphic processes and therefore distinctive landform assemblages regardless of structural and lithological controls has been outlined by such workers as Peltier (1950), Birot (1960), Budel (1963) and Holzner and Weaver (1965). Glaciation was regarded as a 'Climatic Accident' by Davis (1899) and Cotton (1942), in that it represented a departure from the normal processes of landform development under fluvial activity. In that glaciation certainly reflects a change in climatic conditions that causes a new system to develop involving the throughput of ice and water rather than only water, a climatic approach to the interpretation of assemblages of glacial landforms is certainly a valid one.

The developments, during the past twenty years, in our understanding of geomorphological processes in glacierized environments allows the construction of more useful models of glaciation and deglaciation. However, there still remain great gaps in our understanding of the glacial system and the models suggested below will need further modification as more data are obtained. As Bambrough (1964, p. 98) has pointed out, 'The ideal limiting case of representation is reduplication, and a duplicate is too true to be useful. Anything that falls short of the ideal limit of reduplication is too useful to be altogether true'. There is no danger of reduplication in the models suggested below. The data upon which the models are based are a very small sample, often far from random. Their imperfections not only reflect the limitations of the data but the conceptual limitations of the model builder. I cannot escape from the fact that my own field experience is limited to a very small selection of present-day glacierized areas, all of which contain ice at the pressure melting point and all of which are undergoing rapid wastage in temperate maritime conditions. I cannot ignore the fact that in conceptual terms I tend to think of the former British ice sheets rather than the great Laurentide or Scandinavian ice sheets. Despite my awareness of these limitations for which allowances have been made in the pages that follow, the reader should bear in mind the conceptual framework that will inevitably affect the writer's attempts at producing models of glaciation and deglaciation.

TYPES OF GLACIATION

The standard approach to the classification of types of glaciation is two-fold. When the pre-glacial relief maintains an element of control over the extent of the ice and its direction of movement, valley glaciers and therefore valley glaciation are the end-product. When the ice mass completely buries the pre-glacial relief and the extent of the ice mass and the direction of movement of the ice is largely uncontrolled by the underlying relief then an ice cap or ice sheet is the dominant type of glaciation. This dual approach is somewhat misleading. In many instances the two types of glaciation are sequential in that an ice sheet can develop from a series of valley glaciers that over-top their divides and ice sheets can also degenerate into a series of valley glaciers. Local ice caps or ice sheets can also feed 'outlet' valley glaciers and valley glaciers that become piedmont glaciers can have many of the characteristics of ice sheets if the piedmont lobes are large enough.

There are two groups of factors that determine the type of glaciation that may occur in any area. The first group of factors can be called morphological and include all those that determine the form of the land surface upon which the ice mass develops. The term 'pre-glacial relief' must be treated with care, because in areas that have experienced multiple glaciation the form of the surface upon which ice masses develop after the initial glaciation will be determined, not only by non-glacial processes, but also by glacial processes. For all practical purposes land areas upon which the Pleistocene glaciers and ice sheets developed were the product of the interaction of fluvial and mass-wasting processes on varying lithologies and structures. In the last resort the type of morphology upon which the Pleistocene ice masses developed largely reflected the structural history of the regions in which glaciation occurred. In those areas in which Tertiary mountain building or uplift of older mountains had taken place, mountain and valley morphology was dominant. In other areas, where relatively long time periods of structural stability had occurred, lowland plains were the sites for ice sheet development. In terms of macroforms, the structural history and the variation in lithology were primarily responsible for the form of the surface upon which the ice masses developed. The geomorphological work of the ice that subsequently occupied these areas resulted in a modification of their 'pre-glacial' form and only to a limited degree did they create totally new surface forms on a macro scale. No one, for example, has argued that complete mountain ranges have been removed by glacial processes but it has been widely demonstrated that 'pre-glacial' drainage divides have been breached and that valley cross-sectional shapes and valley heads have been extensively modified by glacial action.

The second group of factors that determine the type of glaciation can be referred to as climatological. Some workers may prefer to refer to this group as glaciological rather than climatological but the glaciological characteristics of any ice mass are largely a function of the climatological conditions that prevailed during its period of development and existence. Glaciers and ice sheets develop in situations where snow accumulation exceeds ablation and such situations are primarily determined by temperatures which, in the first instance, must be low enough to allow precipitation to occur in the form of snow and secondly, must have a seasonal range that allows snow to accumulate from one year to the next. The transformation of snow to glacier ice and the glaciological characteristics of valley glaciers and ice sheets have already been dealt with in Chapter II. It is now only necessary to point out how the various glaciological attributes of both valley glaciers and ice sheets affect the development of landforms and deposits within the glacial system. The glaciological attributes of any ice mass largely reflect the climatological conditions under which it developed, although the form of the surface upon which it developed also plays a part. The combination of climatological and morphological factors conducive to snow accumulation and therefore to glacier formation occurs above the annual snow-line either on plateau surfaces or in valley heads.

In the remainder of this chapter and in the whole of Chapter VIII, the climatological factors which determine the type of glaciation that occurs must be regarded as background information. The glaciologists have developed their own models that primarily relate climatic conditions to the build up, movement and wastage of ice masses. The geomorphological models to be constructed below must include data derived from the models developed by glaciologists. The mechanics and rate of ice movement, absolute changes and rate of change in volume and therefore in thickness and extent of ice masses, the temperature of the ice, and the amount and location of meltwater in an ice mass are all of fundamental importance in the interpretation of glacial and fluvioglacial landforms and deposits.

VALLEY GLACIATION

In constructing a model to account for the landforms and deposits that develop as a result of the glaciation of mountain and valley topography, it is assumed at the outset that the area has considerable local relief, that the pre-glacial valley has a fairly uniform gradient, except in the region of the valley head, and that the lithology and structure are uniform. The climate of the area, which previously was moist and mild, begins to deteriorate and snow accumulation begins in the valley head (Fig. 82*a*). The rate of growth of this snow bank and the subsequent development of a small glacier occupying the valley head is a function of the rate of deterioration of the climate and the net accumulation of snow. If annual precipitation amounts are high and there is considerable ablation during the summer, the transformation of snow to ice will be rapid. With a falling snow-line it is likely that a snow bank in the valley head will grow both in extent and thickness and a cirque glacier (Fig. 82*b*) will develop in hundreds rather than thousands of years.

The freeze–thaw weathering and nivation processes associated with net snow accumulation are not the concern of this book, except that they provide a great deal of weathered rock material beneath and around the new ice mass. Once the ice mass becomes so large that movement begins to take place within it, then the term 'valley glacier' can be applied (Fig. 82*c*). In the early stages of its development the shape of the valley glacier is determined by the form of the valley in which it is developing. In general terms the transformation of snow to ice will take place first at the highest altitudes and with a progressively falling snow-line the glacier will extend farther down-valley. Once the glacier has attained a length of some thousands of metres and a thickness of some hundreds of metres it is possible to recognize an area on the glacier surface that experiences net accumulation and an area that experiences net ablation. The mass balance of the glacier will then determine whether it will continue to extend down-valley or retreat. The glacier experiencing a positive mass balance with net accumulation exceeding net ablation will become thicker and the frontal margin will advance. The

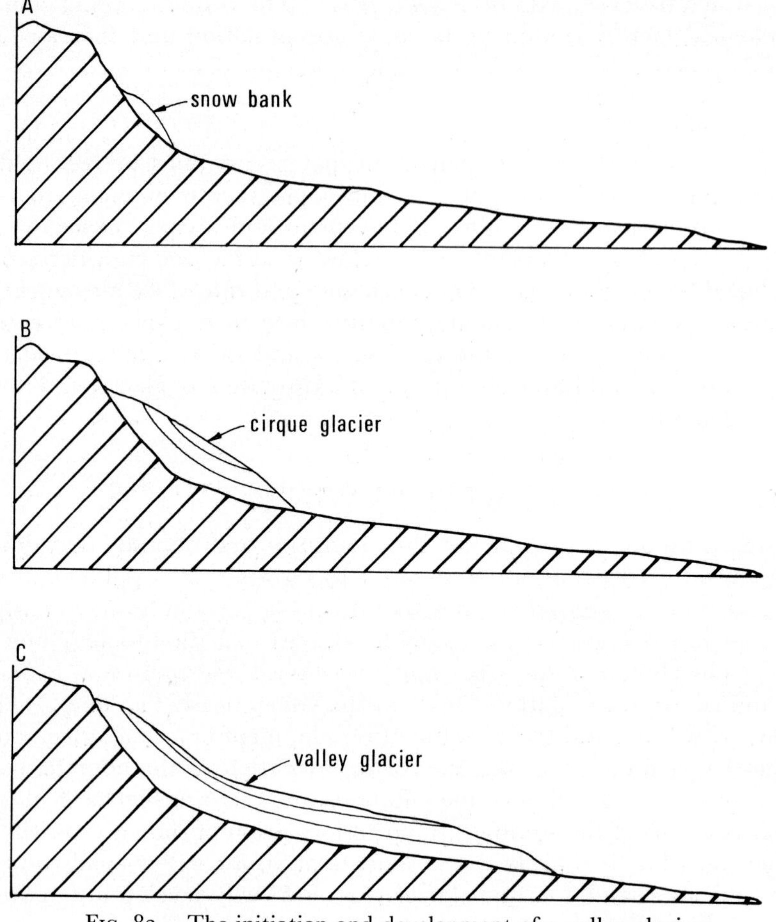

FIG. 82. The initiation and development of a valley glacier.

rate of advance of the fronts of small valley glaciers can be of the order of 5 m to 100 m per year.

Once a valley glacier has become established in the valley head then the normal processes of glacial and fluvioglacial erosion and deposition begin to affect the area covered by the glacier and the area in front of it. The modification of the form of the valley head by the ice mass occupying it results primarily from movement within the ice mass. Nivation processes may well play an important part in preparing the rock surface for glacial erosion, but it is the movement of ice that is responsible for the removal of material from one place to another. The tendency to erode the valley head is related to the flow-lines in the glacier, so that maximum abrasion occurs some distance back from the ice front and minimum abrasion occurs at the ice front. Little or no ice movement occurs at the ice front and material is piled up beneath the ice; this tends to accentuate the basin form of the valley head. The development of semicircular, steep-sided and sometimes over-deepened depressions in valley heads by glaciation is a very characteristic feature of mountain glaciation. The resultant forms are known as cirques and can range from small shallow depressions to very large features, thousands of metres in width and depth. The mechanisms of cirque formation are far from fully understood (Lewis, 1960) but they certainly represent the early phase of mountain glaciation.

The existence of a cirque glacier in a valley head will change the hydrological characteristics of the streams flowing through the rest of the valley. The erosion by the cirque glacier that produces the steep sides and back wall of the cirque provides debris to be transported by the glacier to the frontal margin. This debris is then available for transport by meltwater streams during the ablation season. The cirque glacier acts as a reservoir during the winter, storing precipitation in the solid form. During the summer much higher discharges than during the rest of the year will be produced by ablation and large amounts of debris will be transported and deposited in the valley below the glacier. Any moraine ridges or fluvioglacial deposits that are laid down at the margin of, or in front of, the cirque glacier will be overridden if the cirque glacier continues to have a positive mass balance and to extend down-valley.

If climatic deterioration continues and the annual snow-line descends to lower and lower altitudes then the cirque glacier will expand and develop into a true valley glacier (Fig. 82). In the simple model under discussion the most important geomorphological implications of the extension of a cirque glacier into a valley glacier are the greater area covered by the ice and the increased volume of the ice. Unlike a river channel, which only occupies a very small part of the cross section of a valley, an ice channel occupies a very much larger percentage of the valley cross section. As the ice moves through the valley it is capable of eroding, transporting and depositing at all points of contact between ice and valley sides and floor. The zone of maximum rates of ice movement will, in an ideal glacier, occur approximately at the annual snow-line (the equilibrium line). As the cirque glacier develops into a valley glacier the equilibrium line descends down-valley and therefore the zone of maximum ice movement also proceeds down-valley. Other things being equal, the zone of maximum ice movement can be equated with the zone of maximum glacial erosion so it can be stated that maximum erosion during any one glaciation will occur above the lowest point in a valley reached by the equilibrium line. Above that point glacial erosion will be dominant whereas below it glacial deposition will be dominant. However, it must be remembered that during the retreat phase of a valley glaciation the equilibrium line will proceed up-valley and parts of the valley that had previously experienced erosion will subsequently experience deposition.

A simple model involving the extension of a cirque-glacier into a valley glacier occupying a single valley of uniform gradient produced by fluvial activity on rocks of uniform structure and lithology is a very unsatisfactory generalization. All valley glaciers occupy valleys that are far from regular either in long profile or cross profile. Above the lowest altitude reached by the equilibrium line on a valley glacier there is a tendency for glacial erosion to remove protuberances both on the valley sides and the valley floor. The scratching, scraping, quarrying and plucking of the ice and its included debris result in erosion ranging from minor scratches to the removal of major spurs and rock knobs. The net result of glacial erosion is a reduction in the irregularities both in the cross profile and the long profile of a glaciated valley with one exception. The exception is of course, the development of basins on the floors of glaciated valleys. The over-deepening of cirques to produce rock-cut basins at valley heads has been attributed to rotational slip-line fields of movement in cirque glaciers (Lewis, 1960). The theory of slip-line fields has been applied to relative rates of erosion beneath a valley glacier (Nye and Martin, 1967) and it is demonstrated that the curvature of the bed places a restriction on the relative rates of erosion occurring at different points on the bed. It appears that there is a complex relationship between the surface slope, velocity distribution, structural planes of weakness and the ice-rock interface. The occurrence of rock basins reflects the development of slip-line fields within the glacier which in turn may reflect the pre-glacial shape of the ice-rock interface. It seems likely that glacial erosion may well accentuate irregularities in the surface that existed prior to

FIG. 83. Lateral and basal till, Findelen Glacier, Switzerland.

glaciation. However, in general terms it is the straightness and steep-sided, flat-floored character of glaciated valleys that are the main results of glacial erosion by valley glaciers.

The movement of the ice forming a valley glacier results in the transportation of eroded material down-valley. If the ice is at the pressure melting point throughout then a certain amount of glacial deposition is likely to occur beneath the ice in the ablation zone. Particles entrained in the basal layers of the ice may melt out and a compact subglacial till will occur on the floor and sides of the valley (Fig. 83). The subglacial till may merge into lateral moraines that have developed at the lateral margins of the glacier. Lateral moraines consist either of material that has originated farther up-valley and has been transported and dumped by the glacier, or of weathered material descending the valley sides and being lodged at the glacier margin. Lateral moraines mainly develop below the equilibrium line of a valley glacier where the ice surface tends to be convex.

Glacial deposits at the frontal margin of a valley glacier tend to be derived from three environments. Material being transported on the glacier surface largely results from the merging of two or more lateral moraines when tributary glaciers join the trunk valley glacier. This is an obvious complication to the simple model of a single valley glacier so far discussed. Other material arrives at the terminus either by englacial transport, having been derived from protruding rock knobs, or by transport along flow-lines that tend to curve up towards the surface in the terminal zone. No matter by what method this material arrives on the glacier surface there is a tendency for sliding of the material to take place down the relatively steep ice front, at least during the ablation season. If the frontal ice margin remains stationary for a reasonable period a ridge of glacial deposits will accumulate at the ice front. If, however, the valley glacier is advancing, this material is usually overridden and becomes a part of the subglacial till or it is transported farther down-valley by the glacier. It is also possible for an advancing ice front to act as a 'bull-dozer' and to build up moraine ridges consisting both of material derived from the ice surface and of fluvioglacial materials deposited in the proglacial environment.

During the development of a 'temperate' valley glacier meltwater plays an important part in the conversion of snow to glacier ice. Once a major valley glacier has been established, meltwater may be quite prolific in the ablation area during the summer. Depending on the thickness of the glacier the meltwater may or may not penetrate to the base of the valley. Fluvioglacial erosion often occurs at and below the lateral margins of valley glaciers and suites of marginal, submarginal and subglacial meltwater channels witness the work of meltwater below the equilibrium line of valley glaciers. At the frontal margin of a valley glacier one or more large meltwater streams is usually seen issuing from beneath the ice; this is evidence that the ice, at least in the lower parts of the glacier is at the pressure melting point. The streams are of considerable importance from a geomorphological point of view because they tend to destroy ground moraine and moraine ridges that develop on the valley floors. They are also very important constructional agents in themselves in that they produce valley-sandar by depositing large amounts of fluvioglacial material along the floor of the valley.

The sedimentary sequence produced by an advancing valley glacier is as follows (Fig. 84). In the valley head, periglacial deposits may be succeeded directly by till, but in the middle and lower parts of the valley the basal unit consists of fluvioglacial deposits resulting from deposition by meltwaters derived during the summer from the glacier occupying the upper part of the valley. As the glacier front moves down-valley the former proglacial fluvioglacial deposits are overridden by the ice and unless they are completely removed by glacial erosion the sequence is fluvioglacial sands and gravels

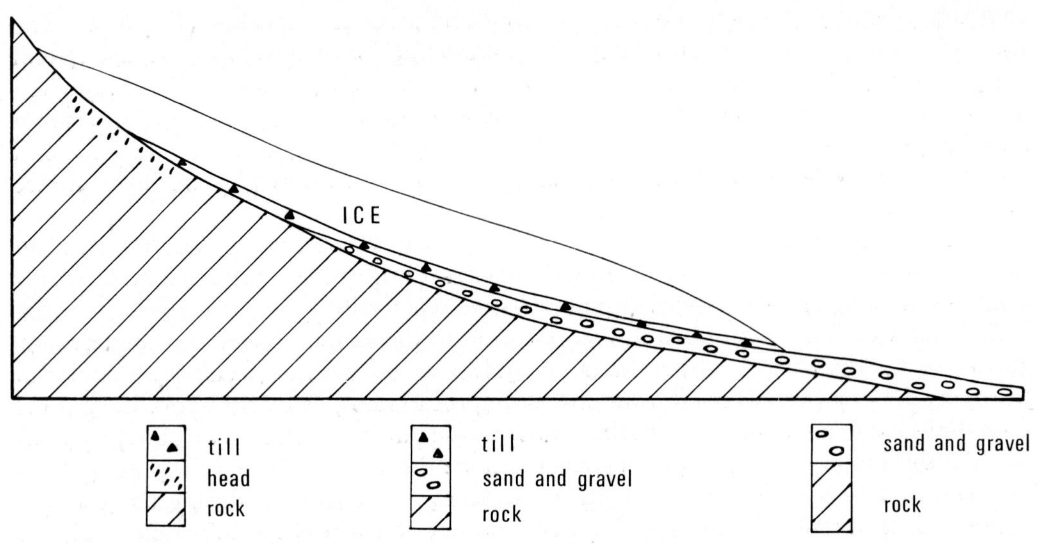

Fig. 84. Stratigraphic sequences associated with an advancing valley glacier.

beneath basal till. Such a sequence is only likely to occur below the lowest altitude reached by the equilibrium line because above that altitude glacial erosion will predominate during the extension of the valley glacier.

Throughout the above discussion a very simple model of the development of a single valley glacier has been used. A basic assumption upon which the model was based was that the glacier resulted from the lowering of the snow-line in an area of humid temperate climate. It has therefore been assumed that the glacier was at the pressure melting point during both its early stages of development and during its extension down-valley. In such circumstances basal sliding would be expected to be a major element in ice movement and therefore basal erosion and deposition could occur. If however, a valley glacier developed either at very high altitudes or very high latitudes it is possible for it to develop and extend down-valley under climatic conditions that would cause the ice in the glacier to be below the pressure melting point. A relatively thin valley glacier developed by net accumulation at low temperatures may well be frozen to its base. In such a situation it is unlikely that any glacial erosion will take place because no basal sliding will occur. Such glaciers presumably are of little geomorphological significance except that they protect the surface beneath them from erosion by other agencies. To complicate the initial model further it must be stated that it is quite possible for a valley glacier to be initiated, and for it to develop, as a temperate ice mass, for it to experience a period when at least part or maybe all of the glacier is below the pressure melting point and then for it to return to the pressure melting point.

Throughout the above discussion a simple, single valley glacier has been the basis of the model. In reality, of course, single valleys and therefore single glaciers are very rare phenomena. Valley systems are characterized by tributary development and once a period of glaciation commences in a mountain area numerous valley heads will contain permanent snow banks that develop into cirque glaciers that in turn will develop into valley glaciers. The former river system is replaced by a glacier system and the various forms and deposits described in the simple model of the single valley glacier may develop in each tributary valley. The modification of a fluvial valley system by glaciation has been the subject of descriptive models in the geomorphological literature since the latter half of the nineteenth century. The erosion of mountains by cirque development to produce arêtes and horns, the greater deepening of trunk valleys compared with minor tributaries to produce hanging valleys, the breaching of divides to produce through valleys and transfluence cols are all elements in the widely accepted model of mountain and valley glaciation.

The importance of the morphology of the area in which valley glaciers develop has already been established. The models discussed so far have assumed that the valley glacier originated by snow accumulation in a pre-existing valley head. Although this is a very common phenomenon it is also possible for valley glaciers to originate on upland surfaces above valleys. These surfaces can be old erosion surfaces and in themselves may be quite irregular and exhibit considerable local relief. On the other hand they may be plateau-like or even basins surrounded by a mountain rim. These upland surfaces, whatever their form, may act as the initial accumulation areas when glaciation is initiated. The development of 'plateau' glaciers or ice caps is usually followed, if the climate continues to deteriorate, by the development of valley glaciers descending the valleys that lead away from the upland plateau. Once again it is the form of the pre-glacial surface that determines the nature of the initial ice form and its subsequent development.

The combination of three variables, namely climate, pre-glacial landform and time, results in various types of glaciation. The relatively greater severity of glacial erosion on the north-east side of mountains in the northern hemisphere is a good example of the importance of these three variables. The existence of a mountain mass in the first place, the favourable climatic conditions on the north-east side of the mass and the relatively longer period during which those climatic conditions are likely to prevail in that situation compared with others, explains the more intense glacial erosion on north-east facing slopes.

The time factor is also important in another sense. If climatic deterioration lasts long enough in a mountain and valley area, it is possible for ice thicknesses to become so great that the pre-glacial landforms become completely buried by the ice; they then cease to control the boundaries of individual ice streams, and the directions of movement of at least the upper ice layers. It is therefore possible for a glaciation which was initially a valley glaciation to develop into an ice sheet. If, after the establishment of such an ice sheet a negative mass balance develops, the ice sheet will once again return to valley glaciation conditions. It must be remembered, therefore, that two periods of valley glaciation can occur in the same area associated with one ice sheet. Such a situation occurred on more than one occasion in northern Britain and in Scandinavia during the Pleistocene. The time periods involved during which valley glaciers rather than an ice sheet existed can only be guessed at. In Scotland for example, the last glaciation is known to have developed after 28000 years BP and the last glaciers disappeared about 10000 years BP. The glaciation reached its maximum southerly extent about 17000 years BP and had certainly been reduced to individual valley glaciers in central Scotland by 13000 years BP. The valley glaciation in Scotland during its build up may have lasted of the order of 6000 to 8000 years before the ice sheet became dominant over large areas, but the break up of the ice sheet and a return to valley glaciation was rapid and was accomplished in a maximum of 4000 years.

ICE SHEET GLACIATION

The great ice sheets of the Pleistocene covered, at their maximum extent, about 30 per cent of the earth's land area. The landforms and deposits associated with ice sheet glaciations are therefore also very extensive. Because of their sheer size and inaccessibility our knowledge of the glaciology of ice sheets is very limited and consists of sample studies undertaken during the last twenty years. However, a great deal of information about the Greenland and Antarctic ice sheets is now available and it is remarkable how much this new data has substantiated the theoretical models suggested by several glaciologists and geophysicists (Paterson, 1969). From a geomorphologist's point of view the data relating to existing ice sheets gives some indication of the surface slope, areal extent, thickness, internal temperature conditions and rates of movement of the Pleistocene ice sheets. In terms of the theoretical models that the glaciologists have developed for ice sheets, certain basic assumptions are common to nearly all of them (Paterson, 1969, p. 151). The fact that it is assumed that the base of the ice sheet is horizontal, that the mass balance is assumed to be constant over all of the ice sheet and that the temperature, roughness of the bed and other factors that might influence rates of ice movement, are constant, are all assumptions that, from the geomorphological point of view, are rather misleading. Not only are the irregularities of the ice-rock interface of great importance in determining the nature of the geomorphological processes that are active, but temperature changes over both short and long distances beneath ice sheets are of fundamental importance to the type of landforms and deposits that develop during ice sheet glaciation.

Before examining the landforms and deposits of ice sheets it is necessary to discuss briefly their glaciology. It is theoretically possible for an ice sheet to develop on a horizontal surface so long as there is net accumulation of snow on that surface. The size, thickness and shape of the resultant ice sheet will be a function of the plastic properties of ice. Paterson (1969, pp. 145-52), has summarized the attempts by various workers to construct theoretical surface profiles of ice sheets (Fig. 85). The theoretical profiles based on equations 7 and 9 in Paterson (1969, p. 148-9) are almost identical on the scale of Fig. 85, which have been plotted as one line. These theoretical profiles are also very similar to an ellipse. A profile of the Antarctic ice sheet measured by a Soviet Antarctic expedition (Vialov, 1958) is also plotted on Fig. 85 and it can be seen that it is very similar to the profiles derived from formulae 7 and 9 in Paterson. Other workers in other areas have also found that an ellipse is a fairly good approximation of the surface slope of ice sheets. Since the surface slopes that have been measured have been from ice sheets with different rates of accumulation and with different degrees of bed roughness it can be argued that the shape of an ice sheet is largely determined by the plastic properties of ice rather than by variations in accumulation, bed roughness or ice temperature.

An ice sheet will become larger when the snowfall on it increases or the melting on its edges decreases. It is likely that all the Pleistocene ice sheets started as mountain glaciers, developed into small ice caps and subsequently developed into continental ice sheets. In this sense many of the assumptions used by glaciologists to develop models of ice sheets are rather unrealistic, because the Pleistocene ice sheets certainly did not develop on horizontal surfaces and it is very unlikely that any one ice sheet had

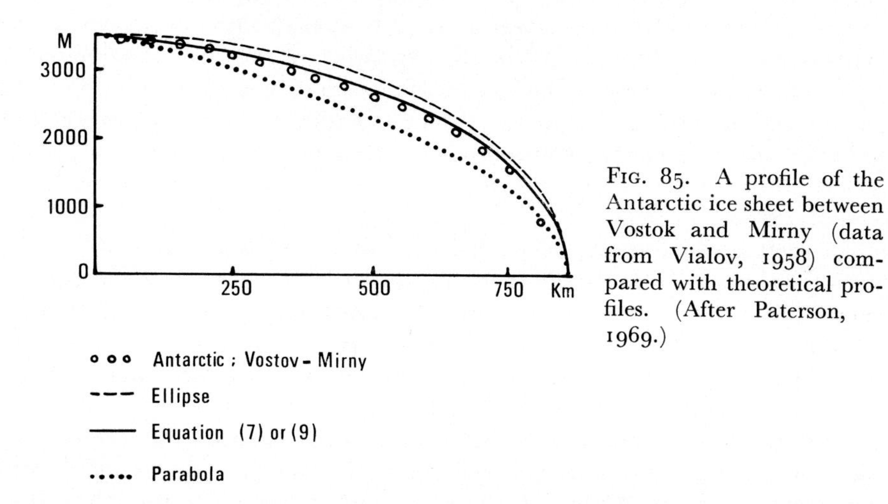

FIG. 85. A profile of the Antarctic ice sheet between Vostok and Mirny (data from Vialov, 1958) compared with theoretical profiles. (After Paterson, 1969.)

uniform rates of accumulation over its surface. Bearing in mind the low annual snow fall on present-day ice sheets it is likely that accumulation on large areas of the Pleistocene ice sheets was also small and therefore their rate of development would have been slow. As Weertman (1961a, p. 3783) has pointed out, 'The accumulation on the ice sheet itself would be expected to decrease as it became bigger, both because the accumulation areas would be at a high elevation and because the cooling of the earth by the presence of the sheet might lead to reduced precipitation.' Another factor to be considered in the development of ice sheets is the ice temperature. If ice sheet development begins under low temperature conditions in high polar areas it is likely that the ice sheet will be frozen to its bed and will remain so even during its expansion. Although small thicknesses and high accumulation rates favour the development of a cold ice mass frozen to its bed, the large thicknesses and low accumulation rates of a more extensive ice sheet increase the probability that the basal layers of the ice sheet will be at the pressure melting point. Because of the geothermal heat flux it is possible for an otherwise cold, large ice sheet to have basal layers at the pressure melting point and once such a condition is produced the basal sliding that takes place also generates heat of an order of magnitude similar to the geothermal heat flux (Weertman, 1957). The temperature conditions at the base of ice sheets are of great importance to the geomorphological work carried out beneath and at the margins of ice sheets and this topic will be returned to later in this chapter.

Weertman (1961*a*) has also made some estimates of the time required by an ice sheet to develop a size similar to that attained by the great Pleistocene ice sheets. On the basis of glaciological arguments he concluded that the large Pleistocene ice sheets would require of the order of 15 000 years to reach their maximum extent but that they could disappear completely in about 5000 years. This theoretical time scale compares very favourably with the time scale deduced by C^{14} dating methods from stratigraphical evidence left by the Pleistocene ice sheets.

Robin (1966), states that once an ice sheet exceeds a minimum size of about 100 km across, it helps to create a climate favourable to its own survival and even to its continued expansion. One of the limits on the expansion of the present Greenland and Antarctic ice sheets is, of course, the extent of the land area, beyond which the ice calves as bergs into the surrounding sea. The climate developed in association with the Greenland and Antarctic ice sheets would have to change quite dramatically, possibly a warming of the order of 10°C or more, before either ice sheet would experience any great reduction in its extent.

From the models of ice sheets developed by glaciologists and the fairly satisfactory comparisons that have been made between these models and existing ice sheets in Greenland and the Antarctic, it is possible to obtain at least a general glaciological model for each of the major Pleistocene ice sheets. Robin (1964, p. 104), states, '. . . provided we know the outline of an ice sheet and have an approximate idea of the underlying topography, we can make a good estimate of its surface form.'

Various authors have attempted reconstructions of the European ice sheets. One of the problems in attempting such reconstructions is the fact that the ice sheet, which at its maximum extent stretched from off the west coast of Ireland to the River Don and from northern Germany to beyond the North Cape, developed from a series of mountain glaciers that eventually coalesced and over-topped their host mountains to produce mountain ice sheets that in turn coalesced to produce the European ice sheet. In 1964, Robin reconstructed the shape and surface profile of the Scandinavian ice sheet as it was about 11 000 years BP (Fig. 86). This certainly was not the maximum extent of that ice mass during the last glaciation but at least by that time the Scandinavian ice sheet was unlikely to have been very much affected by the British ice sheet, which had virtually disappeared by that date. By applying a typical profile for continental ice sheets to a line drawn across the ice sheet from Trondheim to Helsinki, and allowing for the fact that the mountains caused the ice surface to rise more rapidly on the western side, Robin was able to establish the surface contours of the ice sheet. The extra weight of the ice depressed the earth's crust by about one third of the ice thickness.

The build up of the Scandinavian ice sheet has been described by various authors. The lowering of the snow-line produced an expansion of the valley glaciers on the great mountain backbone of the Scandinavian peninsula (Fig. 87*a*). This mountain system, with peaks ranging between 1000 and 2500 m, receives a great deal of precipitation (1000-3000 mm per year) from the moist air masses that track north-eastward across the North Atlantic. Even under present climatic conditions these mountains maintain small ice caps and valley glaciers. With a lowering of the snow-line the accumulation on the west facing slopes increased and rapidly flowing valley glaciers would soon have reached sea level (Fig. 87*b*). From morphological and stratigraphical evidence, however, it is known that at the maximum development of the ice sheet, ice was thickest in a region east of the mountains. The steep western glaciers were relatively short because they soon calved into the sea. The eastern glaciers followed a more gentle gradient

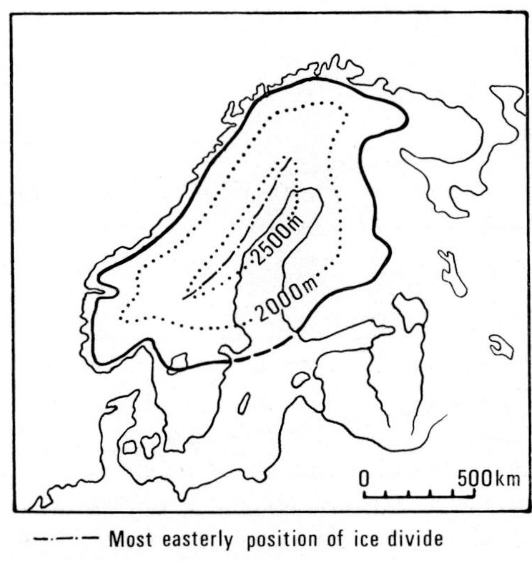

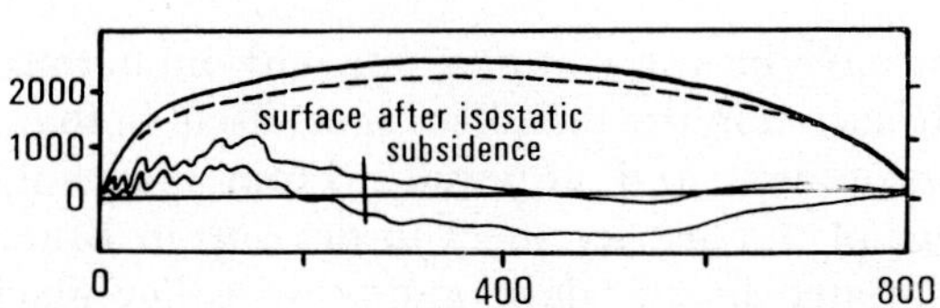

FIG. 86. Approximate surface contours of the Scandinavian ice sheet and an estimated profile across the ice sheet from the site of Trondheim to Helsinki. (After Robin, 1964.)

and were able to expand as piedmont glaciers on the lowlands to the east of the mountains and to eventually build up into an ice sheet (Fig. 87*c*). With the progressive development of the ice sheet the ice-divide moved east of the mountains and at the maximum was probably about 250 km east of the mountain divide. This movement of the ice divide east of the mountain divide beneath the Scandinavian ice sheet is not an isolated phenomenon. It also occurred in the Scottish Highlands and in British Columbia.

It is extremely difficult to determine the maximum thickness of the Scandinavian ice sheet. On theoretical grounds Robin has suggested that at about 11 000 years BP the highest part of the ice sheet was at between 2500 and 3000 m. This would give a maximum thickness of about 2500 m. Holtedahl (1967) quotes a probable thickness of 3000 m over Sognefjord and the 3000 m value occurs repeatedly in the literature. During the Riss glaciation Robin (1966) has estimated that the combined Scandinavian-British ice sheet reached a maximum surface altitude of 3000 m over the Gulf of Finland

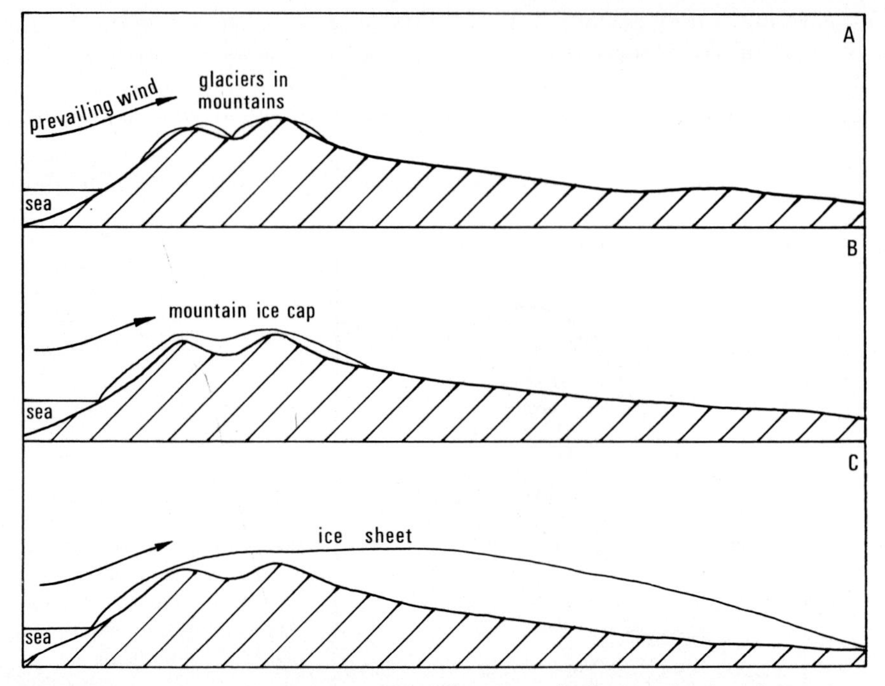

Fig. 87. Idealized development of an ice sheet similar to the ice sheets that developed in Scandinavia, Scotland and Western Canada during the Pleistocene.

but in this case the ice sheet must have been a complex one with numerous centres of dispersal. A similar complexity occurred in the British ice sheet itself, in that there were numerous centres of glacier development and dispersal. From morphological and stratigraphical evidence the extent of the last ice sheet to develop in Britain is fairly well known. Although it was initiated by the development of valley glaciers in the mountains of Scotland, the Lake District, Wales and Ireland, and its western and northern margins terminated in the sea and its eastern margin was affected by Scandinavian ice, it is likely that the ice sheet had a radius of about 250 to 300 km. It is therefore likely that the maximum surface altitudes of the British ice sheet over the Scottish mountains were of the order of 1000 to 1500 m. These statements are supported by the morphological and stratigraphical evidence, in that it is generally believed that even the highest Scottish mountains were submerged beneath the ice. However, it is highly likely that even though an ice sheet developed, the pattern of movement within that ice sheet and its rather crenulated margin reflects the strength of the control of the underlying topography on the development of the ice sheet. There are a few major exceptions to this statement. In the Western Highlands of Scotland the ice divide at the maximum development of the ice sheet was to the east of the mountain divide. In the Southern Uplands a major centre of ice dispersal developed to the west of the centre of the Uplands and in northern England ice from the Lake District over-topped the Pennine ridge and moved down the east side of England.

The period of time during which the last ice sheets in Scandinavia and Britain developed is known in general terms. They both took about 10000 years to develop and reached their maximum extent about 17000 years BP. In Britain the ice sheet had been reduced to a series of valley glaciers by 13000 years BP although a sizeable ice sheet still existed in Scandinavia at that time. As late as 9500 years BP an ice sheet

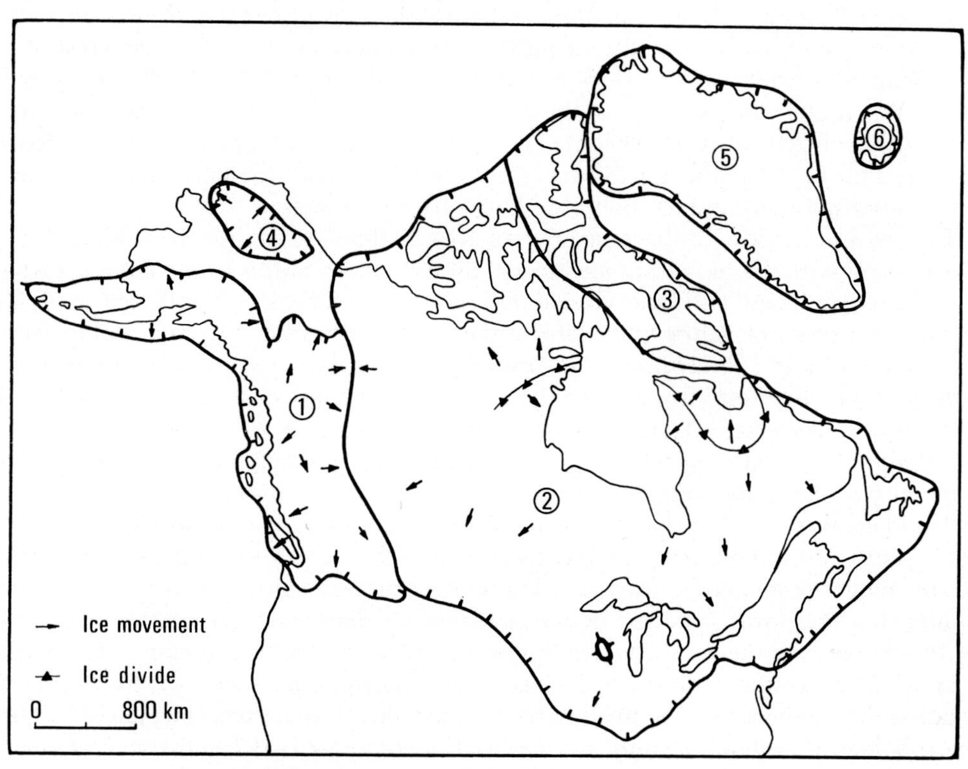

FIG. 88. Pleistocene glaciers and ice sheets: 1—Cordilleran; 2—Laurentide; 3—Baffin Island and Ellesmere Island; 4—Brooks Range; 5—Greenland; 6—Iceland. (After various sources.)

still existed in Scandinavia although it was probably stagnant and wasting rapidly, but all the valley glaciers had disappeared from Britain by that date.

On the North American continent two major ice sheets developed during the last Pleistocene glaciation and are known as the Cordilleran and the Laurentide ice sheets. The Cordilleran ice sheet (Fig. 88) was similar to the Scandinavian and British ice sheets in that its development was largely controlled by mountain systems. Flint (1971) has suggested that there were four phases in the development of the Cordilleran ice sheet. It was initiated by the growth of valley glaciers on the moist Coastal Ranges and to a lesser extent on the drier Rocky Mountains. The second phase involved the growth and coalescence of valley glaciers and the development of major trunk valley glaciers. During the third phase the continued thickening of both valley and piedmont glaciers resulted in the development of a mountain ice sheet over the Coast Ranges. The fourth phase involved a thickening of the piedmont glaciers between the Coast Ranges and the Rocky Mountains to produce an ice sheet with a thickness of at least 2300 m. Flint (1971, p. 469), states, 'In this general sequence the transition from the third to the fourth phase, which is supported by evidence of reversal of direction of flow in the Coast Ranges, implies that the ice divide, the axis of outflow of the ice, shifted from the mountain crest inward to a position over lower land. From the crest of the Coast Ranges the shift was eastward, but from the lesser crest of the Rockies it was westward. When the two axes of outflow coalesced, the Cordilleran glacier complex temporarily centred in an ice sheet.' The ice sheet only extended east of the Rocky Mountains on to the plains for distances of between 60 and 100 km where, at least in some locations, it was confluent with the Laurentide ice sheet.

The last Laurentide ice sheet covered an area of about 12 million km^2 (Fig. 88). It was confluent with the mountain ice sheet of the Ellesmere-Baffin Island complex which in turn was confluent with the Greenland ice sheet. The Laurentide ice sheet was initiated as a series of valley and plateau glaciers in Labrador, Ungava, Baffin Island and Ellesmere Island, and spread out from this favourable area towards the west and south, that is in directions from which the moisture that contributed to the growth of the ice sheet was derived. At least during the late stages of its existence, ice-divides unrelated to the underlying relief developed within the ice sheet (Fig. 88) that separated ice moving to the south-west, south and south-east from ice moving to the north-west, north and north-east. The location of these ice divides suggests that profiles from north-west to south-east or north-east to south-west across the ice sheet would be asymmetrical with the longer more gentle surfaces being on the southern side of the ice sheet. If one assumes that there was only one major centre of ice dispersal over northern Hudson's Bay this would give the ice sheet a radius of the order of 1800 to 2000 km. From what we know of the gradients along flow-lines of the Antarctic ice sheet it is likely that the surface of the ice sheet in the central parts attained altitudes in excess of 3500 m. There is morphological evidence to indicate that ice thicknesses of between 800 m and 1600 m were attained in several parts of the ice sheet and that the altitude of the ice surface on Mt Washington in New Hampshire was 1900 m, in the Adirondack Mountains, 1600 m and in the Torngat Mountains in Labrador, 1500 m.

The time period over which the Laurentide ice sheet was built up appears to coincide with that for the build up of the Scandinavian and British ice sheets, namely about 10000 years. It reached its most southerly position about 18000 years BP.

The above discussion of the European and North American ice sheets in terms of their glaciological characteristics has omitted one very important characteristic in terms of the geomorphological significance of ice sheet glaciation. In earlier chapters of this book the temperature profile of the ice constituting either valley glaciers or ice sheets has been shown to be of fundamental importance in determining the geomorphological processes that will be active beneath or at the margins of any ice mass. There are two extreme conditions as far as ice temperature is concerned, either the ice mass is at the pressure melting point throughout so that meltwater can penetrate to the base of the ice sheet and sliding of the ice over the ground beneath can occur, or the ice mass is below the pressure melting point and is frozen to its base. These two extremes are separated by parts of the basal layers of some ice sheets being at the pressure melting point while other parts are frozen to the base. To what extent and over what periods of time the great Pleistocene ice sheets exhibited each of these possible temperature conditions is unknown. The thermal regime of an ice sheet is largely determined by the climate that affects the mass balance and the temperature of the surface layers of ice, the ice thickness, the geothermal heat flux and the heat generated by ice movement. Since there is little accurate information available about the extent and nature of climatic deterioration that produced the Pleistocene ice sheets, no information about the climate that those ice sheets themselves created over their own surfaces, and no information about their mass balance characteristics, it is impossible to produce a useful model of their thermal characteristics at any particular stage in their development.

However, it is highly likely that during the build up of the mountain ice sheets in Scandinavia, Britain and the Cordilleran ice sheet in North America where high accumulation rates would have prevailed along with rapid rates of movement, that the valley glaciers and maybe even the ice sheets themselves were at the pressure melting point throughout. This is to some extent borne out by the intensive erosion that took place in the mountain systems that were buried beneath each ice sheet. Once the ice sheets had attained their maximum extent, however, they may have developed their own more severe continental climates in which accumulation would have been much less and surface ice temperatures much lower. Even with the geothermal heat flux at the base a negative temperature profile could develop, and parts, if not all of the ice sheet would have been reduced to temperatures below the pressure melting point. On the other hand, the ice thicknesses could have been sufficiently great that the ice sheets could have maintained pressure melting temperatures in their basal layers.

The Laurentide ice sheet developed in rather different circumstances from the other great Pleistocene ice sheets. Its initiation and growth took place under relatively continental conditions, with much lower rates of accumulation and ablation and it may have existed as a true 'cold' ice mass for a relatively longer period. However, even the Laurentide ice sheet became 'temperate' in character during its wastage, as is evidenced by the very extensive occurrence of subglacial fluvioglacial landforms and deposits produced during its withdrawal.

It can be concluded from the above discussion of ice sheets that the form of the surface on which they developed, the climatic conditions under which they were initiated and subsequently developed, the plastic properties of ice, the period of time over which they developed and their changing glaciological characteristics over both time and space, determined the size and shape of the ice sheets and the geomorphological processes that were associated with them. An ice sheet differs from a valley glacier in its geomorphological activity in that its boundaries and direction of movement are not entirely determined by the form of the surface on which it develops. It must be remembered, however, that in those situations in which an ice sheet develops, as a result of the growth of a valley glacier system, into a mountain ice sheet, that it may be difficult to determine which landforms developed during the valley glacier stage and which at the ice sheet stage.

The landforms produced by glacial erosion beneath an ice sheet only differ from the landforms associated with valley glaciers in their size and location and in the absence of certain distinctive features. Cirques, for example, are less likely to be associated with ice sheets than with valley glaciers except where mountain ridges remain above the surface of an ice sheet and allow cirque glaciers to develop on their slopes.

The small scale features of glacial erosion, such as striae, chatter marks and *roche moutonnées* will be controlled largely by the surface gradient of the ice sheet which in turn will affect the general direction of movement within the ice sheet. Under ice sheet conditions these features will occur in a much greater variety of positions compared with a valley glaciation. They may be found on ridge crests, on upland plateau surfaces and on lowland bedrock surfaces. Under a mountain ice sheet the direction of ice movement indicated by such features beyond the confines of valley systems may indicate directions of ice movement that are significantly different from similar features within the valleys. On piedmont plains it is possible for cross-striae to occur as a result of the dominance of valley-controlled ice movements at one stage being superceded, at a later ice sheet stage, by movements being controlled by the gradient of the surface of the ice sheet.

During the development of an ice sheet over a lowland, or even an upland plain, erosion at the ice-rock interface, assuming that the basal ice is at or near the pressure melting point, will tend to produce certain irregularities in some locations while in others protuberances will be removed. On a small scale, rock knobs exhibiting a general smoothing on the proximal side and rough and broken surfaces on their distal sides are common features on plains that have been overridden by ice sheets. Structural alignment in these rock surfaces often assist the development of these features. On a larger scale, major structural features such as faults or shatter zones, where they approximately parallel the direction of ice movement, encourage intensive glacial erosion and significant troughs can develop even in lowland situations. Rock basins have been developed in lowland areas where the local relief is less than 300 m. Linton (1963) has described such features in central Scotland, one of which has been cut to a depth of over 100 m below present sea-level. However, it is the general streamlining and smoothing of exposed rock surfaces that are the major expression of ice erosion on rock plains beneath ice sheets.

During the discussion of the development of the Scandinavian, British and North American Cordilleran ice sheets the migration of the ice-divide away from the mountain divide was discussed in each case. When an ice sheet develops flow lines that are transverse to mountain systems beneath it, one of the most characteristic features of ice sheet erosion is produced. The ice-breached watershed or col is a very common phenomenon in areas that have been buried beneath an ice sheet. In the north-west Highlands of Scotland, Dury (1953) has recognized some thirty ice-breached cols in the former watershed. The effect of the destruction or modification of former watersheds by ice sheet erosion can be very significant to the movement of meltwater through and beneath ice sheets and to the development of post-glacial drainage systems. The development of a series of radiating troughs represents the ultimate adjustment to the surface form of an ice sheet.

The question of glacial deposition during the build up of an ice sheet is full of uncertainty. Some authors have maintained that beneath the interior portions of an ice sheet glacial erosion would be dominant whereas towards the periphery of ice sheets glacial deposition would predominate. This is probably a gross oversimplification because during the development of the ice sheet a particular location that was at one stage peripheral would become relatively close to the centre of ice dispersal as the ice sheet expanded. It is much more likely that pressure and temperature variations at the ice-rock interface are of much greater significance in determining whether glacial deposition will take place beneath any particular part of an ice sheet. Boulton (1972) has defined four boundary conditions at the ice-rock interface: 1. net basal melting; 2. a balance between melting and freezing; 3. sufficient meltwater freezing to the basal ice to maintain it at the melting point; 4. the basal ice below the pressure melting point. Only under conditions 1 and 2 will significant glacial deposition occur. Under condition 3, basal sliding is accompanied by plucking, abrasion and crushing and glacial erosion and transportation is more important than glacial deposition. Under condition 4, the glacier adheres to its bed and movement takes place by internal flow so there will be no subglacial deposition. Where, and for what length of time, each of these conditions prevailed beneath the Pleistocene ice sheets is impossible to determine. From limited evidence of bore holes in the Antarctic and Greenland ice sheets, meltwater and ice at the pressure melting point is known to occur at the base of and beneath more than 2000 m of ice in ice sheets that are generally regarded as cold.

When the basal layers of an ice sheet fulfil the first two boundary conditions indicated by Boulton (1972), then subglacial till can accumulate. Melting of ice that contains debris is the fundamental process in the deposition of subglacial till. According to Boulton (1972) deposition will take place when the frictional drag on debris particles being moved over the bed equals or exceeds the tractional force exerted on it by the ice. It is therefore likely that the rate of deposition will increase with an increase in ice thickness, although the presence of water at the ice-rock interface could reduce friction and mitigate the effect of increased ice thickness. Boulton (1972, p. 12), concludes '. . . lodgement will be favoured under thicker ice by areas where the bed is rough or where it is composed of highly permeable sediments. Deposition will initially be concentrated against obstructions, thus developing drumlinoid features, but will eventually tend to fill up low points, which together with erosion of prominancies, will tend to produce a less irregular bed.' It is highly likely that the internal fabrics of the tills deposited beneath an ice sheet will reflect directions of ice movement in the ice sheet if the tills are subsequently frozen. If they remain unfrozen they are usually disturbed by either shearing produced by the overriding ice, flowage resulting from differential stresses beneath the ice or squeezing. The extent to which till sheets associated with ice sheet glaciation are deposited during the development or during the wastage of ice sheets is not known.

From a geomorphological point of view glacial deposition along the margin of an advancing ice sheet by means of slumping and sliding of either supraglacial debris or of material melted out of the englacial environment at the margin, is of little significance. Such deposits may form moraine ridges and the advancing margin may bulldoze proglacial material into ridges, but such ridges will be subsequently destroyed by the advancing ice margin. Only when the ice sheet reaches its maximum extent does marginal deposition produce significant landforms. The actual mechanisms involved and the landforms produced in that situation are very similar to those that occur during deglaciation and will be discussed in Chapter VIII.

Fluvioglacial erosion and deposition during the build-up of an ice sheet are largely dependent on the temperature profile of the ice mass and the climate at the surface of the ice sheet. If the ice mass is below the pressure melting point at all levels and the ablation season is very short, meltwater streams will develop on the ice surface, along the ice margins and in the proglacial area. The geomorphological activity of these meltwater streams will be of very limited significance not only because of the relatively short periods over which they function in any one year but because any minor channels or deposits they produce will be subsequently overridden by the expanding ice sheet.

As the ice sheet grows, a certain amount of fluvioglacial deposition takes place in the proglacial area. This suggests that the classical stratigraphical sequence of fluvioglacial sands and gravels overlain by subglacial till can be equated with the expansion of an ice sheet. However, the survival of fluvioglacial deposits during ice sheet expansion will depend on the thermal regime at the ice-rock interface, because under certain conditions the overriding ice would erode the fluvioglacial deposits and incorporate them in the subglacial till.

If the whole of an ice sheet is at the pressure melting point during its build-up and expansion, fluvioglacial erosion and deposition will be much more active compared with that associated with a cold ice sheet. Once again, however, the geomorphological activity of the meltwater is likely to be of little significance because of subsequent overriding by the expanding ice sheet. The forms and deposits produced by meltwater activity when the ice sheet attains its maximum extent will be similar to those produced during deglaciation and will be discussed in Chapter VIII. However, it is likely that glacial erosion and transportation will steadily increase in importance during the build-up of an ice sheet and will reach a maximum when the ice sheet reaches its greatest extent. It is therefore likely that the amount of debris reaching the marginal zone will be greatest at the ice sheet's maximum extent and the meltwater streams will carry their greatest loads at that time. In contrast, meltwater discharge will tend to increase after the ice sheet has reached its greatest extent and during the period of recession.

EFFECTS OF MULTIPLE GLACIATION

Throughout the above discussion of valley and ice sheet glaciation it has been generally assumed that only one glaciation has taken place, in order to simplify the models. However, it has been well established that multiple glaciation occurred during the Pleistocene. The landforms and deposits that develop during one glaciation will affect the build-up and expansion of any subsequent ice mass and the landforms and deposits it develops.

In an area that has already experienced a valley glaciation the build-up of any subsequent valley glaciers can be both aided and hindered by the landforms developed during the previous glaciation. Cirques on north-facing slopes in the northern hemisphere may prove to be ideal sites for the initiation of glaciers, and glacial troughs already in existence provide easy routeways for glacier expansion. On the other hand, intensive glacial erosion during a previous glaciation may have destroyed some areas of high ground that may otherwise have acted as accumulation areas during a period of falling snow-line. In an ice sheet glaciation over an area of previous glaciation, former high areas that affected the location of ice divides in the first glaciation may have been sufficiently modified to cause the second ice sheet to develop different centres of ice dispersal and to develop different flow-lines.

In areas of multiple glaciation it is also often difficult to determine to what extent any given landform is solely the product of the last glaciation, or in fact inherited a great deal of its form from previous glaciations and was only modified by the last glaciation. Such a situation can arise when glacial or fluvioglacial deposits initially buried a particular landform that was subsequently exhumed by glacial erosion during the last glaciation.

The interpretation of the stratigraphy of glacial and fluvioglacial deposits in areas of multiple glaciation can also be very difficult. For example, in a lowland area covered by several ice sheets, each ice sheet may have been responsible for a layer of subglacial till. If the direction of movement during the expansion of each ice sheet was identical then the fabric and erratic content of each till sheet could be identical. It would be possible to separate each till sheet if they were separated by proglacial fluvioglacial deposits or non-glacial deposits or soil, but the survival of non-glacial sediments, or of soils, following overriding by an expanding ice sheet must be the exception rather than the rule. It is also possible for large parts of till sheets to be completely removed by subsequent glaciations. Bearing in mind the limited exposures that most areas provide, it is often extremely difficult to interpret stratigraphy in areas of multiple glaciation unless datable horizons are encountered.

VIII MODELS OF DEGLACIATION

INTRODUCTION

THIS chapter is concerned with the glaciological conditions that prevail, and the landforms and deposits that are produced, when an ice mass is at its maximum extent and during the retreat from those limits. A valley glacier or ice sheet begins to decrease in volume when a negative mass balance is achieved. Expressed in another way, the rate of ablation in the ablation area is greater than the rate of replenishment of ice from the accumulation area. The establishment of a negative mass balance on any ice mass involves the migration of the equilibrium line to higher altitudes, a lowering of the ice surface in the ablation area and the retreat of frontal and lateral margins.

It is important to note that forward movement within an ice mass which is decreasing in volume need not cease even at the outer margins. It is possible for rates of movement to be unchanged and yet for volume to decrease if the increase in ablation is sufficiently great. However, it is more likely that when a negative mass balance has been established less ice has to be transmitted through the system and rates of movement will decrease. With decreasing rates of forward movement within the ice and increased rates of ablation then volume decreases and marginal retreat will occur.

There are occasions when parts of ice masses become separated from their source areas and in such situations all forward motion ceases as there is no longer any discharge through the system. In such circumstances the ice mass becomes stagnant and ablation rates are the sole determinant of rates of thinning and marginal retreat.

The form of any ice mass is either strictly controlled by the underlying topography, that is in the case of a valley glacier, or the topography is completely buried by the ice mass. The deglaciation process in each case can be significantly different in the way in which it affects the development of landforms and deposits. In its simplest form, valley deglaciation results in the concentration of geomorphological processes on the valley floor and the energy that produces many of the landforms and deposits in the marginal zone is often responsible for their destruction at a later stage in the deglaciation. In contrast, deglaciation of an area covered by an ice sheet involves a much greater length of frontal margin, a wider variety of environments and a greater dispersal of geomorphological activity. However, it must be stressed that the actual processes responsible for the development of glacial and fluvioglacial deposits and landforms do not differ to any significant extent between valley deglaciation and ice sheet deglaciation.

Of paramount importance to the geomorphological processes that occur during deglaciation is the temperature of the ice in the ablation area. When the ice is at the pressure melting point large volumes of meltwater are produced which descend through the ice mass to the ice-rock interface. The ice also slides over its bed and rates of basal movement can even be increased as a result of the presence of increased amounts of meltwater at the ice-rock interface. When the ice is at the pressure melting point the range and intensity of all the geomorphological processes is at its greatest. In contrast, if the ice is below the pressure melting point during deglaciation then the range of geomorphological activity is at its minimum. It is likely that the ablation season will be short, that only small volumes of meltwater will be discharged, and even these are unable to penetrate into the ice. Geomorphological activity is largely restricted to the zone of contact between the ice and the ice-free land and to the proglacial area.

VALLEY DEGLACIATION

The deglaciation of a valley is a function of the relative importance of the supply of ice and the rate of ablation at the ice front. The shape of the frontal margin is largely controlled by the cross profile of the valley and the rate of retreat of the ice front is strongly controlled by the slope of the ice surface and the long profile of the valley floor. Frontal retreat is simply a reflection of the lowering or downwastage of the ice surface and this in turn reflects the relationship between ablation and replacement of ice by forward movement.

Downwastage in the frontal zone of an active valley glacier can range from as low as a few centimetres per year to as high as 10 m per year, depending upon the climate.

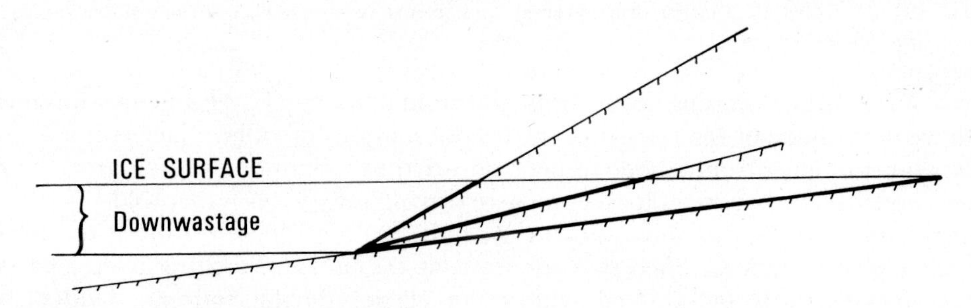

FIG. 89. Relative rates of retreat across horizontal and sloping surfaces produced by the same amount of downwastage of the ice surface.

Assuming a constant ice surface slope and a constant amount of downwastage (Fig. 89), more rapid retreat will take place across a land surface sloping away from the ice front than across a horizontal surface, or a surface sloping towards an ice front. It is possible to construct a series of graphs (Fig. 90) to show the relationship between the three variables of rates of downwastage, ice surface slope and ground slope. Three ice surface slopes of 6, 18, and 27 degrees respectively are matched against three ground surface slopes of 6, 18 and 27 degrees. It can be seen from the graphs that at any given rate of downwastage the rate of horizontal retreat is least on a reverse slope (*A*) and

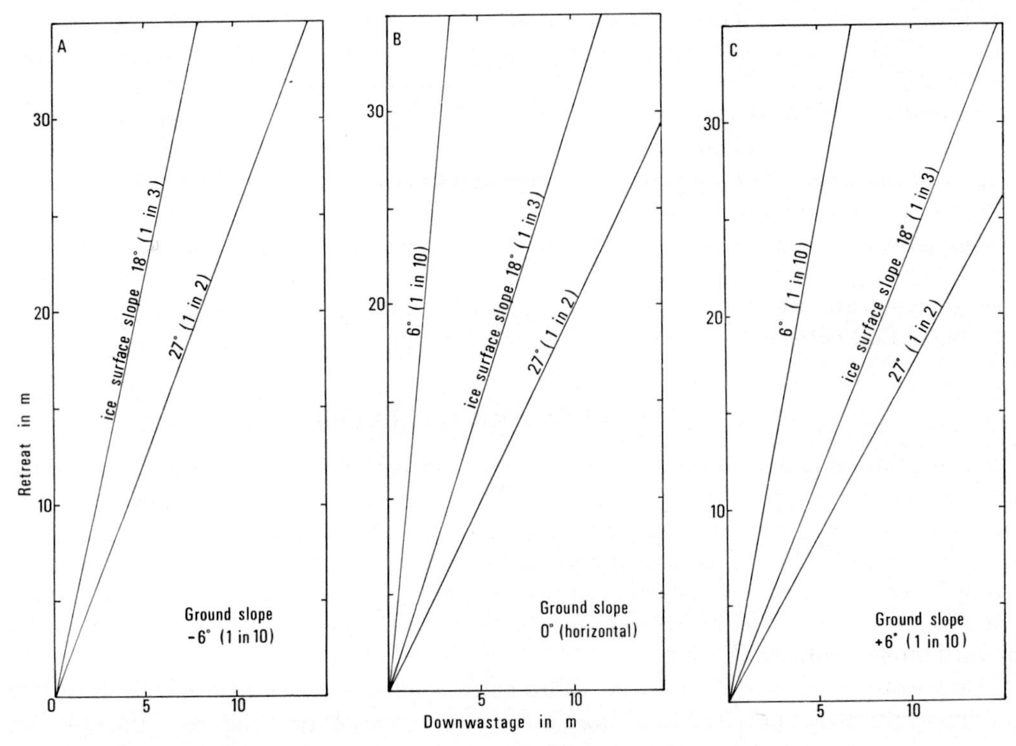

FIG. 90. Graphs relating downwastage and retreat to selected ice surface slopes and ground slopes.

greatest on a surface sloping away from the frontal zone (*C*). From measurements that have been made of the retreat of the frontal margins of valley glaciers it is known that amounts ranging from 10-100 m per year are quite common but that greater rates of frontal retreat are uncommon except where calving into water is involved.

In the discussion of the development of landforms and deposits during the retreat of a valley glacier, it is assumed that the ice is at the pressure melting point and that forward movement is maintained within the glacier during retreat. Under such circumstances glacial erosion and transportation will occur in the ablation zone during the retreat phase. The amount of erosion may not be great, as the ice may well be thin and rates of movement low. However, striations and chatter marks will develop and the general smoothing of rock protuberances may continue. Material that was entrained in the ice and accumulated on the ice surface during the advance of the valley glacier will be concentrated in the ablation zone during the retreat phase. This material is available for deposition both in the marginal and subglacial environments. There is also a tendency for flow-lines at the snout to be inclined upwards towards the surface; and material frequently occurs along these flow-lines and, due to the lowering of the ice surface then accumulates on the ice surface (Fig. 91). The material on the snout of a valley glacier tends to move down-slope under gravity and if the ice margin

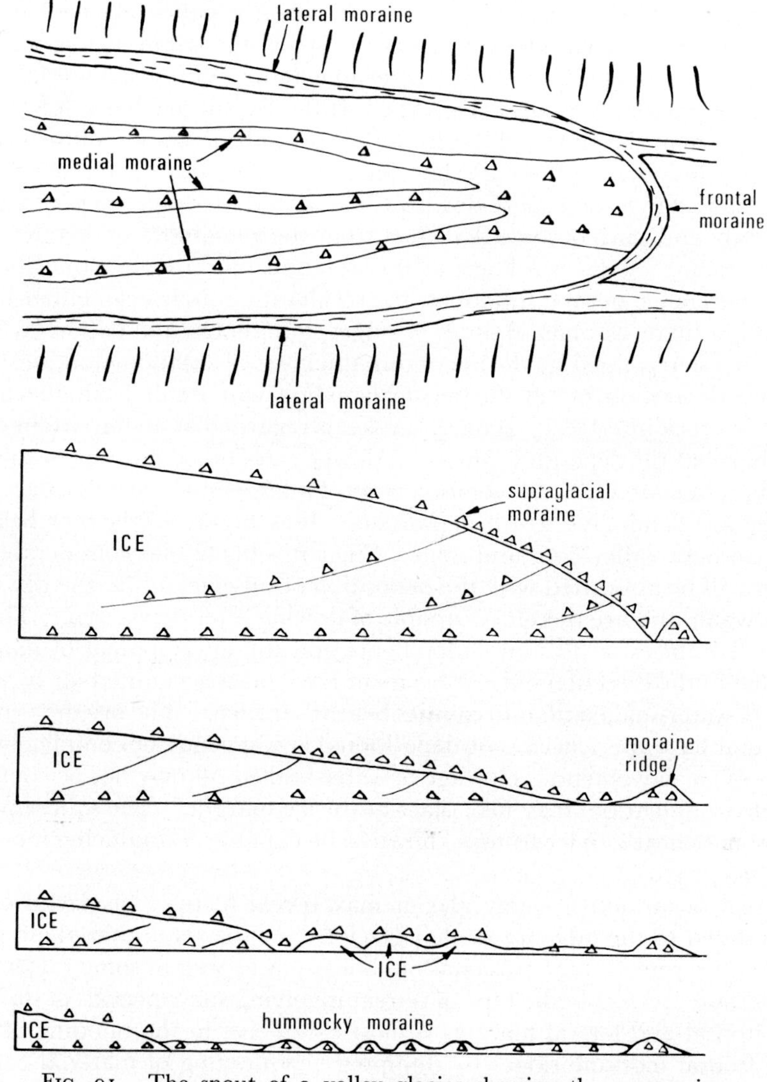

FIG. 91. The snout of a valley glacier showing the progressive wastage of the ice and the associated development of morainic deposits.

remains in one position long enough, ridges of this material will develop when the supporting ice on the proximal side is removed by melting. Ice blocks, which can also be buried by accumulation of this material, melt out leaving depressions. The dumping of material off an ice front therefore tends to produce a somewhat confused morphology of ridges, hummocks and hollows that may have an overall arcuate distribution across the valley floor which approximates to the outline of the glacier front.

In some instances the snout of a valley glacier becomes completely buried by its own ablation moraine. This is most likely to occur when a valley glacier has numerous tributaries each with its own lateral moraines. When the tributaries produce a trunk glacier the numerous lateral moraines provide an extensive cover of material in the form of closely packed medial moraines (Fig. 91). If the ice surface has a low angle slope, very little of this material slides off the frontal slope and thicknesses of the order of 1 to 2 m of ablation moraine can accumulate on the ice surface. Such accumulation of material will slow down the rate of downwastage and quite large areas of ice may survive many years and may become detached from the remainder of a retreating valley glacier. Depending on the thickness of the ablation moraine and the thickness and extent of the buried ice in such situations, the resultant landform can either be simply a sheet of ablation till or a confused series of ridges, hummocks and hollows with a maximum local relief determined by the maximum thickness of ablation moraine.

Subglacial deposition of till during deglaciation can result from the melt-out of debris at the ice-rock interface. If basal friction is regarded as an important controlling factor in subglacial till deposition, then the slower rates of ice movement and the high debris content associated with an ablation zone of a temperate valley glacier during the retreat phase are conducive to till deposition. It is highly likely that basal till will accumulate on both valley floor and sides. Since it is likely that considerable amounts of meltwater will be associated with the deposition of subglacial tills, the tills themselves will contain water and are therefore capable of flowing after deposition. The development of fluted surfaces on tills on valley floors and the development of short moraine ridges parallel to the direction of ice movement have been accounted for by postulating the flowage of water-soaked till into cavities beneath the ice. The orientation of pebbles in material that has experienced post-depositional flowage may be completely unrelated to directions of ice movement. Flowage of water-soaked till may not be confined to the subglacial environment but may take place at the ice margin. The squeezing of water-soaked till from beneath an ice front is known to be capable of producing moraine ridges 3 m high (Price, 1970).

The retreat of an active valley glacier may reveal features of glacial erosion and deposition related to the advance of that glacier. At the same time the retreat phase may also produce some minor erosional modifications as well as some major features of glacial deposition. An episodic type of retreat involving some periods of stability in the position of frontal and lateral margins is most conducive to the building of distinctive lateral and frontal moraine ridges by dumping. Squeezing of material can also produce minor moraine ridges near or at the ice margin. The net result of glacial deposition is the accumulation of basal till or ablation till on the floor and sides of the valley, in either sheets or ridges on top of the fluvioglacial deposits laid down during the advance of the valley glacier. Glacial deposits laid down on the valley floor are very likely to be eroded by meltwaters that are also concentrated in that location when ice front recession continues. Glacial deposits laid down on steep valley sides are unlikely to maintain any distinctive form for very long because of their subsequent movement downslope, *en masse*, or by modification by hill-wash or periglacial processes.

The wastage and retreat of a temperate valley glacier produces large volumes of meltwater. The erosional activity of this meltwater is not simply restricted to the lateral margins but because of crevasses and moulins the meltwater is able to penetrate into and beneath the ice of the ablation area. It has been suggested that active valley glaciers are more likely to sustain long supraglacial and lateral marginal streams than stagnant valley glaciers. This is probably an over-simplification, since stream systems on or at the lateral margins of valley glaciers are likely to be controlled by ice structures and by crevasse systems to a large extent. These features are not simply related to the amount of movement in the glacier but also to the slope of the ice surface, the long and cross profiles of the valley and to the amount of sinuosity in the ice stream. Wherever ice structures allow lateral marginal streams to develop, marginal meltwater channels and benches may be cut either in the drift deposits or in the solid rock. Supraglacial or englacial meltwater streams, whenever they come into contact with spurs or subglacial surfaces, may be superimposed on to them and produce meltwater channels which, when the glacier has retreated, may be observed to occupy rather unusual locations. The general direction of meltwater flow in a valley glacier is controlled largely by the slope of the ice surface and is therefore roughly parallel with the direction of ice movement. Meltwater channels produced by valley deglaciation are therefore usually orientated down-valley and are only distinguished from normal fluvial channels by their unusual locations, which are determined by the presence of structures within the ice.

The concentration of all meltwater associated with a valley glacier on the valley floor at the frontal margin has considerable geomorphological significance. The proglacial meltwater stream can experience large daily and seasonal fluctuation. When such streams experience maximum discharge they are capable of a great deal of erosion and much of the material and many of the landforms previously produced during the advance and retreat of the valley glacier can be removed. If there is any scope for lateral swinging of proglacial meltwater streams then they can destroy even large arcuate moraines and completely clean out glacial deposits and landforms from the floor of a valley. For this reason it is rare to see a series of arcuate moraine ridges linked to lateral moraines and interspersed with fluted ground moraine or ablation moraine on valley floors, if deglaciation has been accompanied by large discharges of meltwater.

Meltwater flow from ablation on valley glaciers is often supplemented by the drainage of ice-dammed lakes. Such lakes may develop in tributary valleys because of differential rates of retreat in different glaciers, or at the junction of two tributary glaciers. These lakes sometimes overflow across the ice surface or along the glacier margin but more frequently they drain by englacial or subglacial routes. Their drainage can be of a catastrophic nature and very large volumes of meltwater can be added to the normal meltwater flow over relatively short time periods. Under such circumstances both subglacial and proglacial meltwater streams can experience massive increases in their discharge rates, that not only have dramatic geomorphological implications but can be very significant in terms of damage to settlements, communications and crops in the proglacial area.

Fluvioglacial deposition associated with valley deglaciation is subject to many of the constraints discussed in connection with fluvioglacial erosion. The same meltwater streams that erode and transport material are capable of deposition. Deposition may occur on lateral margins to produce kame terraces, or in submarginal tunnels producing what Mannerfelt (1945) has called 'subglacially engorged eskers'. Tunnels and caves in and beneath the ice can also become choked with fluvioglacial sediments and would produce eskers and kames when the ice melted away. However, it is unlikely that many ridges and mounds of fluvioglacial deposits laid down on valley floors will survive the deglaciation process. The fact that they are associated with actively moving ice may well, in itself, explain their destruction if ice movement continues after their deposition. On the other hand their deposition on the valley floor also exposes them to the full power of fluvioglacial erosion as deglaciation is continued. Such features are easily destroyed during flood stages of proglacial meltwater streams.

One set of fluvioglacial deposits that is likely to survive deglaciation are the deltaic deposits associated with ice-dammed lakes. Where extra-glacial or meltwater streams enter these lakes, deltas are built up. If the lakes are drained catastrophically, then the deltas tend to survive and are only eroded to the extent that a stream system usually establishes an incised course through the delta form.

A similar situation can develop when the retreat of a valley glacier reveals an over-deepened basin that allows the development of a proglacial lake. Not only will glacial and fluvioglacial deposition be affected by the presence of such a lake but it is likely that the rate of retreat of the ice front will be accelerated as a result of calving of icebergs into the lake. The meltwater streams entering a proglacial lake may produce a delta which in turn may represent the distal part of a valley sandur as retreat progresses.

The most significant result of fluvioglacial deposition during the deglaciation of a valley is the accumulation of sands and gravels as channel deposits in the proglacial area. The development of valley-sandar associated with braided proglacial streams is a very common phenomenon. The infilling of the valley floor with fluvioglacial deposits tends to accentuate the U-shaped cross profile of many glaciated valleys. The subsequent incision of the proglacial fluvioglacial deposits by proglacial streams can also produce terraces that may often be linked to specific moraines and therefore to periods of still-stand in the ice retreat.

The above discussion has been primarily concerned with valley deglaciation in situations where the glacier has retained its forward movement during deglaciation. Local stagnation and separation of relatively small ice masses was discussed but it is now necessary to turn to the effect of ice stagnation on a large scale. Flint (1929, 1930), Lougee (1940), Flint and Demorest (1942), Mannerfelt (1945, 1949), and Hoppe (1950), have all suggested models of deglaciation in which downwastage of ice sheets in areas of mountain topography has produced stagnation of valley glaciers. The lowering of the surface of an ice sheet to such an extent that it is split up into a series of valley glaciers not only changes the climate of the area but also the glaciological characteristics of the ice. It is likely that when the underlying valley system becomes dominant in determining the flow lines within the ice some interfluves will cut off some glaciers from their accumulation areas. In such circumstances forward movement will soon cease, downwastage will become dominant and stagnant ice masses will occupy valley floors. Examples of downwasting, stagnant valley glaciers are known to have existed in the

Scandinavian mountains, the Scottish Southern Uplands and the mountains of New England during the last of the Pleistocene glaciations.

When stagnation of a valley glacier occurs, glacial erosion ceases and glacial deposition is limited in the subglacial environment to the melt-out of debris already entrained in the ice and to the dumping of ablation moraine on the final melting of the stagnant ice. A certain amount of squeezing of basal till into subglacial crevasses and cavities may occur when the large amounts of meltwater that are released by the melting of the stagnant ice penetrate the subglacial materials. In such circumstances moraine ridges and hummocks may be produced as the decaying ice masses sink into the water-soaked subglacial till.

Undoubtedly the most significant geomorphological processes associated with stagnant valley glaciers are fluvioglacial erosion and deposition. The development of lateral marginal, submarginal, supraglacial, englacial and subglacial drainage systems allows fluvioglacial erosion and deposition to produce a wide range of erosional and depositional landforms that characterize the wastage of a stagnant valley glacier. Wherever meltwater streams impinge on the ice-rock interface suites of meltwater channels are cut. These are often linked by deposits in the form of kame terraces or eskers indicating depositional phases of the same meltwaters that cut the channels. Wherever fluvioglacial deposits are washed over detached blocks of ice, kame and kettle topography develops and where similar deposits are accumulated in ice caves, individual kames are formed.

Some authors (e.g. Andersen, 1931) have suggested that meltwaters are so important during the wastage of stagnant valley glaciers that a water table is established within the decaying ice, and this can rise to sufficient heights to cause an irregular streaming of meltwater beneath the ice so that deposits occupy every hollow and depression in the sub-ice surface. Other authors (e.g. Sissons, 1957) have stressed the importance of englacial water tables in separating zones of fluvioglacial erosion from zones of fluvioglacial deposition. Such a situation is most likely to develop when the surface slope of the ice is in the opposite direction to the slope of the land surface beneath the ice. The thicker ice at the lower end of the valley may well assist in the development of an englacial water table within the thinner, more decayed ice in the upper end of the valley (Fig. 92).

In the above discussion of the deglaciation of valleys occupied by active or stagnant glaciers it has been assumed that the ice was at the pressure melting point. It is now necessary to ask in what ways the landforms and deposits would be different if the ice were below the pressure melting point. There appears to have been no systematic study of the differences between landforms and deposits that develop under these two sets of very different ice conditions, and this is an area of investigation that urgently needs the attention of glacial geomorphologists. On the basis of general principles, it can be stated that when a valley glacier is frozen to its bed the subglacial environment is of little significance for landform development, either by erosion or deposition, and most geomorphological activity is concentrated along the lateral and frontal margins. It is also likely that fluvioglacial processes will be much less important than when the ice is at the pressure melting point. Glacial deposition both by the melting out of englacial debris at the frontal margin and by dumping of supraglacial debris will produce moraines at the frontal margin. With only relatively small discharges of meltwater such frontal moraines are more likely to survive than if they were developed at the margin of a temperate valley glacier.

If a valley glacier is below the pressure melting point meltwater will not penetrate below its surface and meltwater streams will occur at the lateral ice margins, on the glacier surface and in the proglacial area during the ablation season. This situation could result in the removal of a lot of debris from the ice surface into the proglacial area to produce a valley-sandur.

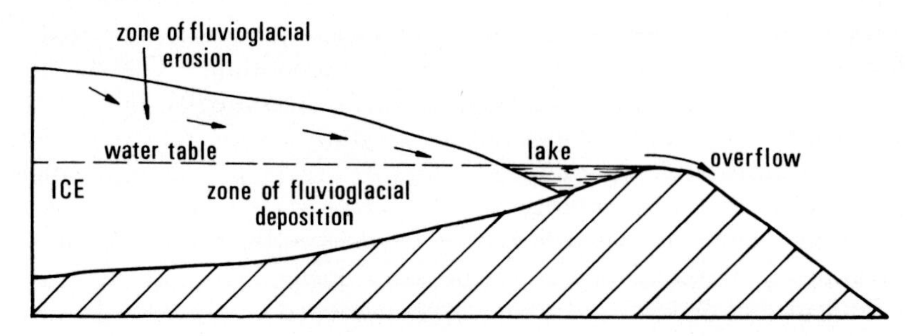

FIG. 92. The development of a water table within an ice mass. The water table will control the extent of fluvioglacial erosion and deposition.

Whether large areas of valley glaciers that are below the pressure melting point become stagnant is not known. Since meltwater activity is likely to be of limited significance in such circumstances it is unlikely that the complex of fluvioglacial erosional and depositional landforms that is associated with wasting, stagnant, temperate ice will develop.

The interpretation of the stratigraphy that is produced during the retreat of a valley glacier is dependent upon the survival of at least remnants of each of the types of deposits associated with deglaciation. If the stratigraphic record is complete then the fluvioglacial and glacial deposits laid down during the advance of a valley glacier are succeeded upwards by further accumulations of basal till followed by further fluvioglacial deposits. If the fluvioglacial deposits were laid down in the subglacial environment then they may have a cover of ablation till. There are two reasons why this simple succession is rarely seen in a valley that has experienced one period of glaciation and one of deglaciation. The deposits are either accumulated on the valley sides where mass movements subsequent to deposition tend to destroy the stratigraphy, or they are accumulated on the valley floor, where proglacial meltwater streams tend to destroy at least significant parts of the sequence.

Even during one period of deglaciation it is unlikely that the retreat of a valley glacier, once started, will continue without either periods of still-stand or readvance. The effects of readvances are in many ways similar to the effects of multiple glaciation except in terms of the size of the area affected. A readvance of a valley glacier will either override or erode the deposits produced during the previous period of deglaciation. If erosion takes place then it may be very difficult to prove readvances on stratigraphic grounds. In the same way, in areas of multiple valley glaciation and deglaciation, the concentration of erosional and depositional processes along the bottom of valleys makes the stratigraphic sequences on the valley floors unreliable indicators of multiple glaciation and deglaciation.

Any model of valley deglaciation must stress the confining effect of the valley walls and the concentration of geomorphic processes within the narrow zone of the valley floor. Landforms and deposits that are produced during one phase of deglaciation are destroyed by later phases, by readvances or by multiple glaciation. For this reason geomorphological studies of deglaciation phenomena are usually much more fruitful along the frontal margins of ice sheets than along the margins of valley glaciers.

ICE SHEET DEGLACIATION

The major distinguishing feature separating ice sheet deglaciation from valley deglaciation is the total length of the marginal zone and therefore the wide variety of marginal environments in which deposits and landforms can develop. Although the same factors that control the retreat of a valley glacier control the retreat of the margin of an ice sheet the effect of the form of the underlying land surface is far less constricting.

Once an ice sheet begins to experience a negative mass balance and its total volume begins to decrease, retreat of its margins will commence. The rate of downwastage in the marginal zone will reflect the relationship between ice replenishment by forward motion and the ablation rates. The actual rate of marginal retreat will be determined by the rate of downwastage, the slope of the ice surface and the slope of the ground surface (Figs 89, 90). Information is available about the nature and rate of retreat of the Laurentide and Scandinavian ice sheets during the last glaciation (Flint, 1971, p. 492 and 608).

The Laurentide ice sheet began to retreat between 18000 and 15000 years BP depending on location, and the ice sheet had ceased to exist by 7000 years BP. Deglaciation was a process accelerating with the passage of time. On the northern and eastern margins calving of ice into the sea constituted a major method of retreat. On its western margins the ice sheet rested against the regional slope and thinning and stagnation accompanied by the development of many proglacial lakes characterized the retreat. Along the southern margins the ice retreated across a land surface with a general slope towards the south until the Great Lakes basins were reached, and proglacial lakes developed and accelerated the retreat rate. Two substantial readvances occurred, after the Great Lakes developed, at about 13000 and 11000 years BP respectively. Flint (1971, p. 493), describes the deglaciation and readvances in the Great Lakes area as follows: 'As examples of the fast timetable we can cite retreat from the Lake Border moraines south of Lake Erie, mostly by calving, to the center of the Lake Huron basin, about 300 km, followed by readvance of about 200 km to build the Port Huron moraine, all within about 1000 years. In the following 2000 years the terminus retreated between 400 and 500 km and then readvanced nearly the same distance (in the Lake Michigan basin) to the Valders maximum.' It appears that maximum rates of retreat of the margin of the Laurentide ice sheet were of the order of 400 to 500 m per year and that rates of between 100 and 200 m per year were quite common. The gradual thinning of the ice sheet is demonstrated by the increased lobation of the ice margin as wastage progressed. It is possible that conditions along the southern margins of the Laurentide ice sheet may have been similar to those that exist today along the western and southern margins of Vatnajökull in Iceland, in as much as the southern side of the Laurentide ice sheet would have been exposed to moist warm air streams originating in the Gulf of Mexico. The final break up of the Laurentide ice sheet occurred when the sea entered Hudson's Bay about 8000 years BP and the bay itself was free of glacial ice by 7000 years BP.

The retreat of the last Scandinavian ice sheet is also fairly well known. Since it originated as a mountain ice sheet, the lowering of its surface produced important areas of stagnation within the mountain area. It also thinned and became stagnant over considerable lengths of its outer margin. This ice sheet reached its maximum extent between 18000 and 20000 years BP. Rates of retreat during the first 6000 years of wastage probably ranged between 40 and 60 m per year. By 8000 years BP the lowland ice sheet had disappeared.

Information about the rates of retreat of the margins of the last British ice sheet is very limited. The break-up of the ice sheet into ice streams largely controlled by the underlying ridges and valleys probably occurred fairly early in the deglaciation. The maximum extent of the ice sheet probably occurred about 18000 years BP and one of the longest flow lines in the ice sheet was from the Firth of Clyde to the east of the Isle of Arran to the southern side of Cardigan Bay, a distance of about 400 km. It is likely that the east coast of Arran was largely ice free by 13000 years BP and that the average rate of frontal retreat through the Irish Sea Basin and the Firth of Clyde was of the order of 100 m per year. The extent to which this retreat occurred on land or in the sea is unknown.

Remembering that a great deal of the retreat of the Scandinavian, British and Laurentide ice sheets occurred at relatively low altitudes (i.e. below 500 m) and that probably the most common marginal environment was one of relatively warm and moist conditions, the best locations to examine comparable retreat phenomena along the margins of existing glaciers are in southern Alaska and Iceland. Both areas contain piedmont ice lobes, the margins of which are at low altitudes, and are experiencing rapid wastage under relatively mild and moist climatic conditions. Rates of retreat of Alaskan glaciers across land surfaces are of the order of 50 to 200 m per year. Much more rapid rates of retreat occur when the ice fronts end in sea water and rates as high as 400 to 800 m per year are known to have occurred in Glacier Bay, Alaska (Price, 1964). In south-east Iceland rates of retreat over land surfaces during the last century commonly range between 40 and 100 m per year.

Variations in the slope of the land surface across which an ice sheet margin retreats are of great significance. If one assumes a constant surface slope direction and therefore direction of movement on the periphery of an ice sheet, the lowering of that ice surface will eventually result in the underlying land surface being intersected. If that land surface has a local relief of the order of hundreds of metres then a stage will be reached when at least the basal layers at the margin of the ice sheet will become stagnant and the ice surface will be intercepted by the emergent land surface. This situation can produce blocks of stagnant ice in depressions that may be a couple of hundred metres thick and several kilometres in length.

The contrast between deglaciation of an area as the result of the retreat of an ice margin associated with active ice and the deglaciation of an area by means of downwasting stagnant ice will be discussed below. From a glaciological point of view it must be stressed that stagnation on the periphery of ice sheets is not an unusual phenomenon. The peripheral zones of ice sheets can be at great distances from the accumulation areas that feed them. They may also experience climatic conditions even at the maximum extent of the ice sheet, and therefore at a time when climatic deterioration is at its greatest, which are only marginally in favour of the extension of the ice sheet. The first amelioration of climatic conditions may mean that the supply of ice to the periphery of the ice sheet will rapidly decrease, and thinning and stagnation of the marginal zones will occur. In such circumstances downwastage rates of between 5 and 10 m per year would not be excessive.

If it is assumed that the marginal zone of an ice sheet at its maximum extent is at the pressure melting point and that ablation is sufficiently great to cause a lowering of the ice surface even though ice is continuing to arrive at the margin from the accumulation zone, the ice margin will retreat even though ice movement is sustained. In such a situation glacial erosion can still occur, although it will not be severe and may be restricted to abrasion of rock surfaces and the moulding of drift deposits beneath the ice. It is possible for fluted surfaces on subglacial till to develop at this stage.

The continued forward movement of the ice means that debris in the basal layers, within the ice or on the ice surface is continually moving towards the ice margin. With ablation taking place, this debris is released on the ice surface or beneath the ice. The surface of an ice sheet can become covered by debris although it is commonly not more than 'one-boulder' thick, but if there are depressions on the ice surface, and meltwater streams or mass movements cause the concentration of the debris in certain areas, it may attain thicknesses of 1-3 m. The continued wastage of the ice will eventually result in this debris cover being let down on to the land surface in the form of ablation moraine consisting of a sheet of angular unsorted boulders and gravel. Many of the finer particles are often removed from ablation moraine by meltwater transport.

Subglacial deposition may be considerable during the retreat phase of a temperate ice sheet. The penetration of meltwater to the base of the ice sheet may accelerate basal melting and therefore debris in the basal zone of the ice sheet may be released to produce basal till. Since forward movement is still taking place, variations in basal friction may allow till accumulation in mound form. The development of drumlins is most likely to take place under such circumstances. The water-soaked till in the subglacial environment, which is also likely to move under differential pressures, will fill any subglacial caves and crevasses and cavities on the lee side of boulders, and move into the proglacial environment along the frontal margin. To what extent subglacial crevasses and caves remain open when ice movement is still taking place is unknown. Ice movement certainly tends to decrease towards the margins of ice sheets and may be as little as a few millimetres per day right at the margin. The slowing down of rates of forward movement near the ice margin may be sufficient for the ice to behave as if it were stagnant and for crevasses, caves and tunnels to remain open.

The formation of moraine ridges along the margin of a retreating, but active ice sheet, is most likely to occur when the ice margin remains stationary long enough for material to accumulate at the margin. Material will slump off the ice surface and be dumped along the ice margin and it will assume a ridge form when the supporting ice on

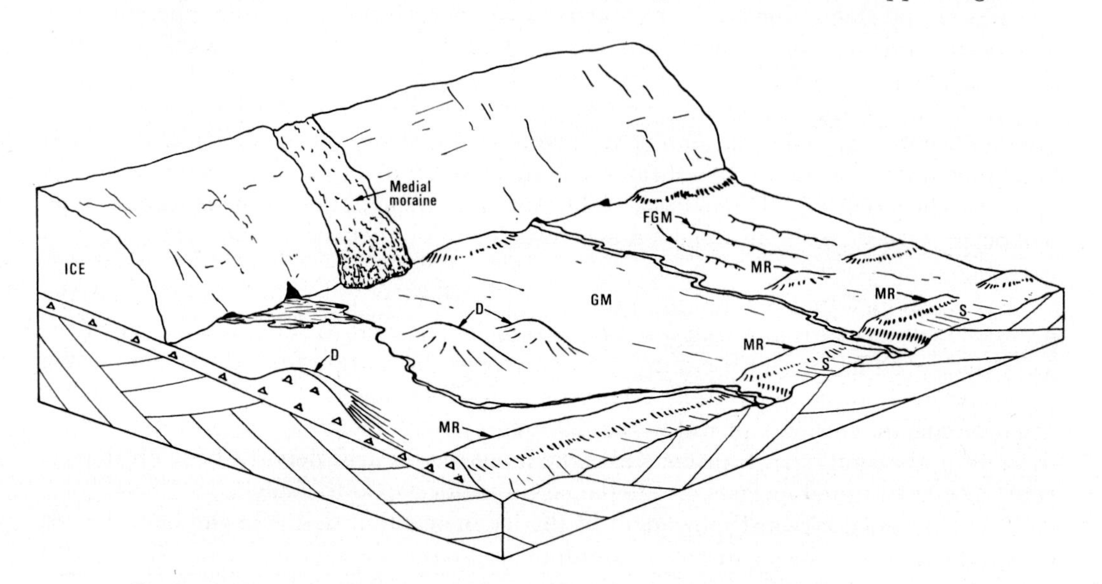

FIG. 93. A proglacial area with forms mainly produced by glacial deposition. *MR*—moraine ridge; *GM*—ground moraine; *FGM*—fluted ground moraine; *D*—drumlin; *S*—sandur.

the proximal side melts away. Other moraines are produced when material actually within 'shear planes' melts out and is let down on to the sub-ice surface. The squeezing of water-soaked till from beneath the ice can also produce moraine ridges at the ice margin.

As the ice margin retreats, glacial deposition is responsible for the development of three types of landforms (Fig. 93). Beneath the ice, a sheet of basal till produces a ground moraine or till plain. This sheet may be only a matter of a metre thick or it may attain a thickness of tens of metres. Its upper surface may be flat or it may have a local relief of a few metres or tens of metres. If the upper surface is fluted this may represent the corrugated nature of the sub-ice surface and the fluted ground moraine will have a local relief of 1-10 m. On the other hand the development of drumlins on the ground moraine may produce a local relief of 50 to 100 m. The flat ground moraine and the fluted or drumlinoid surface are usually also associated with moraine ridges. When these consist solely of supraglacial ablation till and basal till it is unusual for them to attain heights greater than 20 m. Whereas the lineations on ground moraine are generally parallel to the direction of ice movement, moraine ridges are usually transverse to it and frequently outline any crenulations in the ice margin.

For the sake of simplicity the deposits and landforms associated with glacial erosion and deposition have been treated separately from those produced by fluvioglacial erosion and deposition. However, the retreat of the ice sheet is accompanied by the release of meltwater and, at the same time as glacial deposition is taking place, the meltwaters are eroding these glacial deposits both under the ice and in the proglacial area. A major function of fluvioglacial erosion during the retreat of an ice sheet is the removal of glacial deposits, their transportation and redeposition both on, within, beneath and in front of the ice sheet. Since the ice sheet is at the pressure melting point meltwater will flow in channels on the ice surface, in tunnels in and under the ice surface and across the proglacial area.

Supraglacial streams will transport debris into the englacial and subglacial tunnels. Again the effect of the rate of movement on the survival of such tunnels and their deposits is important. So long as movement is not too rapid the tunnels will survive and will allow the accumulation of sediments on their floors. The complicated hydrology, and its associated sedimentary environments, of englacial and subglacial drainage systems, can only be interpreted from the landforms and deposits they produce. There is little doubt that both fluvioglacial erosion and deposition does take place both within and beneath the ice in the peripheral zone of a retreating active ice sheet. Wherever meltwater streams impinge on the sub-ice surface, channels can be cut either in the drift deposits or in solid rock. In other locations the sands and gravels deposited in the supraglacial channels or englacial tunnels and caves can eventually be let down, as the result of continued ice wastage, on to the sub-ice surface to produce eskers and kames. In a similar way the deposits accumulated in subglacial tunnels and caves are revealed by ice retreat and reflect the shape of the tunnels and caves in which they accumulated. The width of the zone in which these channels, tunnels and caves occur presumably reflects the progressive change in rates of ice movement from the interior to the periphery of the ice sheet. This zone will differ very little, in terms of the landforms it generates, from a completely stagnant ice mass.

Meltwaters crossing a proglacial area covered by a till sheet, fluted ground moraine and moraine ridges will develop a drainage network that can either erode the glacial deposits, partially erode them, erode and bury them or simply bury them with fluvioglacial deposits. The complexity of the marginal environment both in terms of the changing conditions over short periods of time and from place to place is very great. Moraine ridges developed along one part of an ice sheet margin over a few weeks may be destroyed in a matter of days by a meltwater stream paralleling the ice front. On the other hand, the development of a particular moraine ridge may direct the course of meltwater movement elsewhere and the moraine ridge will survive and an area of fluted ground moraine may be eroded or buried beneath a spread of sand and gravel. The evolution of marginal and proglacial drainage systems is largely controlled by the detailed form of the recently exposed land surface as a result of deglaciation. Minor changes in slope of the emerging sub-ice surface will direct meltwater flow first in one direction and then another. Channel systems both parallel to and at right angles to the retreating ice margin develop during deglaciation.

Meltwater channels parallel to the retreating ice margin often develop between moraine ridges. They may result in the partial or complete destruction of the moraine ridges and certainly remove traces of ground moraine between them (Fig. 94). Channel systems that result from meltwaters issuing at the ice front and flowing directly away from the ice produce gaps in moraine systems. Beyond these gaps their routes may simply take on a braided pattern and ground moraine is systematically stripped

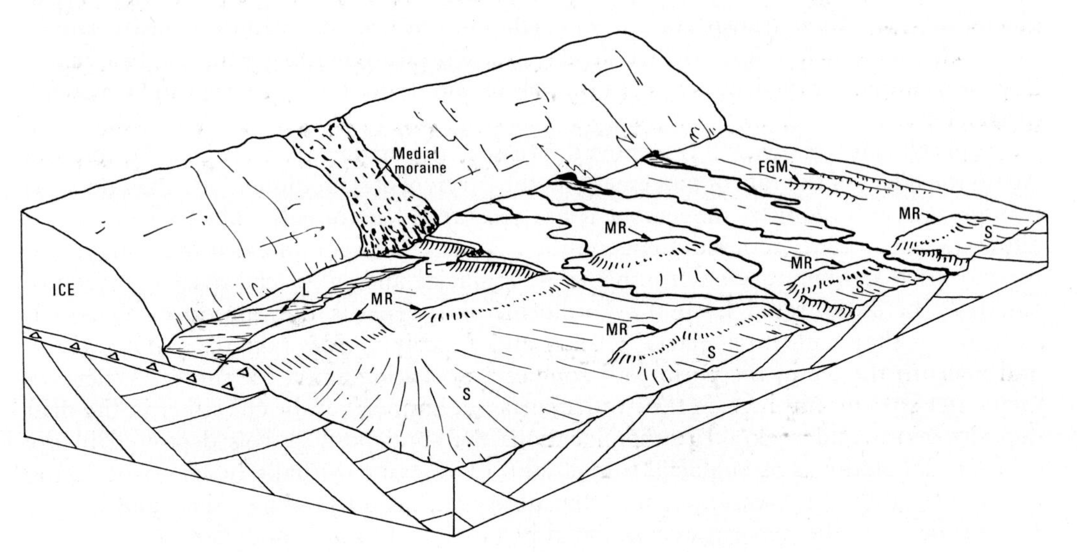

FIG. 94. A proglacial area dominated by fluvioglacial erosion and desposition. *S*—sandur; *E*—esker; *FGM*—fluted ground moraine; *MR*—moraine ridge; *L*—lake.

from the proglacial area and redeposited as a sandur farther down the system. Where ground moraine is strongly fluted, or where drumlins occur, the proglacial drainage pattern will be strongly influenced by the distribution of flutes or drumlins. On a fluted surface the drainage pattern tends to be trellis-like with many right-angle bends.

Whether or not a proglacial drainage system will be primarily responsible for erosion or deposition largely depends on the discharge and load characteristics of the meltwater streams and the slope of the surface across which they develop. It is likely that each proglacial stream will experience periods in which it is dominantly an eroding and transporting agent and other periods when it is dominantly a depositional agent. When a period of deposition has occurred then proglacial sandar develop. In their simplest form they are fan-shaped wedges of fluvioglacial sediments produced by migrating braided streams flowing across the proglacial area. The rapid changes in discharge, load and point of emergence of subglacial and englacial streams mean that the apex of these fans may change location during ice retreat. This means that they are rarely simple fans but consist of overlapping fans with migrating channel systems on their surfaces. Large parts of till plains and even moraine ridges can be buried by sandar development. If an ice margin remains stationary for any length of time, streams issuing at the ice front under hydrostatic pressure or supraglacial or englacial streams can cause fluvioglacial deposition to take place at altitudes well above the sub-ice surface and parts of the marginal ice may become buried by the proximal ends of sandar. In some instances, eskers developed in tunnels within the ice represent continuations of sandur deposits within the ice mass. Both the proximal parts of the sandar and the eskers, in such situations, will be significantly altered when the ice against or upon which they developed melts out.

In the model of deglaciation discussed above it was assumed that the frontal margin was retreating across either a horizontal or low angle normal slope. In such circumstances small scale ponding of water may occur between successive moraine ridges. If, however, ice retreat takes place across a reverse slope (Fig. 95), then the proglacial

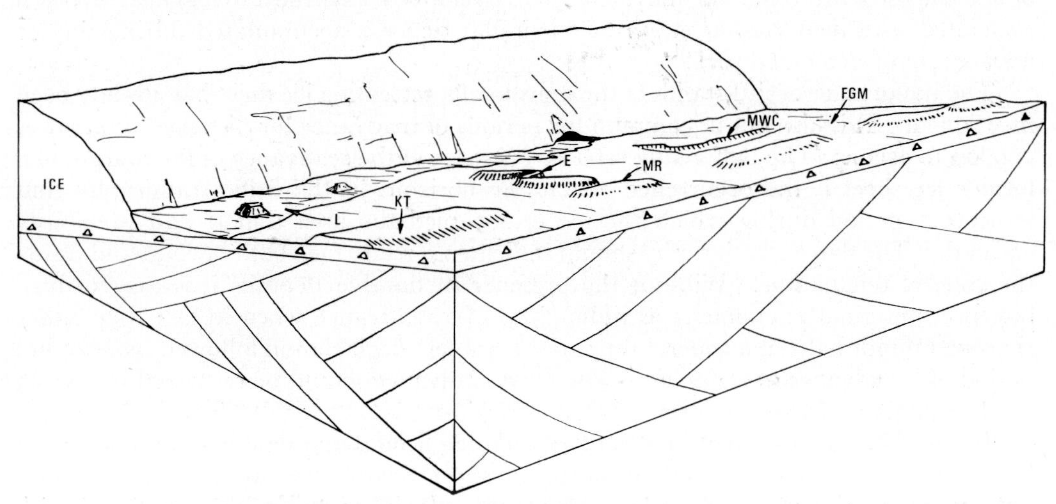

FIG. 95. A proglacial area with a ground surface slope towards the retreating ice margin. *MR*—moraine ridge; *FGM*—fluted ground moraine; *MWC*—meltwater channel; *KT*—kame terrace; *E*—esker; *L*—lake.

system may be entirely different. Meltwater streams flowing parallel to the ice front may even turn back into the ice to take advantage of lower outlets at other points along the margin. The development of ribbon lakes along the ice front will result in the deposition of sands and gravels to produce kame terraces. The expansion of ribbon lakes into major proglacial lakes will introduce the whole range of lacustrine environments for sediment accumulation. Deltas, beaches and bottom deposits will be laid down on top of the newly emerged sub-ice surface. The lake itself may well accelerate the rate of frontal retreat, or if its waters penetrate beneath the ice sheet they can accelerate rates of ice movement. The period of time over which the proglacial lakes exist is determined by the efficiency of the ice as a dam and the maintenance of the reverse slope on the proglacial surface. If the reverse slope is composed entirely of glacial and fluvioglacial deposits then the establishment of an overflow channel through the ridge may soon lead to a lowering and perhaps the drainage of the lake. Whether retreat of an ice front is associated with a lacustrine environment or with a terrestrial one can produce fundamental differences in the range of deposits and landforms that are produced during delgaciation.

The stratigraphy of an area that has experienced one period of deglaciation and in which all the deposits have survived after deposition is as follows. The subglacial till or ground moraine is succeeded upwards by fluvioglacial deposits that were laid down either beyond the ice margin or within the ice. If the fluvioglacial deposits were laid down within the ice they may have a thin cover of ablation moraine. Since it is impossible to distinguish between a subglacial till deposited during ice advance and one produced during retreat the only significant indicator of a period of deglaciation is a series of fluvioglacial deposits. If meltwater erosion occurred prior to the accumulation of the deposits the basal till may have been removed and the fluvioglacial deposits associated with deglaciation may rest on similar deposits accumulated during the ice advance, or even on solid rock.

The importance of still-stands of the margin of a retreating ice sheet has already been mentioned. It is also not uncommon for periods of readvance for distances as great as 200 km to occur. The only really reliable evidence of the readvance of the margin of a former ice sheet is the occurrence of datable horizons in the drift stratigraphy that indicate a period during which the ice re-occupied the area that it had previously vacated. The dated stratigraphy should indicate no readvance both beyond and inside the zone of fluctuation. Without the presence of datable deposits it is easy to misinterpret marginal phenomena as indications of a readvance when in fact they either represent minor halts in a general retreat or complete deglaciation followed by a second period of ice advance and retreat. The term readvance should be restricted to proven local oscillations of an ice margin.

In the above discussion of ice sheet deglaciation, frontal marginal environments have been stressed. When the thinning of an ice sheet occurs over an area that contains isolated mountains then, of course, these areas of higher ground emerge as the ice surface is lowered. The term 'nunatak' is of Eskimo origin and refers to a rock peak appearing above the Greenland ice sheet. It is unfortunate that the English sounds in this term suggests the meaning 'not-attacked' because in many instances such rock peaks were buried beneath an ice cover and then appeared above it as the ice surface was lowered. The landforms and deposits that develop along these nunatak ice margins are similar to those that develop along the frontal margin.

The landforms and deposits that develop as the result of the retreat of the margin of an ice sheet that is below the pressure melting point will mainly differ from those associated with the wastage of a temperate ice sheet in terms of the extent of fluvioglacial activity. It is possible that a greater percentage of marginal glacial deposits will survive because fluvioglacial erosion will be less active. The lack of penetration of meltwater into the retreating ice sheet will restrict both fluvioglacial erosion and deposition to the marginal and proglacial environments.

Ice sheet deglaciation has so far been discussed in terms of active ice with ice margins retreating due to ablation even though the ice itself continued to have forward motion. It has also been suggested that under slow rates of movement englacial and subglacial cavities and their associated deposits can survive. When forward movement ceases completely over extensive peripheral zones of an ice sheet then new and different environments for the development of deposits and landforms are created. Whatever the reasons for stagnation are, and whether they affect the whole of a wasting ice sheet or only peripheral areas, the net result is the decay of ice primarily by downwastage and bottom melting.

No erosional phenomena are known to be created by stagnant ice itself, although the meltwaters produced from the wasting stagnant ice can be significant erosional agents. The most notable landforms developed by the wastage of stagnant ice are depositional in origin and the types of landforms may well be affected by the extent of the stagnant ice, the structures that ice inherited from its active phase and the amount and type of debris on, within and beneath the ice. Some of these characteristics are related to the causes of stagnation. If, for example, stagnation has been induced by thinning of the ice sheet resulting in the introduction of an element of control by irregularities in the sub-ice surface, stagnation may be local. On the other hand stagnation may be very extensive due to a rapid and major change in climatic conditions and what was an active ice sheet may suddenly become stagnant and inherit all the structural features and debris content of the active phase.

The wastage of completely debris-free stagnant ice is unlikely to produce any significant landforms and will simply increase meltwater discharge. When stagnant ice that has a high debris content begins to waste, then the range of deposits and landforms that are produced as it melts away can be large. Debris occurring on and within stagnant ice is released by melting and as the ice surface downwastes it can accumulate to a thickness of several metres. The debris itself can inhibit and produce differential rates of ice wastage. Areas of thin debris cover can, in fact, speed up ice wastage due to the greater absorbing power of thin layers of dark coloured material, and in such situations depressions tend to develop on the ice surface. Ice covered by thicker debris layers, is protected from melting at such a fast rate as clean ice, and areas of upstanding debris-covered ice develop. There is a tendency for an irregular surface to develop and for mass-movements of debris to occur across the steeper slopes. It is possible for inversion of relief to take place when debris concentrated in hollows provides protection for the ice beneath the hollow so that as general downwastage continues hollows are converted into mounds. The redistribution of supraglacial debris in this way can produce a very confused ice surface morphology, and an equally confused series of mounds and hollows consisting of ablation till, when all of the ice has melted away. If the wasting ice mass has a strongly developed crevasse system, some surface debris migrates into the crevasses and will assume a linear pattern of ridges reflecting the former crevasse pattern on final deglaciation. There may be a certain amount of reworking of supraglacial debris by meltwater on the downwasting ice surface. This may simply be limited to the removal of the finer materials, leaving a coarse deposit of angular fragments. If surface meltwater activity is more widespread the supraglacial material may experience considerable water transport and fluvioglacial deposits will occur in hollows and open crevasses on the ice surface.

Beneath masses of stagnant ice englacial debris is released by bottom melting and subglacial tills can be formed in this way. The large amounts of meltwater that occur in association with wasting, stagnant, temperate ice cause basal tills to become soaked with water. Any caves or basal crevasses tend to become filled with water-soaked till which flows into these cavities under pressure from the ice above, and as a result of differential pore-water pressures in the till. The movement of water-soaked till in the subglacial environment is capable of producing a wide range of landforms, ranging from simple ridges and mounds to complex areas of what has been described by some Scandinavian workers as hummocky moraine. These complex areas may contain single ridges tens of metres high, extensive plateaux with or without rim ridges, or an assortment of mounds and hollows. If the alignment of ridges and mounds produced in this way is determined by structures within the melting ice, the resultant forms are known as 'controlled disintegration features'. The supraglacial debris, in the form either of a simple sheet or a complex of hummocks or ridges may be let down on top of the subglacial features to produce a very complicated sequence of landforms and deposits.

The wastage of a large area of stagnant ice can produce a great deal of meltwater. The movement of this meltwater on, through and beneath the ice produces a complicated network of channels. The evolution of the meltwater drainage system associated with wasting stagnant ice has been likened to drainage systems on limestone areas. The erosion of the ice is largely by melting, but tunnels, caves and surface depressions develop along with collapse forms such as gorges. Wherever the drainage system impinges on the ground surface beneath the wasting ice meltwater channels are cut. The patterns exhibited by meltwater drainage systems in stagnant ice tend to be controlled by the slope of the ice surface and in some cases by ice structures. The superimposition of meltwater streams from the stagnant ice on to the land surface beneath the ice tends to produce meltwater channel systems transverse to any spurs or ridges that are not parallel to the ice surface slope.

The proglacial drainage system associated with wasting stagnant ice may experience rapid changes in discharge. Any individual proglacial stream may have its drainage area enlarged or reduced from time to time as collapse of tunnels or caves in the ice result in the re-routing of meltwater flow. Quite large supraglacial, englacial or subglacial chambers can develop and these sometimes drain quite rapidly, adding large volumes to the stream that they feed.

Associated with the meltwater drainage system are extensive fluvioglacial deposits. The meltwater streams accumulate sediments in channels on the ice surface, in tunnels and caves within and beneath the ice and in the proglacial area. The lack of forward movement in the ice and the relatively slow lowering of these deposits as the ice wastes away permits the preservation of these deposits so that they appear on the land surface as the casts of the channels or cave systems in which they were deposited. Some of the deposits are destroyed by subsequent erosion by the same streams that deposited them. Those deposits that are not destroyed maintain the sedimentary characteristics of the environments in which they were deposited and have surface forms that reflect the shape of ice walls bounding the sedimentary basins or channels. The sedimentary characteristics of the deposits vary considerably. All of the deposits reflect their fluvial origins but some indicate deposition in high energy environments whereas others are fine grained and laminated and indicate deposition in standing water.

The form of the deposits ranges from single ridges or mounds to complex anastomosing ridge systems and chaotic mounds and hollows. If a complex englacial or subglacial tunnel system becomes completely choked with sediment, possibly as a result of a flood of meltwaters, the resulting landform when all the ice melts away is a ridge system that duplicates the original tunnel system. The removal of ice support from the sides of, and beneath fluvioglacial deposits results in a certain amount of collapsing along the sides of ridges or mounds that in turn can cause at least partial destruction of the original structures within the sediments.

The landforms that result from the accumulation of fluvioglacial sediments in the channels, depressions and caves that develop during the wastage of stagnant ice produce a distinctive assemblage. The ridge forms are known as eskers and the mounds are kames. The complexities in the methods of formation and the classification of these landforms have been discussed in Chapter VI. The fact that these forms are associated with stagnant ice means that blocks of ice can become buried by the deposits and can survive long after the wastage of the main ice mass. When these buried ice blocks eventually melt out, kettle holes develop, and although kettle hole development is not restricted solely to areas of stagnant ice their development in large numbers in association with eskers and kames is usually a strong indication of the former presence of stagnant ice.

Fluvioglacial deposition at and in front of the margin of a retreating stagnant ice sheet occurs in two very distinctive environments. Deposition takes place either in proglacial lakes or in channel systems in an otherwise terrestrial environment. An additional complication is that the line indicating the margin of a stagnant ice sheet is often very difficult to determine. Buried ice beyond the visible ice margin and the extension of lake waters or channel systems from the proglacial area into and beneath the ice means that there is often a continuum of depositional environments across the apparent ice margin.

The streams that issue along a stagnant ice margin into a proglacial lake are responsible for the usual range of sediments in lacustrine environments. Fine, deep water sediments, coarser beach deposits, and typical delta sequences are laid down in the proglacial lakes. What makes this environment distinctive is the possibility of the incorporation of ice blocks or sheets of stagnant ice within the deposits. This means, for example, that deltas can develop on top of ice shelves which project into lakes. When the ice on which these deltas have developed eventually melts away, kettle holes will be produced. Blocks of stagnant ice can also float across the lake allowing the transport of coarse material into the central parts of the proglacial lake. When these ice blocks melt the coarse material they contain is deposited on the lake floor on to otherwise fine material.

Meltwater streams issuing along the retreating margin of a stagnant ice sheet tend to carry large debris loads and to experience variable discharges. Under these circumstances it is not surprising that they develop braided courses. The migration of these braided courses produces fan-shaped accumulations of fluvioglacial sediments that merge together to produce great outwash plains or sandar. Sandar tend to have steeper gradients if they are of limited length and the longer sandar are steeper at their proximal than at their distal ends. The detailed surface morphology of sandar consists of abandoned anastomosing channel systems. If during retreat, one meltwater stream becomes dominant then it may incise its course into the sandur surface, and terrace development and the abandonment of the old sandur surface will result.

The proximal part of a sandur may be built against or even on top of the retreating ice margin. This will cause blocks or even extensive sheets of stagnant ice to be buried by the sandur deposits. The melting out of such blocks causes kettle holes to develop, so producing a pitted-sandur. The increased number of kettle holes towards the proximal part of a sandur can result in the virtual destruction of a sandur surface and the development of a chaotic assemblage of mounds, ridges and hollows, very similar to kame and kettle topography. The meltwater streams that developed the sandur may be an extension of an englacial or subglacial drainage system and if eskers were produced by that system they may, on deglaciation, be joined to the proximal, ice-contact slope of a sandur. The linkage between landforms produced within or beneath the ice and those formed in front of the ice may be clearly defined or may be very confused if the marginal environment is dominated by the burial of ice beneath the accumulating sediments.

No reference has been made to the possibility of large areas of stagnant ice developing during the retreat of an ice sheet that is below the pressure melting point. The writer is unaware of any descriptions of such conditions on the margins of existing ice sheets. If such conditions ever do exist then the absence of any large volumes of meltwater would mitigate the development of many of the landforms and deposits that are usually associated with the wastage of stagnant ice.

CONCLUSIONS

The models of glaciation and deglaciation discussed above represent the selection of elements of reality and their combination to produce simple structures with certain built-in assumptions. Any attempt to achieve simplicity denies the very complexity of the real world. The first six chapters of this book emphasized the limitations of current knowledge regarding landform genesis in areas of existing glaciers. Until more is known about the actual mechanics of landform development and the nature of glaciological and sedimentological environments in areas of existing glaciers the development by analogy of models of glaciation and deglaciation must be of limited value.

To the geomorphologist, the interaction between the specialists concerned with the materials and environments involved in landform development is of fundamental importance. Only when more information is available about the behaviour of glaciers and ice sheets, about the hydrology of glaciers and proglacial areas and about the sediments and sedimentary environments, will it be possible to develop more useful models that can be applied to areas affected by the Pleistocene glaciers and ice sheets. The data upon which the models in Chapters VII and VIII have been based are not only limited by the small sample of existing glaciers and ice sheets that have been studied but by the relative short periods over which any particular sample area has been examined. The variations over both space and time, of glacier and ice sheet environments are immense. When the task of the glacial geomorphologist is to apply these inadequate models to Pleistocene environments which, at their minimum involve extrapolation over 10 000 years and may be over hundreds of thousands of years, the inadequacy of the models themselves is compounded by the limitations of the data to which the models are to be applied. The glacial system poses a great many problems, the answers to which lie in renewed efforts in collecting data from existing glaciers and ice sheets, the development of more refined models of landform development and the rigorous application of these models to areas affected by Pleistocene glaciers and ice sheets.

BIBLIOGRAPHY

Agassiz, J. L. R. 1840a. *Études sur les glaciers.* Privately published, Neuchâtel.

Agassiz, J. L. R. 1840b. On glaciers and boulders in Switzerland. *Rep. Br. Ass., 10th Meeting (Glasgow), Sec.* 2, 113-4.

Agassiz, J. L. R. 1840c. Glaciers, and the evidence of their having once existed in Scotland, Ireland and England. *Proc. Geol. Soc.*, **3**, 321-2.

Ahlmann, H. W. 1935a. Scientific results of the Norwegian-Swedish Spitzbergen expedition 1934. Part V: The Fourteenth of July glacier. *Geogr. Annlr.*, **17**, 167-218.

Ahlmann, H. W. 1935b. Contribution to the physics of glaciers. *Geogr. J.*, **86**, 97-113.

Ahlmann, H. W. 1948. Glaciological research on the North Atlantic coasts. *R. Geogr. Soc. Res. Ser.* 1, 83.

Ahlmann, H. W., and Thorarinsson, S. 1939. Vatnajökull. Scientific results of the Swedish-Icelandic investigations (1936-38). *Geogr. Annlr.*, **21**, 39-66.

Alden, W. C. 1918. The Quaternary geology of southeastern Wisconsin with a chapter on the older rock formations. *U.S. Geol. Surv. Prof. Pap.*, 106.

Allen, J. R. 1970. *Physical processes of sedimentation.* Allen and Unwin, London.

Altmann, J. G. 1751. *Versuch einer historischen und physischen Beschreibung der helvetischen Eisgebirge.* Zurich.

Andersen, S. A. 1931. The waning of the last continental glacier in Denmark as illustrated by varved clays and eskers. *J. Geol.*, **39**, 609-24.

Anderson, R. C. 1957. Pebble and sand lithology of the major Wisconsin glacial lobes of the central lowland. *Bull. Geol. Soc. Am.*, **68**, 1415-50.

Andrews, J. T. 1963. Cross-valley moraines of the Rimrock and Isotoq river valleys, Baffin Island. A descriptive analysis. *Geogr. Bull.*, **19**, 49-77.

Andrews, J. T. 1965. The corries of the northern Nainokak section of Labrador. *Geogr. Bull.*, **7**, 129-36.

Andrews, J. T. 1970. A geomorphological study of postglacial uplift with particular reference to Arctic Canada. *Inst. Br. Geogr. Spec. Pub.* **2**.

Andrews, J. T., and King, C. A. M. 1968. Comparative till fabrics and till fabric variability in a till sheet and a drumlin, a small-scale study. *Proc. York. Geol. Soc.*, **36**, 435-61.

Andrews, J. T., and Smithson, B. B. 1966. Till fabrics of the cross-valley moraines of north central Baffin Island. *Bull. Geol. Soc. Am.*, **77**, 271-90.

Antevs, E. 1925. Retreat of the last ice sheet in eastern Canada. *Geol. Surv. Can. Mem.*, 146.

Antevs, E. 1957. Geological tests of the varve and radiocarbon chronologies. *J. Geol.*, **65**, 129-48.

Arnborg, L. 1955a. Hydrology of the glacial river Austurfljot. *Geogr. Annlr.*, **37**, 185-201.

Arnborg, L. 1955b. Ice-marginal lakes at Hoffellsjökull. *Geogr. Annlr.*, **37**, 202-28.

Askelsson, J. 1936. On the last eruptions in Vatnajökull. *Societas Scientiarium Islandica* XVIII.

Bambrough, R. 1964. Principia metaphysica. *Philosophy*, **39**, 97-109.

Barry, R. G. 1966. Meteorological aspects of the glacial history of Labrador-Ungava with special reference to atmospheric vapour transport. *Geogr. Bull.*, **8**, 319-40.

Battey, M. H. 1960. Geological factors in the development of Veslgjuv-botn and Vesl-Skautbotn. In Norwegian cirque glaciers. Ed. W. V. Lewis. *R. Geogr. Soc. Res. Ser. 4*, 5-10.

Battle, W. R. B. 1960. Temperature observations in bergschrunds and their relationship to frost shattering. In Norwegian cirque glaciers. Ed. W. V. Lewis. *R. Geogr. Soc. Res. Ser. 4*, 83-95.

Benson, C. S. 1961. Stratigraphic studies in the snow and firn of the Greenland ice sheet. *Folia Geographica Danica*, **9**, 13-37.

Bernhardi, A. 1832. Wie kamen die aus dem Norden stammenden Felsbruchsücke und Geschiebe . . . an ihre gegenwärtigen Fundorte? *Heidelb. Jb. Minert. Geogn. Petrefaktenk.*, **3**, 257-67.

Birot, P. 1960. *Le cycle d'érosion sous les*

différents climats. Centro des Pesguisas de Geografia do Brazil. Faculdade Nacional de Felosofia, University of Brazil.

Bishop, B. C. 1957. Shear moraines in Thule area, north-west Greenland. *U.S. Snow, Ice and Permafrost Research Establishment Res. Rep.*, 17.

Boer, G. de 1949. Ice-margin features, Leirbreen, Norway. *J. Glaciol.*, **1**, 332-6.

Bordier, A. C. 1773. *Voyage pittoresque aux glacières de Savoie Suisse.* Geneva.

Boulton, G. S. 1967. The development of a complex supraglacial moraine at the margin of Sorbreen, Ny Friesland, Vestspitzbergen. *J. Glaciol.*, **6**, 717-36.

Boulton, G. S. 1968. Flow tills and related deposits on some Vestspitzbergen glaciers. *J. Glaciol.*, **7**, 391-412.

Boulton, G. S. 1970a. On the origin and transport of englacial debris in Svalbard glaciers. *J. Glaciol.*, **9**, 213-29.

Boulton, G. S. 1970b. On the deposition of subglacial and melt-out tills at the margins of certain Svalbard glaciers. *J. Glaciol.*, **9**, 231-46.

Boulton, G. S. 1972. The role of thermal regime in glacial sedimentation—a general theory. *Inst. Br. Geogr. Spec. Pub.*, **4** (in press).

Bowen, D. Q. 1964-66. On the supposed ice-dammed lakes of South Wales. *Trans. Cardiff Nat. Soc.*, **93**, 4-17.

Bowen, D. Q., and Gregory, K. J. 1965. A glacial drainage system near Fishguard, Pembrokeshire. *Proc. Geol. Ass.*, **76**, 275-82.

Bretz, J. H. 1959. The double Calumet stage of Lake Chicago. *J. Geol.*, **67**, 675-84.

Brooks, C. E. P. 1949. *Climate through the ages.* Ernest Benn, London.

Brown, T. C. 1931. Kames and kame terraces of central Massachusetts. *Bull. Geol. Soc. Am.*, **42**, 467-79.

Buckland, W. 1840-1. On the evidences of glaciers in Scotland and the north of England. *Proc. Geol. Soc.*, **3**, 332-7 and 345-8.

Budd, W., Jenssen, D., and Radok, U. 1970. The extent of basal melting in Antarctica. *Polarfarschung*, **1**, 293-306.

Büdel, J. 1963. Klima-genetische geomorphologie. *Geogr. Rundaschau*, **15**, 269-85.

Carey, S. W., and Ahmad, N. 1961. Glacial and marine sedimentation. *Proc. 1st Int. Symp. Arctic Geol.*, **2**, 865-94.

Carol, H. 1947. Formation of roches moutoneés. *J. Glaciol.*, **1**, 57-9.

Carruthers, R. G. 1947-48. The secret of the glacial drifts. *Proc. York. Geol. Soc.*, **27**, 43-57 and 129-72.

Carruthers, R. G. 1953. *Glacial drifts and the undermelt theory.* Priv. Pub. Newcastle upon Tyne.

Chadwick, G. H. 1928. Adirondack eskers. *Bull. Geol. Soc. Am.*, **39**, 923-9.

Chamberlin, T. C. 1883. Terminal moraine of the second glacial epoch. *U.S. Geol. Surv. 3rd Ann. Rep.*, 291-402.

Chamberlin, T. C. 1888. The rock scorings of the great ice invasions. *U.S. Geol. Surv. 7th Ann. Rep.*, 155-248.

Charlesworth, J. K. 1926a. Glacial geology of the Southern Uplands, west of Annandale and upper Clydesdale. *Trans. R. Soc. Edinb.*, **55**, 1-23.

Charlesworth, J. K. 1926b. The re-advance marginal kame moraine of the south of Scotland and some later stages of the retreat. *Trans. R. Soc. Edinb.*, **55**, 25-50.

Charlesworth, J. K. 1955. Late glacial history of the Highlands and Islands of Scotland. *Trans. R. Soc. Edinb.* **62**, 769-928.

Charlesworth, J. K. 1957. *The Quaternary era.* Vols. I and II. Edward Arnold, London.

Charpentier, J. de 1823. *Essai sur la constitution géognostique des Pyrénées.* Paris.

Charpentier, J. de 1835. Notice sur la cause probable de transport des blocs erratiques de la Suisse. *Annls. des Mines, Paris*, 8.

Chorley, R. J. 1959. The shape of drumlins. *J. Glaciol.*, **3**, 339-44.

Chorley, R. J. 1967. Models in geomorphology. In *Physical and information models in geography.* Ed. Chorley and Hagget, 57-96. University Paperback, Methuen, London.

Chorley, R. J., Dunn, A. J., and Beckinsale, R. P. 1964. *The history of the study of landforms.* Vol. I. Methuen, London.

Clapperton, C. M. 1968. Channels formed by the superimposition of glacial meltwater streams, with special reference to the East Cheviot Hills, northeast England. *Geogr. Annlr.*, **50**, 207-20.

Clapperton, C. M. 1971. The location and origin of glacial meltwater phenomena in the eastern Cheviot Hills. *Proc. York. Geol. Soc.*, **38**, 361-80.

CLAYTON, K. M. 1965. Glacial erosion in the Finger Lakes region. (New York State, U.S.A.). *Zeit. F. Geomorph.*, **9**, 50-62.

CLAYTON, L. 1964. Karst topography on stagnant glaciers. *J. Glaciol.*, **5**, 107-12.

COMMON, R. 1957. Variations in the Cheviot meltwater channels. *Geogr. Stud.*, **4**, 90-103.

CONWAY, M. 1898. An exploration in 1897 of some of the glaciers of Spitzbergen. *Geogr. J.*, **12**, 137-58.

COOK, J. H. 1946a. Ice contacts and the melting of ice below a water level. *Am. J. Sci.*, **244**, 502-12.

COOK, J. H. 1946b. Kame complexes and perforation deposits. *Am. J. Sci.*, **244**, 573-83.

COTTON, C. A. 1942. *Climatic accidents in landscape-making.* Whitcombe and Tombs, Christchurch.

CROSBY, W. O. 1902. Origin of eskers. *Am. Geol.*, **30**, 1-38.

DAHL, R. 1965. Plastically sculptured detail forms on rock surfaces in northern Nordland. *Geogr. Annlr.*, **47**, 83-140.

DANSGAARD, W. 1961. The isotopic composition of natural waters with special reference to the Greenland ice cap. *Meddr. om Grønland*, **165** (2), 7-166.

DAVIS, W. M. 1899. The geographical cycle. *Geogr. J.*, **14**, 481-504.

DAVIS, W. M. 1900. Glacial erosion in France, Switzerland and Norway. *Proc. Boston Soc. Nat. Hist.*, **29**, 303-32.

DEMOREST, M. 1938. Ice flowage as revealed by glacial striae. *J. Geol.*, **46**, 700-25.

DERBYSHIRE, E. 1961. Subglacial col gullies and the deglaciation of the north-east Cheviots. *Trans. Inst. Br. Geogr.*, **29**, 31-46.

DERBYSHIRE, E. 1962. Fluvioglacial erosion near Knob Lake central Quebec-Labrador, Canada. *Bull. Geol. Soc. Am.*, **73**, 1111-26.

DESIO, A. 1954. An exceptional glacier advance in the Karakorum-Ladakl region. *J. Glaciol.*, **2**, 383-5.

DONN, W. L., FARRAND, W. R., and EWING, M. 1962. Pleistocene ice volumes and sea lowering. *J. Geol.*, **70**, 206-14.

DOORNKAMP, J. C., and KING, C. A. M. 1971. *Numerical analysis in geomorphology.* Arnold, London.

DREIMANIS, A. *et al.* 1957. Heavy mineral studies in tills in Ontario and adjacent areas. *J. Sed. Pet.*, **27**, 48-161.

DREIMANIS, A. and VAGNERS, U. J. 1969. Lithologic relation of till to bedrock. In *Quaternary geology and climate.* Ed. H. E. Wright. **16** Proc. VII INQUA Cong., Pub. Nat. Acad. Sci., Washington D.C., 93-8.

DRYER, C. R. 1901. Certain peculiar eskers and esker lakes of north-eastern Indiana. *J. Geol.*, **9**, 123-9.

DURY, G. H. 1953. A glacial breach in the north-western Highlands. *Scott. Geogr. Mag.*, **69**, 106-17.

DWERRYHOUSE, A. R. 1902. The glaciation of Teesdale, Weardale and the Tyne Valley, and their tributary valleys. *Q. J. Geol. Soc. Lond.*, **58**, 572-608.

DYSON, J. L. 1952. Ice-ridged moraines and their relation to glaciers. *Am. J. Sci.*, **250**, 204-11.

EASTERBROOK, J. 1964. Void ratios and bulk densities as means of identifying Pleistocene tills. *Bull. Geol. Soc. Am.*, **75**, 745-50.

EBERS, E. 1961. Die Gletscherschliffe und-rinnen. In Der Gletscherschliffe von Fischback am Inn. Ed. H. Fehn. *Landesk. Forsch., Munchen*, 40.

EDELMAN, N. 1949. Some morphological details of the roches moutonnées in the archipelago of S.W. Finland. *Bull. Comm. Geol. Finl.*, **144**, 129-37.

EDELMAN, N. 1951. Glacial abrasion and ice movement in th earea of Rosala-Noto, S.W. Finland. *Bull. Comm. Geol. Finl.*, **154**, 157-69.

ELLISTON, G. 1963. In discussion following J. Weertman, Catastrophic glacier advances. *Bull. Int. Ass. Sci. Hydrol.*, **8**, 65-6.

ELSON, J. A. 1957. Lake Agassiz and the Mankato-Valders problem. *Sci.*, **126**, 999-1002.

EMBLETON, C. 1964. Subglacial drainage and supposed ice-dammed lakes in north-east Wales. *Proc. Geol. Ass.*, **75**, 31-8.

EMBLETON, C., and KING, C. A. M. 1968. *Glacial and periglacial geomorphology.* Arnold, London.

ENGELN, O. D. von 1912. Phenomena associated with glacier drainage and wastage, with special reference to observations in the Yakutat Bay region, Alaska. *Zeit. f. Gletscher.*, **6**, 104-50.

ESMARK, J. 1827. Remarks tending to explain

the geological history of the earth. *Edinb. New Phil. J.*, **2**, 107-21.

Fahnestock, R. K. 1963. Morphology and hydrology of a glacial stream—White River, Mount Ranier Washington. *U.S. Geol. Surv. Prof. Pap.*, 422-A.

Fairchild, H. L. 1929. New York drumlins. *Rochester Acad. Sci. Proc.*, **7**, 1-37.

Fisher, J. E. 1963. Two tunnels in cold ice at 4000 m on the Breithorn. *J. Glaciol.*, **4**, 513-20.

Flint, R. F. 1928. Eskers and crevasse fillings. *Am. J. Sci.*, **215**, 410-6.

Flint, R. F. 1929. The stagnation and dissipation of the last ice sheet. *Geogr. Rev.*, **19**, 256-89.

Flint, R. F. 1930. The origin of the Irish eskers. *Geogr. Rev.*, **20**, 615-30.

Flint, R. F. 1947. *Glacial geology and the Pleistocene epoch.* Wiley, New York.

Flint, R. F. 1957. *Glacial and Pleistocene geology.* Wiley, New York.

Flint, R. F. 1971. *Glacial and Quaternary geology.* Wiley, New York.

Flint, R. F., and Demorest, M., 1942. Glacier thinning during deglaciation. *Am. J. Sci.*, **240**, 29-66 and 113-36.

Forbes, J. D. 1843. *Travels through the Alps of Savoy.* Edinburgh.

Forbes, J. D. 1845. Notes on the topography and geology of the Cuillin Hills in Skye, and on the traces of ancient glaciers which they present. *Edinb. New Phil. J.*, **40**, 76-99.

Forbes, J. D. 1859. *Occasional papers on the theory of glaciers.* Black, Edinburgh.

Fristrup, B. 1964. Further investigations of the Greenland ice cap. *Collected papers, Denmark, 20th Int. Geogr. Congr.*, 59-67.

Frodin, G. 1954. The distribution of late glacial subfossil sandar in northern Sweden. *Geogr. Annlr.*, **36**, 112-34.

Frye, J. C., Glass, H. D., Kempton, J., and Willman, H. 1969. Glacial tills of north-western Illinois. *Ill. Geol. Surv. Circ.*, 437.

Frye, J. C., Glass, H. D., and Willman, H. B. 1968. Mineral zonation of Woodfordian loesses of Illinois. *Ill. Geol. Surv. Circ.*, 427.

Gage, M. 1961. On the definition, date and character of the Ross glaciation, early Pleistocene, New Zealand. *Trans. R. Soc. New Zealand*, **88**, 631-7.

Garwood, E. J. 1899. Additional notes on the glacial phenomena of Spitzbergen. *Q. J. Geol. Soc. Lond.*, **55**, 681-91.

Garwood, E. J., and Gregory, J. W. 1898. Contributions to the glacial geology of Spitzbergen. *Q. J. Geol. Soc. Lond.*, **54**, 197-227

Geer, G. de 1897. Om rullstensåsarns bildnigssatt. *Geol. Fören Stockn. Förh.*, **19**, 366-88.

Geer, G. de 1912. A geochronology of the last 12,000 years. *11th Int. Geol. Congr,. Stockholm 1910, Compte Rendu*, VI, 241-58.

Geer, G. de 1940. *Geochronologia Suecica principles.* K. Svenska Vetensk, Handl., 3d ser., **18**, (6), text and atlas. Amquist and Wiksells, Stockholm.

Geikie, A. 1862. On the date of the last elevation of central Scotland. *Q. J. Geol. Soc. Lond.*, **18**, 218-32.

Geikie, A. 1863. On the phenomena of glacial drift in Scotland. *Trans. Geol. Soc. Glasgow*, **1**, 1-171.

Geikie, A. 1865. *The scenery of Scotland.* 1st Edn., McMillan, London.

Geikie, J. 1869. Explanation of sheet 24—Peeblesshire with part of Lanark, Edinburgh and Selkirk. *Mem. Geol. Surv. Scott.*, 1-24.

Geikie, J. 1873. *The great ice age.* 1st Edn., Edward Stanford, London.

Giles, A. W. 1918. Eskers in the vicinity of Rochester, N.Y. *Rochester Acad. Sci. Proc.*, **5**, 161-240.

Gillberg, G. 1956. Den glaciala utvecklingen inom Sydsvenska hoglandets vastrarandzon issjor och isavsmaltning. *Geol. Fören. Stockh. Förh.*, **78**, 357-458.

Gjessing, J. 1960. *The drainage of the deglaciation period, its trends amd morphogenetic activity in northern Atnedalen—with comparative studies from northern Gudbrandsdalen and northern Østerdalen.* Ad Novas, 3, Oslo. (Eng. Sum.)

Gjessing, J. 1967. On plastic scouring and subglacial erosion. *Norsk Geogr. Tidsskr.*, **20**, 1-37.

Glen, A. R. 1941. The latest map of North East Land. *Geogr. J.*, **98**, 206-7.

Glen, J. W. 1952. Experiments on the deformation of ice. *J. Glaciol.*, **2**, 111-4.

Glen, J. W. 1954. The stability of ice-dammed lakes and other water-filled holes in glaciers. *J. Glaciol.*, **2**, 316-8.

Glen, J. W. 1955. The creep of polycrystalline ice. *Proc. R. Soc. Ser. A.*, **228**, 519-38.

GLEN, J. W. 1956. Measurement of the deformation of ice in a tunnel at the foot of an ice fall. *J. Glaciol.*, **2**, 735-45.

GLEN, J. W., and LEWIS, W. V. 1961. Measurements of side-slip at Austerdalsbreen, 1959. *J. Glaciol.*, **3**, 1109-22.

GLENTWORTH, R., MITCHELL, W. A., and MITCHELL, B. D. 1964. The red glacial drift deposits of north-east Scotland. *Clay Mineral Bull.*, **5**, 373-81.

GOLDTHWAIT, J. W. 1924. Physiography of Nova Scotia. *Can. Geol. Surv. Mem.*, 140.

GOLDTHWAIT, R. P. 1951. Development of end moraines in east central Baffin Island. *J. Geol.*, **59**, 567-77.

GOLDTHWAIT, R. P. 1963. Dating the little ice age in Glacier Bay, Alaska. *Rep. Int. Geol. Congr. XXI* 1960, *Pt.* 37.

GOODYEAR, J. 1962. X-Ray examination of some east Yorkshire boulder-clays. *Clay Minerals Bull.*, **5**, 43-4.

GOW, A. J. 1965. The ice sheet. In *Antarctica*. Ed. T. Hatherton, 221-58. Methuen, London.

GRAF, W. L. 1970. The geomorphology of the glacial valley cross section. *J. Arct. Alp. Res.*, **2**, 303-12.

GRAVENOR, C. P. 1953. The origin of drumlins. *Am. J. Sci.*, **251**, 674-81.

GRAVENOR, C. P., and KUPSCH, W. O. 1959. Ice disintegration features in western Canada. *J. Geol.*, **67**, 48-64.

GREGORY, J. W. 1912. The relation of eskers and kames. *Geogr. J.*, **40**, 169-75.

GREGORY, J. W. 1926. The Scottish kames and their evidence on the glaciation of Scotland. *Trans. R. Soc. Edinb.*, **54**, 395-432.

GREGORY, K. J. 1965. Proglacial Lake Eskdale after sixty years. *Trans. Inst. Br. Geogr.*, **36**, 140-62.

GRIPP, K. 1929. Glaciologische und geologische ergebnisse der Hamburgischen Spitzbergen expedition 1927. *Abh. nw. V Hamb.*, **22**, 145-249.

GWYNNE, C. S. 1942. Swell and swale pattern of the Mankato lobe of the Wisconsin drift plain in Iowa. *J. Geol.*, **50**, 200-8.

HAEFELI, R. 1951. Some observations on glacier flow. *J. Glaciol.*, **1**, 496-500.

HAGGET, P., and CHORLEY, R. J. 1967. Models, paradigms and the new geography. In *Physical and information models in geography*. Ed. Chorley and Hagget, 19-41. University Paperbacks, Methuen, London.

HANSEN, B. L., and LANGWAY, C. C. 1966. Deep core drilling in ice and core analysis at Camp Century, Greenland, 1961-66. *Antarctic J. of the U.S.*, **1**, 207-8.

HANSON, G. F. 1943. A contribution to experimental geology, the origin of eskers. *Am. J. Sci.*, **241**, 447-52.

HARMER, F. W. 1907. On the origin of certain canyon-like valleys associated with lake-like areas of depression. *Q. J. Geol. Soc. Lond.*, **63**, 470-514.

HARRISON, P. W. 1957. A clay till fabric; its characteristics and origin. *J. Geol.*, **65**, 275-308.

HASELTON, G. M. 1966. Glacial geology of Muir Inlet, south-east Alaska. *Ohio State Univ. Inst. Polar Stud. Rep.* 18.

HATTERSLEY-SMITH, G. 1964. Rapid advance of glacier in northern Ellesmere Island. *Nature*, **201**, 176.

HELLAAKOSKI, A. 1931. On the transportation of materials in the esker of Laitila. *Fennia*, **52**, 1-41.

HENDERSON, E. P. 1958. A glacial study of central Quebec-Labrador. *Geol. Surv. Can. Bull.*, 50.

HERSHEY, O. H. 1897. Eskers indicating stages of glacial recession in the Kansan Epoch in Northern Illinois. *Am. Geol.*, **19**, 197-209 and 237-53.

HEWITT, K. 1967. Ice-front deposition and the seasonal effect: a Himalayan example. *Trans. Inst. Br. Geogr.*, **42**, 93-106.

HILL, A. R. 1971. The internal composition and structure of drumlins in north Down and south Antrim, Northern Ireland. *Geogr. Annlr.*, **53**, 14-31.

HIND, H. Y. 1859. A preliminary and general report on the Assiniboine and Saskatchewan exploring expedition. *Canada Legislative Assembly Journal*, **19**, Appendix 36.

HJULSTRÖM, F. 1935. *Studies of the morphological activity of rivers as illustrated by the River Fyris*. Uppsala.

HJULSTRÖM, F. 1954. The Hoffellsandur. *Geogr. Annlr.*, **36**, 135-45.

HOBBS, P. V., and MASON, B. J. 1964. The sintering and adhesion of ice. *Phil. Mag.*, **9**, 181.

HOLDAR, C. G. 1957. Deglaciationsforloppet i Tornetraskomradet efter senaste nedisningsperioden, med vissa tillbakablinkar och regionalo Jamforelser. *Geol. Fören. Stockl. Förh.*, **79**, 291-528.

HOLLIN, J. T. 1962. On the glacial history of Antarctica. *J. Glaciol.*, **4**, 173-95.

HOLLINGWORTH, S. E. 1931. The glaciation of W. Edenside and adjoining areas. *Q.J. Geol. Soc. Lond.*, **87**, 281-359.

HOLLINGWORTH, S. E. 1952. A note on the use of marginal drainage channels in the recognition of unglaciated enclaves. *J. Glaciol.*, **2**, 107-8.

HOLMES, C. D. 1941. Till fabric. *Bull Geol. Soc. Am.*, **52**, 1299-1354.

HOLMES, C. D. 1947. Kames. *Am. J. Sci.*, **245**, 240-9.

HOLMES, C. D. 1960. Evolution of till-stone shapes, central New York. *Bull. Geol. Soc. Am.*, **71**, 1645-60.

HOLTEDAHL, H. 1967. Notes on the formation of fjords and fjord valleys. *Geogr. Annlr.*, **49**, 188-203.

HOLZNER, L., and WEAVER, G. D. 1965. Geographic evaluation of climatic and climate-genetic geomorphology. *Ann. Ass. Am. Geogr.*, **55**, 592-602.

HOOKE, R. LEB. 1970. Morphology of the ice-sheet margin near Thule, Greenland, *J. Glaciol.*, **9**, 303-24.

HOPPE, G. 1950. Nagra exempel pa glacifluvial dranering fran det Inre Norrbotten. *Geogr. Annlr.*, **32**, 37-59.

HOPPE, G. 1952. Hummocky moraine regions with special reference to the interior of Norbotten. *Geogr. Annlr.*, **34**, 1-72.

HOPPE, G. 1957. Problems of glacial morphology and the ice age. *Geogr. Annlr.*, **39**, 1-18.

HOPPE, G. 1959. Glacial morphology and the inland ice recession in northern Sweden. *Geogr. Annlr.*, **41**, 193-212.

HOPPE, G. 1963. Subglacial sedimentation with examples from northern Sweden. *Geogr. Annlr.*, **45**, 41-9.

HOPPE, G., and SCHYTT, V. 1953. Some observations on fluted moraine surfaces. *Geogr. Annlr.*, **35**, 105-15.

HOWARTH, P. J. 1966. An esker, Breidamerkurjökull, Iceland. *Br. Geom. Res. Group, Occ. Pap.*, **3**, 6-9.

HOWARTH, P. J. 1968a. *Geomorphological and glaciological studies, eastern Breidamerkurjökull, Iceland.* Unpub. Ph.D. thesis, University of Glasgow.

HOWARTH, P. J. 1968b. A supraglacial extension of an ice-dammed lake, Tunsbergdalsbreen, Norway. *J. Glaciol.*, **7**, 413-9.

HOWARTH, P. J. 1971. Investigations of two eskers at eastern Breidamerkurjökull, Iceland. *J. Arct. Alp. Res.*, **3**, 305-318.

HOWARTH, P. J., and PRICE, R. J. 1969. The proglacial lakes of Breidamerkurjökull and Fjallsjökull, Iceland. *Geogr. J.*, **135**, 573-81.

IVES, J. D. 1958. Glacial drainage channels as indicators of late-glacial conditions in Labrador-Ungava: a discussion. *Cah. de Geogr. de Queb.*, **5**, 57-72.

IVES, J. D. 1960. Former ice-dammed lakes and the deglaciation of the middle reaches of the George River, Labrador-Ungava. *Geogr. Bull.*, **14**, 44-70.

IVES, J. D., and KIRBY, R. P. 1964. Fluvioglacial erosion near Knob Lake, central Quebec-Labrador, Canada: discussion. *Bull. Geol. Soc. Am.*, **75**, 917-22.

JAMIESON, T. F. 1860. On the drift and rolled gravel of the north of Scotland. *Q. J. Geol. Soc. Lond.*, **16**, 347-71.

JAMIESON, T. F. 1862. On the ice-worn rocks of Scotland. *Q.J. Geol. Soc. Lond.*, **18**, 164-84.

JAMIESON, T. F. 1863. On the parallel roads of Glen Roy and their place in the history of the glacial period. *Q.J. Geol. Soc. Lond.*, **19**, 235-59.

JAMIESON, T. F. 1865. On the history of the last geological changes in Scotland. *Q.J. Geol. Soc. Lond.*, **21**, 161-203.

JAMIESON, T. F. 1892. Supplementary remarks on Glen Roy. *Q.J. Geol. Soc. Lond.*, **48**, 4-28.

JARNEFORS, B. 1952. A sediment—petrographic study of glacial till from Pajala district, northern Sweden. *Geol. Fören. Stockh. Förh.*, **74**, 185-211.

JOHNSON, W. H. 1964. Stratigraphy and petrography of Illinoian and Kansan drift in central Illinois. *Ill. Geol. Surv. Circ.*, 378.

JOHNSSON, G. 1956. Glacialmorfologiska studier i Södra Sverige. *Meddn. Lunds. Geogr. Instn.*, **30**, 1-407.

KAMB, W. B., and LACHAPELLE, E. 1964. Direct observation of the mechanism of glacier sliding over bedrock. *J. Glaciol.*, **5**, 159-72.

KAMB, W. B., and SHREVE, R. L. 1963.

Texture and fabric of ice at depth in a temperate glacier. *Trans. Am. Geophys. Un.*, **44**, 103. (Abs. only).

KELLY, T. E., and BAKER, C. H. 1966. Colour variations within glacial till—east central North Dakota. *J. Sed. Pet.*, **31**, 75-86.

KEMPTON, J. P. 1963. Subsurface stratigraphy of the Pleistocene deposits of central northern Illinois. *Ill. Geol. Surv. Circ.*, 356.

KEMPTON, J. P., and HACKETT, J. E. 1968a. The Late-Altonian (Wisconsinan) glacial sequence in northern Illinois. In *Means of correlation of Quaternary successions.* 535-46. Proc. VII Congr. INQUA, Univ. Utah Press.

KEMPTON, J. P., and HACKETT, J. E. 1968b. Stratigraphy of the Woodfordian and Altonian drifts of central northern Illinois. In the Quaternary of Illinois. *Univ. of Ill. College of Agriculture Spec. Pub.*, **14**, 27-34.

KENDALL, P. F. 1902. A system of glacier-lakes in the Cleveland Hills. *Q.J. Geol. Soc. Lond.*, **58**, 471-571.

KILBURN, C., PITCHER, W. S., and SHACKLETON R. M. 1965. The stratigraphy and origin of the Portaskaig boulder bed series (Dalradian). *J. Geol.*, **4**, 343-60.

KINDLE, E. M. 1930. Sedimentation in a glacial lake. *J. Geol.*, **38**, 81-7.

KING, C. A. M., and BUCKLEY, J. T. 1968. The analysis of stone size and shape in Arctic environments. *J. Sed. Pet.*, **38**, 200-14.

KING, C. A. M., and GAGE, M. 1961. Note on the extent of glaciation in part of west Kerry. *Ir. Geogr.*, **4**, 202-8.

KINGERY, W. D. (ed.) 1963. *Ice and snow.* M.I.T. Press, Cambridge, Mass.

KIRBY, R. P. 1969. Variation in glacial deposition in a subglacial environment: an example from Midlothian. *Scott. J. Geol.*, **5**, 49-53.

KOCH, J., and WEGNER, A. 1911. Die glaziologischen beobachtungen der Denmark-expedition. *Meddr. om Grønland*, **46**, 1-467.

KRIGSTROM, A. 1962. Geomorphological studies of sandar plains and their braided rivers in Iceland. *Geogr. Annlr.*, **44**, 328-46.

KUENEN, P. H. 1951. The mechanics of varve formation and the action of turbidity currents. *Geol. Fören. Stockh. Förh.*, **6**, 149-62.

LAMPLUGH, G. W. 1911. On the shelly moraine of the Sefström glacier and other Spitzbergen phenomena illustrative of British glaciological conditions. *Proc. York. Geol. Soc.*, **17**, 216-41.

LEE, H. A. 1965. Investigation of eskers for mineral exploration. *Geol. Surv. Can. Pap.*, 65-14, 1-17.

LEMKE, R. W. 1958. Narrow linear drumlins near Velva, North Dakota. *Am. J. Sci.*, **256**, 270-83.

LEOPOLD, L. B., WOLMAN, M. G., and MILLER, J. P. 1964. *Fluvial processes in geomorphology.* Freeman, San Francisco.

LEVERETT, F., and TAYLOR, F. B. 1915. The Pleistocene of Indiana and Michigan and the history of the Great Lakes. *U.S. Geol. Surv. Mon.*, 53.

LEWIS, W. V. 1938. A meltwater hypothesis of cirque formation. *Geol. Mag.*, **75**, 249-65.

LEWIS, W. V. 1949. An esker in process of formation, Böverbreen, Jotunheim, 1947. *J. Glaciol.*, **1**, 314-9.

LEWIS, W. V. 1954. Pressure release and glacial erosion. *J. Glaciol.*, **2**, 417-22.

LEWIS, W. V. (ed.) 1960. Norwegian cirque glaciers. *R. Geogr. Soc. Res. Ser. 4.*

LIESTØL, O. 1955. Glacier dammed lakes in Norway. *Norsk Geogr. Tidsskr.*, **15**, 122-49.

LINDSAY, J. F. 1966. Observations on the level of a self-draining lake on the Casement Glacier, Alaska. *J. Glaciol.*, **6**, 443-5.

LINTON, D. L. 1933. The 'Tinto Glacier' and some glacial features in Clydesdale. *Geol. Mag.*, **70**, 549-54.

LINTON, D. L. 1963. The forms of glacial erosion. *Trans. Inst. Br. Geogr.*, **33**, 1-28.

LISTER, H., PENDLINGTON, A., and CHARLTON, J. 1967. Laboratory experiments on abrasion of sandstones by ice. *Int. Ass. Sci. Hydrol., Comm. Snow and Ice, Bern*, 98-106.

LLIBOUTRY, L. 1958. Studies of the shrinkage after a sudden advance, blue bands and wave ogives on Glacier Universidad (central Chilean Andes). *J. Glaciol.*, **3**, 261-8.

LLIBOUTRY, L. 1964. *Traité de glaciologie.* Tome 1 and 2, Masson, Paris.

LOEWE, F. 1966. The temperature of the Sukkertoppen ice cap. *J. Glaciol.*, **6**, 179.

LOUGEE, R. J. 1940. Deglaciation of New England. *J. Geomorph.*, **3**, 189-217.

LOUGEE, R. J. 1954. The role of up-warping in the post-glacial history of Canada. *Rev. Canad. Geogr.*, **8**, 1-14.

LYELL, C. 1840-41. On the geological

evidence of the former existence of glaciers in Forfarshire. *Proc. Geol. Soc.*, **3**, 337-45.

McCabe, L. H. 1939. Nivation and corrie erosion in west Spitzbergen. *Geogr. J.*, **94**, 447-65.

McCall, J. G. 1952. The internal structure of a cirque glacier: report on studies of the englacial movements and temperatures. *J. Glaciol.*, **2**, 122-31.

McCall, J. G. 1960. The flow characteristics of a cirque glacier and their effect on glacial structure and cirque formation. In Norwegian cirque glaciers. Ed. W. V. Lewis. *R. Geogr. Soc. Res. Ser. 4*, 39-62.

McKenzie, G. D. 1969. Observations on a collapsing kame terrace in Glacier Bay National Monument, S. E. Alaska. *J. Glaciol.*, **8**, 413-25.

Manley, G. 1959. The late-glacial climate of north-west England. *Liverpool and Manchester Geol. J.*, **2**, 188-215.

Mannerfelt, C. M. 1945. Nagra glacialmorfologiska formelement. *Geogr. Annlr.*, **27**, 1-239.

Mannerfelt, C. M. 1949. Marginal drainage channels as indicators of the gradients of Quaternary ice caps. *Geogr. Annlr.*, **31**, 194-9.

Martel, P. 1744. *An account of the glaciers or ice Alps in Savoy*. London.

Mathews, W. H. 1963. Discharge of a glacial stream. *Int. Ass. Sci. Hydrol.*, **63**, 290-300.

Mathews, W. H. 1964. Water pressure under a glacier. *J. Glaciol.*, **5**, 235-40.

Meir, M. F. 1954. Recent eskers in the Wind River Mountains of Wyoming. *Iowa Acad. Sci.*, **58**, 291-4.

Mickelson, D. M. 1971. Glacial geology of the Burroughs Glacier area, southeastern Alaska. *Ohio State Univ. Inst. Polar Stud. Rep.* 40.

Miller, H. 1884. On boulder glaciation. *Proc. R. Phys. Soc. Edinb.*, **8**, 156-89.

Miller, M. M. 1954. In discussion on 'The mechanics of glacier flow.' *J. Glaciol.*, **2**, 339-41.

Millis, J. 1911. What caused the drumlins? *Sci.*, 34, 60-2.

Morisawa, M. 1968. *Streams their dynamics and morphology*. McGraw-Hill, New York.

Morrison, A. 1966. *Glacial geomorphology of the Churchill Falls area, Labrador*. Unpub. Ph.D. thesis, McGill University.

Muir, J. 1915. *Travels in Alaska*. Houghton Mifflin Co., Boston.

Müller, F. 1962. Zonation of the accumulation areas of the glaciers of Axel Heiberg Island, N.W.T. Canada. *J. Glaciol.*, **4**, 302-11.

Nichols, R. L., and Miller, M. M. 1952. The Moreno Glacier, Lago Argentino, Patagonia. Advancing glaciers and nearly simultaneously retreating glaciers. *J. Glaciol.*, **2**, 41-7.

North, F. J. 1943. Centenary of the glacial theory. *Proc. Geol. Ass.*, **54**, 1-28.

Nye, J. F. 1951. The flow of glaciers and ice-sheets as a problem in plasticity. *Proc. R. Soc. Ser. A.*, **207**, 554-72.

Nye, J. F. 1952. The mechanics of glacier flow. *J. Glaciol.*, **2**, 82-93.

Nye, J. F. 1957. The distribution of stress and velocity in glaciers and ice sheets. *Proc. R. Soc. Ser. A.*, **239**, 113-33.

Nye, J. F. 1959a. The deformation of a glacier below an ice fall. *J. Glaciol.*, **3**, 387-408.

Nye, J. F. 1959b. The motion of ice sheets and glaciers. *J. Glaciol.*, **2**, 493-507.

Nye, J. F., and Martin, P. C. S. 1967. Glacial erosion. *Int. Ass. Sci. Hydrol., Comm. Snow and Ice, Bern* 78-83.

Ogilvie, I. H. 1904. The effect of superglacial debris on the advance and retreat of some Canadian glaciers. *J. Geol.*, **12**, 722-43.

Okko, V. 1955. Glacial drift in Iceland, its origin and morphology. *Bull. Comm. Geol. Finl.*, **170**, 1-133.

Østrem, G. 1963. Comparative crystallographic studies on ice from ice-cored moraine, snow banks and glaciers. *Geogr. Annlr.*, **45**, 210-40.

Østrem, G. 1964. Ice-cored moraines. *Geogr. Annlr.*, **46**, 282-337.

Paterson, W. S. B. 1969. *The physics of glaciers*. Pergamon Press, London.

Peel, R. F. 1951. The study of two Northumbrian spillways. *Trans. Inst. Br. Geogr.*, **15**, 73-89.

Peel, R. F. 1956. The profiles of glacial drainage channels. *Geogr. J.*, **122**, 483-7.

Peltier, L. C. 1950. The geomorphic cycle in periglacial regions as it is related to climatic geomorphology. *Ann. Ass. Am. Geogr.*, **40**, 214-36.

PETRIE, G., and PRICE, R. J. 1966. Photogrammetric measurements of the ice wastage and morphological changes near the Casement Glacier, Alaska. *Can. J. Earth Sci.*, **3**, 827-40.

PLAYFAIR, J. *Illustrations of the Huttonian theory of the earth.* Edinburgh.

POST, A. S. 1960. The exceptional advances of the Muldrow, Black Rapids and Susitna Glaciers. *J. Geophys. Res.*, **65**, 3703-12.

POST, A. S. 1965. Alaskan glaciers; recent observations in respect to the earthquake advance theory. *Sci.*, **148**, 366-8.

POUNDER, E. R. 1965. *Physics of ice.* Pergamon Press, London.

PREST, V. K. 1968. Nomenclature of moraines and ice-flow features as applied to the glacial map of Canada. *Geol. Surv. Can. Pap.*, 67-57.

PRESTWICH, J. 1879. On the origin of the parallel roads of Lochaber and their bearing on other phenomena of the glacial period. *Phil. Trans. R. Soc. Lond. B.*, **170**, 663-726.

PRICE, R. J. 1960. Glacial meltwater channels in the upper Tweed drainage basin. *Geogr. J.*, **126**, 485-89.

PRICE, R. J. 1961. *The deglaciation of the Tweed drainage area west of Innerleithen.* Unpub. Ph.D. thesis, University of Edinburgh, 2 vols.

PRICE, R. J. 1963a. A glacial meltwater drainage system in Peeblesshire, Scotland. *Scott. Geogr. Mag.*, **79**, 133-41.

PRICE, R. J. 1963b. The glaciation of a part of Peeblesshire, Scotland. *Trans. Edinb. Geol. Soc.*, **19**, 326-48.

PRICE, R. J. 1964. Landforms produced by the wastage of the Casement Glacier, southeast Alaska. *Ohio State Univ. Inst. Polar Stud. Rep.*, 9.

PRICE, R. J. 1965. The changing proglacial environment of the Casement Glacier, Glacier Bay, Alaska. *Trans. Inst. Br. Geogr.*, **36**, 107-16.

PRICE, R. J. 1966. Eskers near the Casement glacier, Alaska. *Geogr. Annlr.*, **48**, 111-25.

PRICE, R. J. 1969. Moraines, sandar, kames and eskers near Breidamerkurjökull, Iceland. *Trans. Inst. Br. Geogr.*, **46**, 17-43.

PRICE, R. J. 1970. Moraines at Fjallsjökull, Iceland. *J. Arct. Alp. Res.*, **2**, 27-42.

PRICE, R. J. 1971. The development and destruction of a sandur, Breidamerkurjökull, Iceland. *J. Arct. Alp. Res.*, **3**, 225-37.

PRICE, R. J. and HOWARTH, P. J. 1970. The evolution of the drainage system (1904-1965) in front of Breidamerkurjökull, Iceland. *Jökull*, **20**, 27-37.

RABOT, C. 1905. Glacier reservoirs and their outbursts. *Geogr. J.*, **25**, 534-48.

RAMSAY, A. C. 1862. On the glacial origin of certain lakes. *Q.J. Geol. Soc. Lond.*, **18**, 185-204.

RANKAMA, K. (ed.) 1965. *The Quaternary.* Vol. 1, (Denmark, Norway, Sweden, Finland). Interscience Publishers, London.

RANKAMA, K. (ed.) 1967. *The Quaternary.* Vol. 2, (British Isles, France, Germany, Netherlands). Interscience Publishers, London.

RAY, L. L. 1935. Some minor features of valley glaciers and valley glaciation. *J. Geol.*, **43**, 297-322.

REED, B., GALVIN, C. J., and MILLER, J. P. 1962. Some aspects of drumlin geometry. *Am. J. Sci.*, **260**, 200-10.

REID, H. F. 1892. Studies of Muir Glacier, Alaska. *Nat. Geogr. Mag.*, **4**, 19-84.

REID, H. F. 1896. Glacier Bay and its glaciers. *U.S. Geol. Surv. 16th Ann. Rep.*, 415-61.

REID, J. R., and CALLENDER, E. 1965. Origin of debris-covered icebergs and mode of flow of ice into 'Miller Lake,' Martin River Glacier, Alaska. *J. Glaciol.*, **5**, 497-503.

REID, J. R., and CLAYTON, L. 1963. Observations of rapid water-level fluctuations in ice sink-hole lakes, Martin River Glacier, Alaska. *J. Glaciol.*, **4**, 650-2.

RICH, J. L. 1908. Marginal glacial drainage features in the Finger Lakes region. *J. Geol.*, 16, 527-48.

RICH, J. L. 1943. Buried stagnant ice as a normal product of a progressively retreating glacier in a hilly region. *Am. J. Sci*, **241**, 95-9.

RICHTER, K. 1936. Gefugestudien im Engeabrae, Fondalsbrae, und ihren Vorlandsedimenten. *Zeit. f. Gletscher.*, **24**, 22-30.

RIGSBY, G. P. 1960. Crystal orientation in glacier and in experimentally deformed ice. *J. Glaciol.*, **3**, 589-606.

ROBIN, G. de Q. 1964. Glaciology. *Endeavour*, **23**, 102-7.

ROBIN, G. de Q. 1966. Origin of the Ice Ages. *Science Journal*, June, 3-8.

RUSSELL, I. C. 1892. Mt. St. Elias and its glaciers. *Am. J. Sci.*, **43**, 169-82.

RUSSELL, I. C. 1893. Malaspina Glacier. *J. Geol.*, **1**, 219-45.

RUSSELL, I. C. 1895. The influence of debris on the flow of glaciers. *J. Geol.*, **3**, 823-32.

SAVAGE, J. C., and PATERSON, W. S. B. 1963. Borehole measurements in the Athabasca Glacier. *J. Geophys. Res.*, **68**, 4521-36.

SCHEUCHZER, J. J. 1723. *Itinera per Helvetiae Alpinas regiones facta annis* 1702-11. Collected Ed., Leyden.

SCHIMPER, K. 1837. Über die Eiszeit. *Mém. Soc. Helv. Sci. Nat.*, **5**, 38-51.

SCHNEIDER, A. F. 1961. Pleistocene geology of the Randall region, central Minnesota. *Bull. Minnesota Geol. Surv.*, 40.

SCHYTT, V. 1956. Lateral drainage channels along the northern side of the Moltka Glacier, northwest Greenland. *Geogr. Annlr.*, **38**, 64-77.

SCHYTT, V. 1959. The glaciers of the Kebnekajse massif. *Geogr. Annlr.*, **41**, 213-27.

SCOTT, J. S., and ST. ONGE, D. A. 1969. Guide to the description of till. *Geol. Surv. Can. Pap.*, 68-6.

SEDDON, B. 1957. Late-glacial cwm glaciers in Wales. *J. Glaciol.*, **3**, 94-9.

SHALER, N. S. 1889. The geology of Cape Ann, Massachusetts. *U.S. Geol. Surv. 9th Ann. Rep.*, 526-611.

SHARP, R. P. 1947. The Wolf Creek Glaciers, St. Elias Range, Yukon Territory. *Geogr. Rev.*, **37**, 26-52.

SHARP, R. P. 1949. Studies of supraglacial debris on valley glaciers. *Am. J. Sci.*, **247**, 289-315.

SHARP, R. P. 1951a. Accumulation and ablation on the Seward Malaspina Glacier. *Bull. Geol. Soc. Am.*, **62**, 725-44.

SHARP, R. P. 1951b. Features of the firn on upper Seward Glacier, St. Elias Mountains, Canada. *J. Geol.*, **59**, 599-621.

SHARP, R. P. 1960. *Glaciers.* Condon Lecture Publs., University of Oregon Press, Eugene.

SHARP, R. P. 1969. Semiquantitative differentiation of glacial moraines near Convict Lake, Sierra Nevada, California. *J. Geol.*, **77**, 68-91.

SHOTTON, F. W. 1953. The Pleistocene deposits of the area between Coventry, Rugby and Leamington, and their bearing on the topographic development of the Midlands. *Phil. Trans. R. Soc. Lond. B.*, **237**, 209-60.

SHUMSKII, P. A. 1964. *Principles of structural glaciology.* Dover, New York.

SIMPSON, J. B. 1933. The late-glacial readvance moraines of the Highland border west of the River Tay. *Trans. R. Soc. Edinb.*, **57**, 633-46.

SISSONS, J. B. 1958a. The deglaciation of part of East Lothian. *Trans. Inst. Br. Geogr.*, **25**, 59-77.

SISSONS, J. B. 1958b. Supposed ice-dammed lakes in Britain with particular reference to the Eddleston Valley. *Geogr. Annlr.*, **40**, 159-87.

SISSONS, J. B. 1958c. Subglacial stream erosion in southern Northumberland. *Scott. Geogr. Mag.*, **74**, 163-74.

SISSONS, J. B. 1960a. Subglacial, marginal and other glacial drainage in the Syracuse-Oneida areas, New York. *Bull. Geol. Soc. Am.*, **71**, 1575-88.

SISSONS, J. B. 1960b. Some aspects of glacial drainage channels in Britain, Part I. *Scott. Geogr. Mag.*, **76**, 131-46.

SISSONS, J. B. 1961a. Some aspects of glacial drainage channels in Britain, Part II. *Scott. Geogr. Mag.*, **77**, 15-36.

SISSONS, J. B. 1961b. A subglacial drainage system by the Tinto Hills, Lanarkshire. *Trans. Edinb. Geol. Soc.*, **18**, 175-93.

SISSONS, J. B. 1961c. The central and eastern parts of the Lammermuir-Stranraer moraine. *Geol. Mag.*, **98**, 380-92.

SISSONS, J. B. 1963. The glacial drainage system around Carlops, Peeblesshire. *Trans. Inst. Br. Geogr.*, **32**, 95-111.

SISSONS, J. B. 1967. *The evolution of Scotland's scenery.* Oliver and Boyd, Edinburgh.

SLATER, G. 1929. The structure of drumlins exposed on the south shore of Lake Ontario. *N.Y. St. Mus. Bull.*, **281**, 3-19.

SMALLEY, I. J., and UNWIN, D. J. 1968. The formation and shape of drumlins and their distribution and orientation in drumlin fields. *J. Glaciol.*, **7**, 377-90.

SMITH, H. T. U. 1948. Giant glacial grooves in northwest Canada. *Am. J. Sci.*, **246**, 503-14.

SOUCHEZ, R. A. 1966. The origin of morainic deposits and the characteristics of glacial erosion in the western Sør Rondane, Antarctica. *J. Glaciol.*, **6**, 249-54.

SPEIGHT, R. 1940. Ice wasting and glacier retreat in New Zealand. *J. Geomorph.*, **3**, 131-43.

STALKER, A. MACS. 1960. Ice-pressed drift forms and associated deposits in Alberta. *Geol. Surv. Can. Bull.*, 57.

STENBORG, T. 1968. Glacier drainage connected with ice structures. *Geogr. Annlr.*, **50**, 25-53.

STENBORG, T. 1969. Studies of the internal drainage of glaciers. *Geogr. Annlr.*, **51**, 13-41.

STOKES, J. C. 1958. An esker-like ridge in process of formation, Flåtisen, Norway. *J. Glaciol.*, **3**, 286-90.

STONE, G. H. 1893. The osar gravels of the coast of Maine. *J. Geol.*, **1**, 246-54.

STONE, G. H. 1899. The glacial gravels of Maine. *U.S. Geol. Surv. Mon.*, 34.

STONE, K. H. 1963. Alaskan ice-dammed lakes. *Ann. Ass. Am. Geogr.*, **53**, 332-47.

SUGDEN, D. E. 1968. The selectivity of glacial erosion in the Cairngorm Mountains, Scotland. *Trans. Inst. Br. Geogr.*, **45**, 79-92.

SVENSSON, H. 1959. Is the cross-section of a glacial valley a parabola? *J. Glaciol.*, **3**, 362-3.

SWEETING, M. M. 1966. The weathering of limestones with particular reference to the Carboniferous limestones of northern England. In *Essays in geomorphology*. Ed. G. H. Dury. 177-210.

SYNGE, F. M. 1970. The Irish Quaternary: current views 1969. In *Irish Geographical Studies in honour of E. Estyn Evans*. Ed. N. Stephens and R. E. Glassock. Pub. Queens Univ. Belfast.

TARR, R. S. 1894. The origin of drumlins. *Am. Geol.*, **13**, 393-407.

TARR, R. S. 1908. Glacial erosion in the Scottish Highlands. *Scott. Geogr. Mag.*, **24**, 575-87.

TARR, R. S. 1909. Some phenomena of the glacier margins in the Yakutat Bay region, Alaska. *Zeit f. Gletscher.*, **3**, 81-110.

TARR, R. S., and BUTLER, B. S. 1909. The Yakutat Bay region, Alaska. *U.S. Geol. Surv. Prof. Pap.*, **64**, 1-178.

TARR, R. S., and MARTIN L. 1906. Glaciers and glaciation of Yakutat Bay, Alaska. *Bull., Am. Geogr. Soc.*, **38**, 145-67.

TARR, R. S., and MARTIN L. 1914. *Alaskan glacier studies*. National Geographic Society, Washington.

TEMPLE, P. H. 1965. Some aspects of cirque distribution in the west-central Lake District, northern England. *Geogr. Annlr.*, **47**, 185-93.

THORARINSSON, S. 1939a. Hoffellsjökull, its movements and drainage., *Geogr. Annlr.*, **21**, 189-215.

THORARINSSON, S. 1939b. The ice dammed lakes of Iceland with particular reference to their values as indicators of glacier oscillations. *Geogr. Annlr.*, **21**, 216-42.

THWAITES, F. T. 1926. The origin and significance of pitted-outwash. *J. Geol.*, **34**, 308-19.

THWAITES, F. T. 1956. *Outline of glacial geology*. University of Wisconsin Press, Madison.

TODTMANN, E. 1932. Glazialgeologische studien am Sudrand des Vatna-Jökull. *Forsch and Fortschr.*, 8.

TREFETHEN, J., and TREFETHEN, H. 1945. Lithology of the Kennebec Valley esker. *Am. J. Sci.*, **242**, 521-7.

TROTTER, F. M. 1929. The glaciation of Eastern Edenside, the Alston Block and the Carlisle Plain. *Q. J. Geol. Soc. Lond.*, **85**, 549-612.

TROWBRIDGE, A. C. 1914. The formation of eskers. *Iowa Acad. Sci.*, **21**, 211-8.

TWIDALE, E. R. 1956a. Longitudinal profiles of some glacial overflow channels. *Geogr. J.*, **122**, 88-92.

TWIDALE, E. R. 1956b. Glacial overflow channels in north Lincolnshire. *Trans. Inst. Br. Geogr.*, **22**, 47-55.

UPHAM, W. 1894. The Madison type of drumlins. *Am. Geol.*, **14**, 69-83.

UPHAM, W. 1896. The glacial lake Agassiz. *U.S. Geol. Surv. Mon.*, 25.

VENETZ, J. 1833. Mémoire sur les variations de la température des Alpes de la Suisse. *Mém. Soc. Helv. Sci. Nat.*, **1**, pt. 2.

VERNON, P. 1966. Drumlins and Pleistocene ice flow over the Ards Peninsula. *J. Glaciol.*, **6**, 401-9.

VIALOV, S. S. 1958. Regularities of glacial shield movements and the theory of plastic viscous flow. *Int. Ass. Sci. Hydrol.*, **47**, 266-75.

WEERTMAN, J. 1957. On the sliding of glaciers. *J. Glaciol.*, **3**, 33-8.

WEERTMAN, J. 1961a. Stability of Ice Age ice sheets. *J. Geophys. Res.*, **66**, 3783-92.

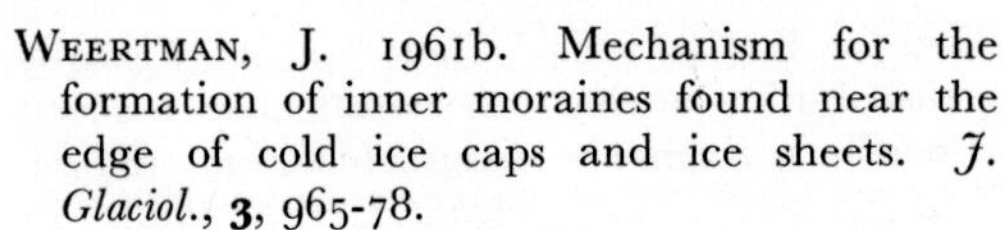

WEERTMAN, J. 1961b. Mechanism for the formation of inner moraines found near the edge of cold ice caps and ice sheets. *J. Glaciol.*, **3**, 965-78.

WEERTMAN, J. 1964a. The theory of glacier sliding. *J. Glaciol.*, **5**, 287-303.

WEERTMAN, J. 1964b. Rate of growth of nonequilibrium ice sheets. *J. Glaciol.*, **5**, 145-58.

WEERTMAN, J. 1966. Effect of a basal water layer on the dimensions of ice sheets. *J. Glaciol.*, **6**, 191-207.

WEERTMAN, J. 1968. Comparison between measured and theoretical temperature profiles of the Camp Century, Greenland, borehole. *J. Geophys. Res.*, **73**, 2691-700.

WELCH, R. 1967. *The application of aerial photography to the study of a glacial area. Breidamerkur, Iceland.* Unpub. Ph.D. thesis, University of Glasgow.

WELCH, R., and HOWARTH, P. J. 1968. Photogrammetric measurement of glacial landforms. *Photogram. Rec.*, **6**, 75-96.

WEST, R. G. 1968. *Pleistocene geology and biology*. Longmans, London.

WOLDSTEDT, P. 1954. Die klimakurve des Tertiärs und Quartärs in Mitteleuropa. *Eiszeitalter Gegenw.*, **4**, 5-9.

WOODWARD, H. B. 1907. *A history of the Geological Society of London*. London.

WRIGHT, G. F. 1889. *The ice age of North America*. Kegan Paul, Trench & Co., London.

WRIGHT, H. E. 1957. Stone orientation in the Wadena drumlin field, Minnesota. *Geogr. Annlr.*, **39**, 19-31.

WRIGHT, H. E., and FREY, D. G. 1965. *The Quaternary of the United States*. Princeton University Press, New Jersey.

YOUNG, J. 1864. On the former existence of glaciers in the high grounds of the south of Scotland. *Q. J. Geol. Soc. Lond.*, **20**, 452-62.

YOUNG, J. A. T. 1969. Variations in till macrafabric over very short distances. *Bull. Geol. Soc. Am.*, **80**, 2343-52.

ZOTIKOV, I. A. 1963. Bottom melting of the central zone of the ice shield on the Antarctic continent. *Bull. Int. Ass. Sci. Hydrol.*, **8**, 36-43.

INDEX

b
18.8.05

BEHAVIORAL ECOLOGY AND CONSERVATION BIOLOGY